60 Excellent Inventions in Metal Forming

A. Erman Tekkaya · Werner Homberg ·
Alexander Brosius
Editors

60 Excellent Inventions in Metal Forming

Springer Vieweg

Editors

A. Erman Tekkaya
Institut für Umformtechnik und Leichtbau
Technische Universität Dortmund
Dortmund, Germany

Alexander Brosius
Institut für Fertigungstechnik
Technische Universität Dresden
Dresden, Germany

Werner Homberg
Lehrstuhl für Umformende und Spanende
Fertigungstechnik
Universität Paderborn
Paderborn, Germany

ISBN 978-3-662-46311-6 ISBN 978-3-662-46312-3 (eBook)
DOI 10.1007/978-3-662-46312-3

Library of Congress Control Number: 2015938597

Springer Vieweg

Printed on acid-free paper

Springer-Verlag GmbH Berlin Heidelberg is part of Springer Science+Business Media
(www.springer.com)

Preface

This book is dedicated to the 60th birthday of Matthias Kleiner by his friends from the international forming community. The title and content of this book "60 Excellent Inventions in Metal Forming" corresponds to the essence of the scientific and technical contribution of Matthias Kleiner, who is very well known for his creativity, out-of-the-box thinking, and innovations. He is not an engineer who is only interested in his specialist area and deals with the optimization of individual technical details. Instead, he is steadily interested in distinct topics and open to new ideas and opportunities. Matthias Kleiner is not a scientist who is doing research behind closed doors. Rather, he seeks contact and exchange with engineers, scientists from other disciplines, and every other people. By this, he defines the notion 'open minded' in a new and personal way. He appreciates the inspirations and the opportunities that arise from this contact and exchange and understands how to collate them to create something fundamentally new. He always understands be the multiplicator as well as the catalyzer in one person and motivates other scientists and his own team in a participating and enjoyable way. Thus, he created many innovative, unconventional ideas, which were often ahead of their time. Numerous patents, publications, and doctoral theses are clear indicators for his success.

The scientific career in academia offered him the necessary environment for developing and trying out unconventional ideas. He initiated several new research activities that achieved international recognition such as Curved Profile Extrusion, Flexible Manufacturing Chains for Lightweight Structures, Hydroforming, Bending, Advancement and Modelling of Electromagnetic Forming Processes, and others. Besides his engineering work, Matthias Kleiner has been involved for many years with the system of academic research. He has developed clear-sighted visions for the improvement of the frame conditions for research while never losing sight of solutions to obvious problems. Matthias Kleiner is a member of numerous local and foreign academies and scientific institutions, including the German Academy of Natural Sciences Leopoldina, the Berlin-Brandenburg Academy of Sciences, the Academia Europaea, the German Academy of Engineering Sciences (acatech), the International Academy for Production Engineering and the Scientific Society of Production Technology (WGP), and the International Academy for Production Engineering (CIRP).

The 60th birthday of Matthias Kleiner is a wonderful opportunity to recognize his impressed and unique national and international contributions to the metal forming community and the research community in general. As the editors, we are very much impressed about the spontaneous and sincere readiness of all colleagues worldwide to accept our invitation to contribute to this book. This is obviously another clear indicator of the international and national recognition of Matthias Kleiner as a scientist, a colleague, and friend.

We would like to thank all colleagues of the CIRP, the colleagues of the Japanese Society for Technology of Plasticity (JSTP) – especially Professor Kozo Osakada for organizing and editing the Japanese contributions –, the colleagues of the German Association of Metal Forming (AGU), Mr. Thomas Lehnert (Springer-Verlag) and his team for supporting our idea and accepting to print this unique book through the distinguished publisher Springer-Verlag, Dr. Nooman Ben Khalifa (Chief Engineer for Research, IUL) for organizing the structure of the book, Dr. Frauke Maevus (IUL) and Dr. Ramona Hölker (IUL) for their contribution to the preface and introduction, Mr. Lars Hiegemann (research assistant at the IUL) for preparing the print of the book and keeping the contact with all authors and the publisher, and all research assistants at the IUL for reviewing the contributions.

The Editors, January 2015

A. Erman Tekkaya
Werner Homberg
Alexander Brosius

Matthias Kleiner wrote his doctoral dissertation at Dortmund University in 1987 on the topic "Multiprocessor Control in Metal Forming". He habilitated in 1991 on the topic "Process Simulation in Metal Forming". From 1994 to 1998, Matthias Kleiner built the Chair of Design and Manufacturing at the newly founded Brandenburg Technical University of Cottbus (BTU Cottbus) as a full professor. In 1997 Matthias Kleiner was awarded the Gottfried Wilhelm Leibniz Prize by the DFG (Deutsche Forschungsgemeinschaft – German Research Funding Organization). He was appointed as the Head of the Chair of Forming Technology of Dortmund University in 1998. In 2004 he transformed the chair into today's Institute of Forming Technology and Lightweight Construction (IUL – Institut für Umformtechnik und Leichtbau). 2007 he was elected, as the first engineer in history, as president of the DFG for a term of 6 years. Since 2014 Matthias Kleiner has been the president of the Leibniz association, where he can contribute with his excellent skills to lead a large and successful research association, his understanding of the problems of the scientific research in Germany, and his visions towards a European research landscape.

Introduction

A. Erman Tekkaya, Nooman Ben Khalifa, Ramona Hölker, and Lars Hiegemann

Metal forming is a symbiosis of tradition and innovation driven by scientific and technological inventions. Comprehensive developments in materials, tooling and machinery lead permanently to new or improved metal forming processes and products. The driving motor in metal forming are the **technological innovations** which lead to new processes or process combinations resulting in novel or improved products. The enormous boost in innovations that occurred in the last decades was only possible by a concurrent development in **analysis methods and measurement techniques**, which act as the enabling tool for metal forming innovations by providing a physical insight to the processes. This is different than in other manufacturing technologies, for instance as in machining, since the effect of actuators in metal forming covers large parts of the workpiece volume based on complicated physical response functions.

Figure 1 shows the new process families that have been invented in the last 60 years. Most technological innovations in this period are related to sheet metal forming, except

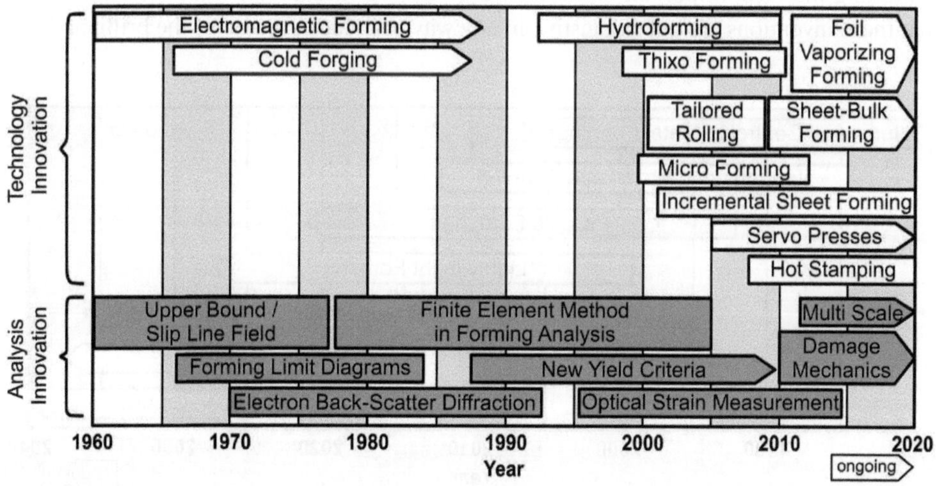

Fig. 1 60 years of inventions in metal forming – overview of the main active development period in selected areas

cold forging, tailored rolling, and parts of the micro forming technology. The analysis innovation can be located in two fields: analytical methods and measurements (see Fig. 1). Upper bound methods and slip line field methods are even today valuable tools for developing insight in complicated metal forming processes. Impressive developments in analyzing the microstructure enabled new and powerful computational methods, the so-called multi-scale methods. These trigger the transition of the phenomenological approach of plasticity to a physics-based approach today.

The basic drivers of these process innovations came in various streams of focused research and development, as shown in Fig. 2. The stream of near-net shape forming boosted the field of cold forging. The increasing product variance (especially in the automotive industry) initiated the flexible forming process development and inventions. Incremental sheet forming methods were basically developed in this era. The awareness of the anthropogenic greenhouse effect enforced energy savings in all branches of design and manufacturing, leading to environmentally benign forming processes. Today, these activities include most of the lightweight efforts, but also cover dry forming processes or low acoustic emission processes. A natural extension of the environmentally conscious design of products is the deterministic setting and utilization of product properties altered in forming processes. This recent development will be one of the main research trends in future decades. Developments in damage mechanics and multi-scale modelling will be the basis of this stream. Finally, it is believed that the precise setting of parameters, despite of several uncertainties in the material, process, environment etc., will be enabled by the closed-loop control of forming processes and forming process chains including the heat treatment processes. This last trend is just starting initiated by the collaborative work of various CIRP colleagues.

This book aims at demonstrating the enormous innovation affinity of metal forming by compiling the most important 60 inventions of the last decades. It must be emphasized that these inventions are not exhaustive in any way. Obviously, some of the brilliant ideas

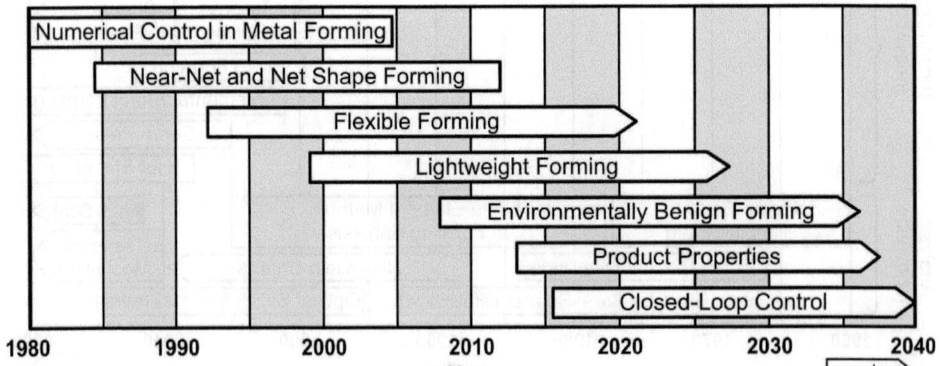

Fig. 2 Main trends in metal forming with a future projection

have been left unmentioned in this book. Nevertheless, we think that this work still can serve the aim mentioned before.

The book covers the 60 new ideas in 9 parts:

- Part 1: Material Characterization and Tribology (6 inventions)
- Part 2: Modelling (5 inventions)
- Part 3: Sheet Metal Forming (9 inventions)
- Part 4: Incremental Forming (7 inventions)
- Part 5: Shear Cutting (5 inventions)
- Part 6: Rolling (6 inventions)
- Part 7: Extrusion and Hot Forging (6 inventions)
- Part 8: Cold Forging (9 inventions)
- Part 9: Tube and Profile Forming (7 inventions)

Part 10 collects further developments covering 7 additional inventions and ideas.

The inventions in this book originate from 10 countries: Denmark, France, Germany, Italy, Japan, Poland, Portugal, Romania, United Kingdom, and the United States. It is noteworthy that several authors from many more countries are involved in these inventions, demonstrating that the forming innovation is internationally widespread.

The authors have confidence that the next decades will bring out many additional inventions and ideas that will improve the technology of metal forming even further and, hence, serve the wealth generation for the society in the same manner as they have in the past decades. The authors also hope that this book will stimulate young students, young researchers, and young engineers in the field of metal forming to recognize the huge opportunity for creativity of this vivid field of manufacturing engineering. At the same time, they hope that it is also recognized how metal forming technology goes hand in hand with analysis methods in mechanics, metal science, chemistry as well as measuring techniques. The terrific developments in mechatronic systems (including sensors and actuators), digital technology, and tailoring materials on nano-scale are wonderful opportunities for the future.

The authors also hope that this impressive collection of inventions in technology and science inspires national funding agencies and university administrations to recognize the huge added value of metal forming research for the society. Metal forming is obviously not an ancient technology that has exceeded its shelf-life for research. In fact, it is an ever-developing technology utilizing all available scientific knowledge and methodologies. Metal forming is an enabling technology without which it would be not possible to realize many of the innovations in biotechnology (e. g. implants, stens), in clean energy generation and transport (e. g. wind mills, super-conductors), space missions (e. g. boosters, structures), and many more. Metal forming could not enable all these technologies without evolving itself as presented in this book.

Table of Contents

Part IV Incremental Forming

Part V Shear Cutting

Part VI Rolling

Part VII Extrusion and Hot Forging

Part VIII Cold Forging

Part IX Tube and Profile Forming

Part I
Material Characterization and Tribology

Novel Method for Combined Tension and Shear Loading of Thin-Walled Tubes

Christopher P. Dick and Yannis P. Korkolis

1 Motivation

The bulk deformation processes that are used to produce thin sheets and tubes introduce preferred crystallographic orientations, or texture, to them. As a result of that, the sheets and tubes exhibit plastic anisotropy, which affects the way that they flow plastically and, consequently, their failure limits. Establishing the plastic anisotropy and calibrating appropriate material models is critical for accurate numerical simulations of forming processes and often determines whether the simulation agrees with the experiment or not [1]. Typical experiments for characterizing the anisotropy of sheet materials are tension tests at different angles to the rolling direction, biaxial testing of cruciform specimens, hydraulic bulge tests as well as a variety of shear tests [2]. For the case of tubes, the most flexible and well-controlled experiments are the combined tension and torsion, or combined tension and internal pressure loading [3, 4]. However, both require sophisticated equipment and are relatively complex to perform.

Two experiments for tubes, which are simple to perform and require only a universal testing machine are described in this paper: a) the Ring Hoop Tension Test, or RHTT [5], and b) the Ring Plane-Strain Tension test, or RPST. In both experiments, a ring is extracted from the parent tube and a test-section is machined on it. The resulting specimen is then placed over two lubricated, close-fitting, D-shaped mandrels. The mandrels are parted in a universal testing machine. Care is taken so that the test-section of the RHTT or RPST specimen remains on one of the mandrels throughout the experiment. As a result of that, the curvature of test-section does not change and hence it experiences only stretching.

Christopher P. Dick · Yannis P. Korkolis ✉
University of New Hampshire, Durham, USA
e-mail: Yannis.Korkolis@unh.edu

A. E. Tekkaya et al. (eds.), *60 Excellent Inventions in Metal Forming*,
DOI 10.1007/978-3-662-46312-3_1

2 Ring Hoop Tension Test

A close-up of the RHTT specimen is shown in Fig. 1. A sandwich of oil and Teflon tape has been used to reduce the friction between the specimen and the mandrels. Also, the specimen has been painted with a random speckle for Digital Image Correlation (DIC) purposes. A valid concern is to what extent does the contact with the mandrel and the resulting friction affect the stress distribution inside the test-section, and make it deviate from the ideal uniaxial case. This was investigated with the aid of finite element analysis (FEA) using solid elements, to capture the stress fields as accurately as possible. The material was modeled as a finitely-deforming, rate-independent elastoplastic solid, using the J_2 flow theory of plasticity with an associated flow-rule. The hardening curve of Al-6061-T4 was input to the simulations. The FEA results for the hoop strain are shown in Fig. 2. The hoop stress is uniform inside the test section and remains elastic in the rest of the specimen. The two regions that are between the parting mandrels experience both stretching and (un-)bending, and develop a through-thickness stress gradient that is visible in Fig. 2. A very detailed probing of this model [6] showed that the contact pressure at the test-section area is very close to being uniform, and that the stresses that are induced by it are negligible, provided that the specimen-mandrel friction remains below 0.15–0.2. The investigation also established that the contact pressure in a well-lubricated RHTT experiment does not have a very different effect on the specimen from that of internal pressure in an inflated tube.

In RHTT experiments on extruded Al-6061-T4 tubes (OD = 60 mm, t = 3 mm), the apparent hoop stress-strain response was measured and found to be different than the axial one. After accounting for the tube eccentricity, specimen-mandrel friction and specimen preparation, the remaining difference between the axial and hoop responses was attributed to material anisotropy (Fig. 3). The DIC technique was used [6] to probe the local strain fields during RHTT, which enabled the determination of the R-values, or Lankford coeffi-

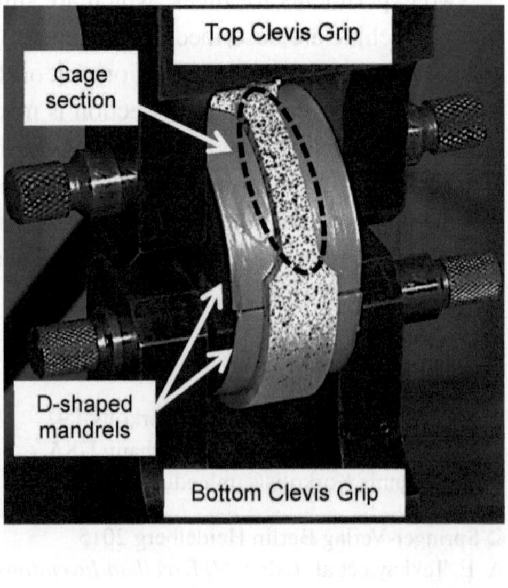

Fig. 1 The Ring Hoop Tension Test [6]

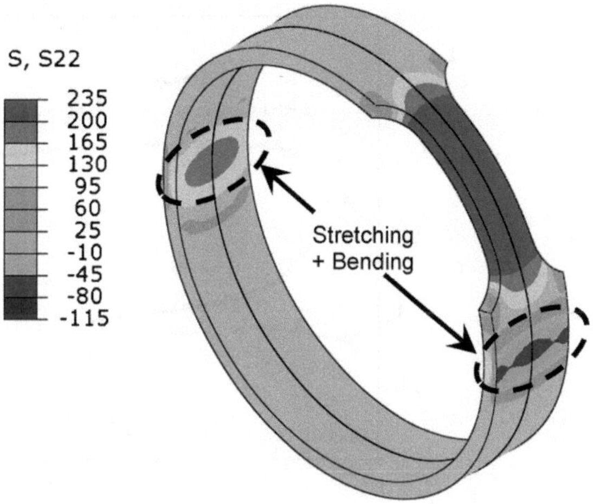

Fig. 2 Finite element prediction of the hoop stress in the RHTT (Units: MPa) [6]

cients of this material. The R-values are seen to evolve with plastic deformation (Fig. 4). The RHTT experiment established that the R-value in the hoop direction is greater than 2.

The RHTT experiment can also be used to probe the response of welds in either electrical-resistance-welded (ERW) or extruded through porthole-die tubes [7]. In contrast to tube inflation experiments, the RHTT allows the direct loading of the weld in uniaxial tension transversely to it.

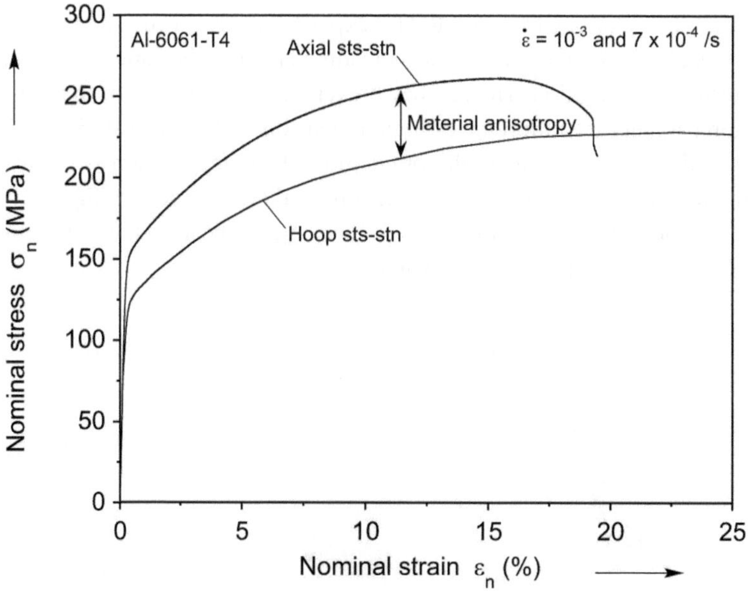

Fig. 3 Nominal stress-strain responses in the axial and hoop direction of the tube [6]

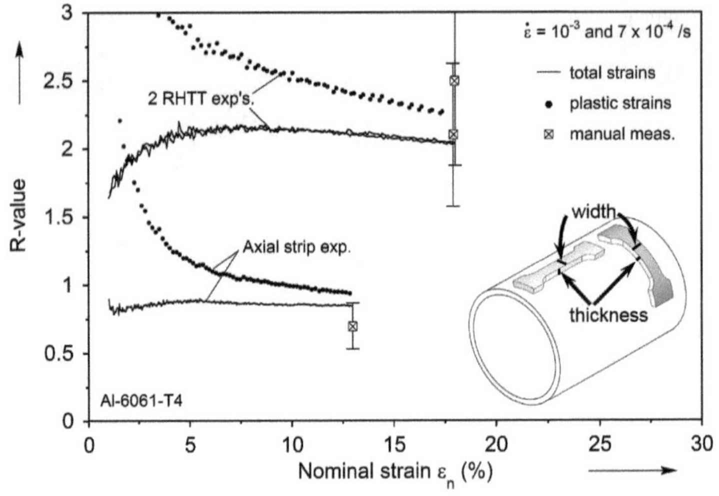

Fig. 4 Evolution of the R-values during deformation, showing axial-hoop anisotropy [6]

3 Ring Plane-Strain Tension Test

The RHTT experiment can be used to determine the hoop stress-strain response of a tube. The plastic response along additional loading paths can be probed with the RPST experiment [8]: a relatively wide and short test-section is prepared on a ring, which is then tested on the same equipment as the RHTT experiment. The wide shoulders above and below the test-section remain elastic during loading and thus they are stiffer than the plastically deforming test-section. Hence, they prevent the deformation of the test-section in the direction parallel to it, enforcing essentially tension-under-plane-strain conditions. To generate additional loading paths, the orientation of the test-section is rotated with respect to the tube generatrix, to extend along a helical arc. By controlling the inclination of the test-section, different combinations of tension and shear on the test-section can be induced. Three such specimens are shown in Fig. 5.

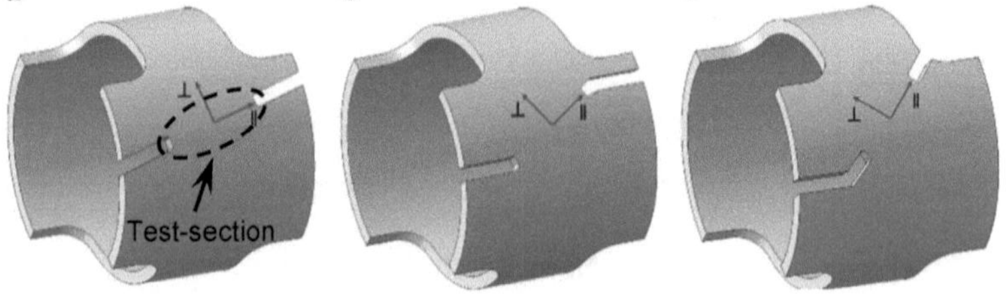

Fig. 5 Ring Plane-Strain Tension specimens with inclined test-sections at **a** 15°, **b** 30° and **c** 45° to the tube axis [8]

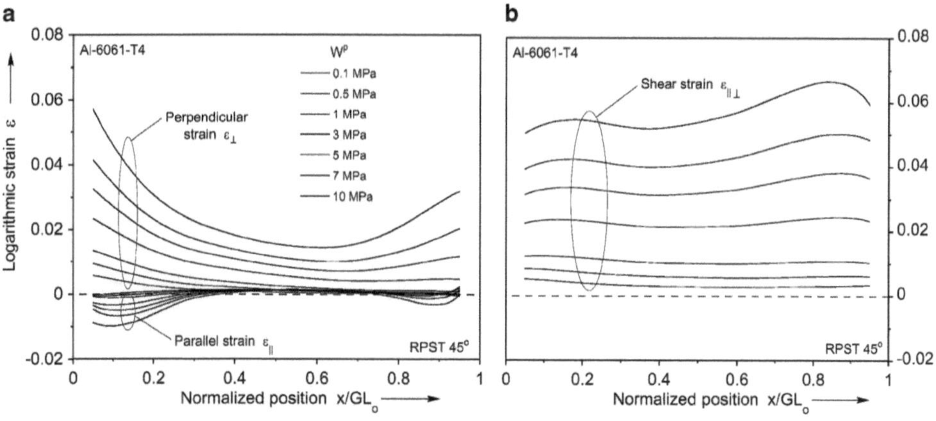

Fig. 6 Evolution of log. strain fields along the test-section, showing that throughout most of the experiment plane-strain conditions prevail at the central portion of test-section. **a** Extensional and **b** shear strains [8]

The DIC technique was used to confirm that plane-strain conditions exist during testing, as seen in Fig. 6 for a specimen where the test-section was inclined by 45°. The DIC results reveal that these conditions are met only at the central portion of the test-section. The strain parallel to the test-section deviates from zero towards the two edges, indicating a transition from plane-strain to uniaxial tension. Interestingly, the shear strain (Fig. 6b) remains relatively constant along the test-section.

The non-uniform strain distribution introduces two complications in analyzing the results of the RPST experiment: a) the net force that the test-section is carrying is not simply half of the load-cell reading, as in the RHTT experiment, and b) the stresses in the test-section cannot be simply assumed to be "force/area". Both of these issues were addressed with the aid of FEA of the RPST specimens [8].

The results can be used to plot contours of constant plastic work, shown in Figs. 7 and 8. In Fig. 7, the plastic work contour is plotted in the axial-hoop plane-stress space, at constant levels of shear stress. The Yld2000-2D yield function by Barlat et al. [9] was calibrated to this data and is also included in this figure. The function captures the experiments very accurately. To examine the evolution of the plastic work contours, the data is plotted in the π-plane, shown in Fig. 8. Also included are the von Mises and Yld2000-2D yield functions. While the former cannot capture either the anisotropy or its evolution with plastic deformation, the latter can be calibrated to fit the experiments very closely.

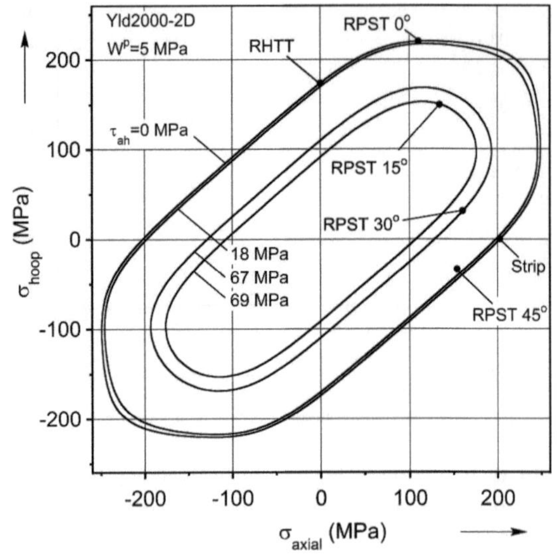

Fig. 7 Plane-stress yield locus of Al-6061-T4 tube, including the RPST experiments and the Yld2000-2D material model [8]

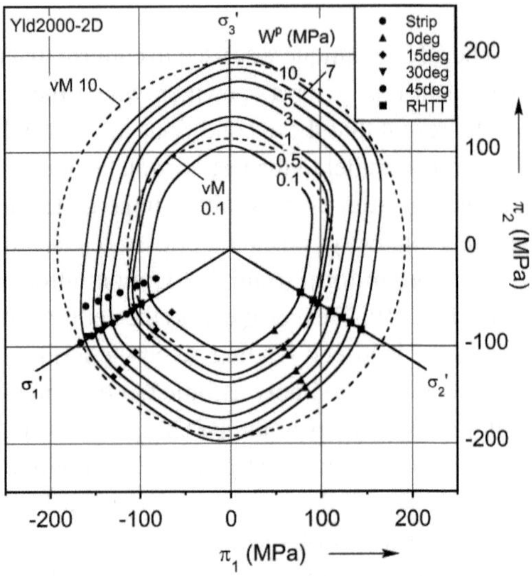

Fig. 8 Yield locus of the Al-6061-T4 tube on the π-plane, including the RPST experiments and the Yld2000-2D material model [8]

4 Conclusions

The RHTT and RPST experiments can be used to probe the hoop response and the yield locus of a tube material, respectively. The experiments require simple, general purpose testing equipment. The RHTT experiment can be used to assess the response of weld seams in ERW or porthole-die-extruded tubes. The RPST experiment provides data under tension and shear and can be used in conjunction with the tube inflation experiments to provide a more complete picture of the tube material anisotropy.

5 Acknowledgement

This work was supported by the U.S. National Science Foundation under the GOALI Grant CMMI 1031169. The extruded Al-6061-T4 tubes were provided by Dr. Cedric Xia of Ford Motor Company.

References

1. Kuwabara, T., Hashimoto, K., Iizuka, E., and Yoon, J.W., 2011, Effect of anisotropic yield functions on the accuracy of hole expansion simulations. J. Mat's Processing Tech., 211, 475–481.
2. Kuwabara, T.: Biaxial Stress Testing Methods for Sheet Metals. In Comprehensive Materials Processing; Van Tyne, C. J., Ed.; Elsevier Ltd., 2014; Vol. 1, pp 95–111.
3. Korkolis, Y.P. and Kyriakides, S., 2008, Inflation and burst of anisotropic aluminum tubes for hydroforming applications. Int'l J. Plasticity 24/3, 509–543.
4. Lee, M.G., Korkolis, Y.P., and Kim, J.H., 2014, Recent developments in hydroforming technology. Proc. Inst. Mech. Eng., Part B: J. Eng. Manuf., 0954405414548463.
5. Arsene, S. and Bai, J., 1996. A New Approach to Measuring Transverse Properties of Structural Tubing by a Ring Test, J. Test. & Eval. 24, 386–391.
6. Dick, C.P. and Korkolis, Y.P., 2014. Mechanics and full-field deformation study of the ring hoop tension test, Int'l J. Solids & Struct, 51, 3042–3057
7. Dick, C.P. and Korkolis, Y.P., 2015, Strength and ductility evaluation of cold-welded seams in aluminum tubes extruded through porthole dies, Materials & Design, 67, 631–636
8. Dick, C.P. and Korkolis, Y.P. Anisotropy of thin-walled tubes by a new method of combined tension and shear loading, (submitted)
9. Barlat, F., Brem, J.C., Yoon, J.W., Chung, K., Dick, R.E., Lege, D.J., Pourboghrat, F., Choi, S.-H., and Chu, E., 2003, Plane Stress Yield Function for Aluminum Alloy Sheets-Part I: Theory, Int'l J. Plasticity, 19, 1297–1319.

5 Acknowledgement

An Innovative Procedure for the Experimental Determination of the Forming Limit Curves

Dorel Banabic, Lucian Lazarescu, and Dan-Sorin Comsa

1 Motivation

The Forming Limit Curve (FLC) is an instrument widely used for the quantitative description of the sheet metal formability [1, 2]. Various methodologies have been proposed for the experimental determination of the FLCs. The FLC should cover the entire deformation domain specific to sheet metal forming processes [3]. In general, the strain combinations span between those induced by uniaxial and biaxial surface loads [4]. The experimental methods commonly used for investigating the deformation domain of the FLCs are presented in the following section.

The uniaxial tension of flat specimens having circular notches (proposed by Brozzo and de Lucca [5]) allows the exploration of the tension-compression range (left branch of the FLC). By using relatively wide specimens, it is also possible to reach the plane strain point. The positive-positive region (right branch) of the FLC can be reproduced in a hydraulic bulging device equipped with dies having circular or elliptic apertures. Different load paths belonging to the tension-tension domain result by varying the eccentricity of the elliptic aperture [6]. Other procedures used for the experimental determination of the FLCs are those based on the punch stretching principle. Keeler [7] used circular specimens and spherical punches with different radii in order to modify the load path. In general, the punch stretching test developed by Keeler is able to investigate only the right end of the tension-tension FLC branch. Hecker [8] extended Keeler's methodology to the whole tension-tension domain by improving the lubrication on the contact surface between punch and specimen. A notable development of this experimental procedure is due to Nakazima [9]. He used a hemispherical punch having a constant radius in combination with rectangular specimens with different widths. In this way, Nakazima was able to

Dorel Banabic ✉ · Lucian Lazarescu · Dan-Sorin Comsa
Technical University of Cluj-Napoca, Cluj Napoca, Romania
e-mail: banabic@tcm.utcluj.ro

A. E. Tekkaya et al. (eds.), *60 Excellent Inventions in Metal Forming*,
DOI 10.1007/978-3-662-46312-3_2

explore both the tension-compression and the tension-tension domains of the FLC. By using circular specimens with lateral notches, Hasek [10] removed the main disadvantage of the Nakazima test, namely the wrinkling of wide specimens. In order to reduce the frictional effects in the case of the flat punch drawing test, Marciniak [11] developed the so-called double blank method (specimen placed on the top of a carrier blank). He was able to obtain different load paths by modifying the cross section of the punch (circular, elliptic or rectangular). Grosnostajski [12] improved Marciniak's test by changing the geometry of the specimen and carrier blank. One may notice that none of those procedures are able to reproduce the whole strain domain of the FLC.

Therefore, an innovative procedure for the experimental determination of entire deformation range of the FLCs has been proposed [13]. The procedure is based on the hydraulic bulging of a double specimen.

2 New Procedure

The formability test proposed by the authors is based on the hydraulic bulging principle. It is well known the fact that, in its standard version, hydraulic bulging is only able to reproduce a biaxial tension in the polar region of the specimen. The capabilities of this test can be extended if the specimen has a pair of holes pierced in symmetric positions with respect to the pole. Of course, the presence of the holes creates a technical problem, namely the need of sealing the hydraulic chamber of the experimental device. The solution of this problem consists in placing a carrier blank under the pierced specimen. The carrier acts both as a transmitter of the increasing pressure developed by the hydraulic agent and a deformable punch.

Figure 1 presents the principle of the new formability test. One may notice that the pierced specimen and the carrier blank are firmly clamped between the die and the blank

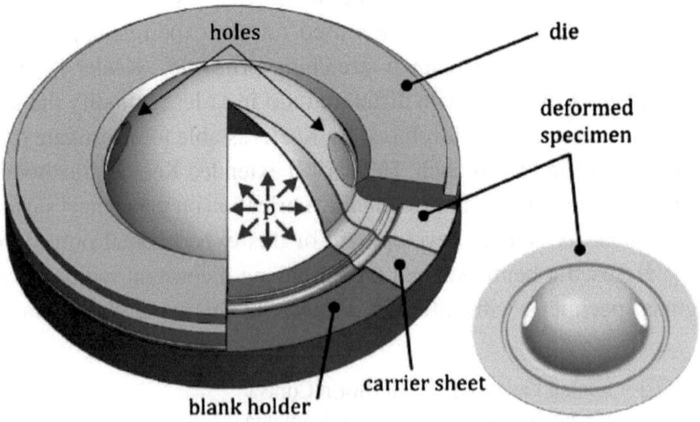

Fig. 1 Schematic view of the new formability test [13] (Copyright of Figures: Elsevier 2013)

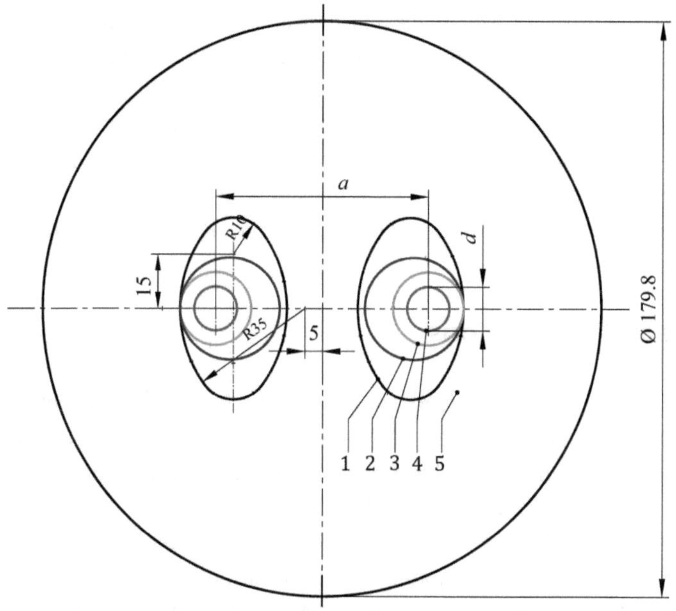

Fig. 2 Strain paths obtained in the hydraulic bulge tests: comparison between the numerical simulation and experimental data [13] (Copyright of Figures: Elsevier 2013)

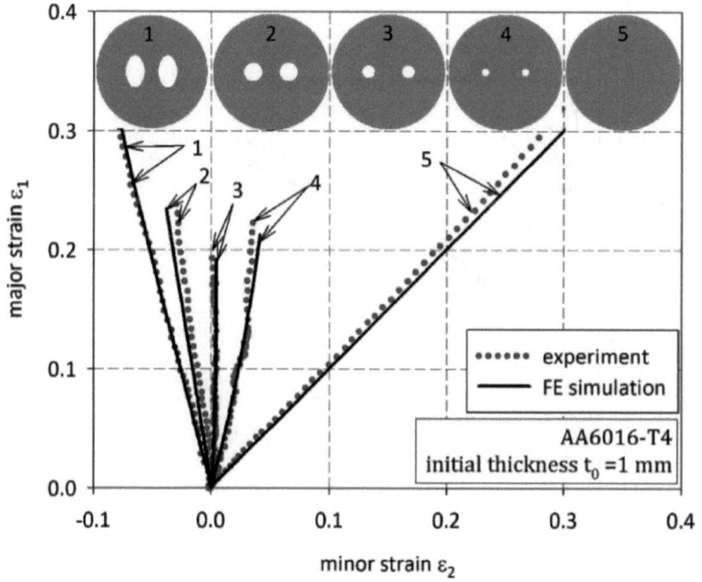

Fig. 3 Geometric parameters of the specimens [13] (Copyright of Figures: Elsevier 2013)

holder. The bulging of the specimen and carrier is caused by the increasing pressure applied on the lower surface of the carrier. The geometric characteristics of the pierced specimen are the dimensions and the reciprocal distance of the holes. By varying these parameters as shown in Fig. 3, it is possible to obtain different load paths during the hydraulic bulge test and, consequently, to investigate the whole deformation range of an FLC. The methodology proposed in [13] allows determining at least five different points on the FLC, in accordance with the specification of the standard ISO 12004-2 (see Fig. 2).

3 Results and Applications

Figure 4 compares the FLCs obtained using the methodology proposed by the authors and the Nakazima test (according to the specifications of the international standard ISO 12004-2). In both cases, the limit strains have been measured using the ARAMIS system. Each measuring point represents the mean value of three specimens. The limit strains have been determined according to the standard ISO 12004-2 methodology implemented in the ARAMIS system. One may notice that the limit points obtained in the plane-strain case are almost the same for both methodologies. In the uniaxial and biaxial regions, the FLC obtained when using the new methodology is slightly translated to lower values of the major principal strain. This fact is in agreement with the theoretical considerations presented in the literature (see, for example, Fig. 9.7 in [14]). Because the fracture takes

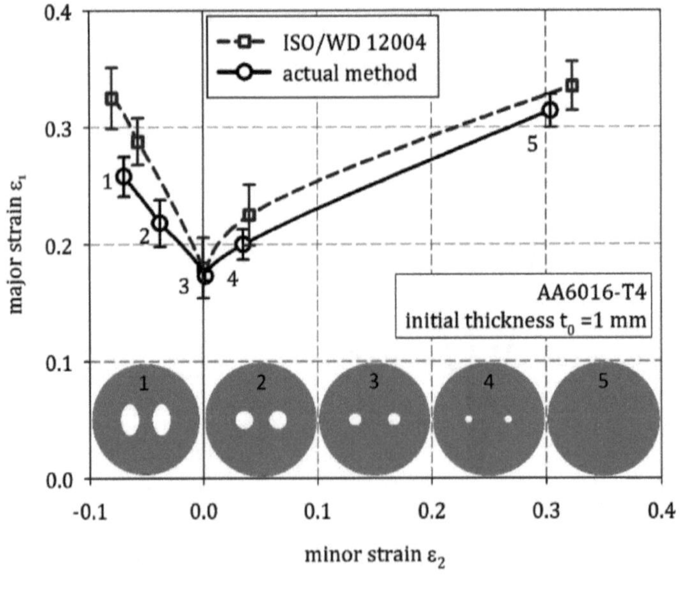

Fig. 4 Forming Limit Diagram of the AA6016-T4 sheet metal [13] (Copyright of Figure: Elsevier 2013)

place at the pole in the case of the hydraulic bulge test, the corresponding limit strains are smaller than in the case of the hemispherical punch stretching. When the rigid punch is used (Nakazima test), the frictional interactions reduce the strain level in the polar region and distribute the strain over a larger area. This leads to a better formability of sheet metals subjected to hemispherical punch stretching.

4 Potential

The new method has numerous potentialities, qualifying it as a testing procedure for the FLC determination:

1. Capability of investigating the whole strain range specific to the sheet metal forming processes
2. Simplicity of the equipment and of the specimen configuration
3. Reduction of the parasitic effects induced by the frictional interactions between the specimen and other components of the experimental device
4. Occurrence of the necking and fracture in the polar region of the specimen

5 Conclusion

An innovative procedure for the experimental determination of the FLCs has been proposed. The methodology is based on the hydraulic bulging of a double specimen. The upper blank has a pair of holes pierced in symmetric positions with respect to the center, while the lower blank acts both as a carrier and a deformable punch. By modifying the dimensions and reciprocal position of the holes, it is possible to investigate the whole strain range specific to the sheet metal forming processes.

6 Acknowledgement

The work has been kindly supported by the Romanian National Research Council (CNCS) under the grant number PCCE 100/2009.

References

1. Banabic D., Bünge H.J., Pöhlandt K., Tekkaya A.E., 2000, Formability of Metallic Materials. Plastic Anisotropy, Formability Testing, Forming Limits, Springer, Berlin Heidelberg.
2. Banabic D., 2010, Sheet Metal Forming Processes, Springer, Berlin Heidelberg.
3. Banabic D., Barlat F., Cazacu O., Kuwabara T., 2010, Advances in anisotropy and formability, International Journal of Material Forming, 3 (2), 165–189.

4. Bruschi S., Altan T., Banabic D., Bariani P.F., Brosius A., Cao J., Ghiotti A., Khraisheh M., Merklein M., Tekkaya E., 2014, Testing and Modeling of Material Behavior and Formability in Sheet Metal Forming Processes, Annales of CIRP, 63 (1), 727–749
5. Brozzo P., de Lucca B., 1971, On the interpretation of the formability limits of metals sheets and their evaluation by means of elementary tests, Proc. ICSTIS, Tokyo, 966–988.
6. Ranta-Eskola A. J., 1979, Use of the hydraulic bulge test in biaxial tensile testing, International Journal of Mechanical Sciences, 21 (8), 457–465.
7. Keeler S.P., 1964, Plastic instability and fracture in sheet stretched over rigid punches, Trans. of the ASM, (56), 25–48.
8. Hecker S. S., 1972, A simple FLC technique and results on some aluminium alloy, Proc. of the 7th IDDRG Congress, Amsterdam, 51–71.
9. Nakazima K., Kikuma T., Kasuka K., 1968, Study on the formability of steel sheets, Yawata Technical Report, Nr. 264, 141–154.
10. Hasek V., 1978, Untersuchung und theoretische Beschreibung wichtiger Einflussgrossen auf das Grenzformaenderungschaubild, Blech, 25 (5,6,10,12), 213–220, 285–292, 493–499, 619–627.
11. Marciniak Z., Kuczynski K., Pokora T., 1973, Influence of the plastic properties of material on the FLD for sheet metal in tension, International Journal of Mechanical Sciences, 15 (10), 789–805.
12. Grosnostajski J., Dolny A., 1980, Determination of FLC by means of Marciniak punch, Mem. Sci. Rev. Met., 4, 570–576.
13. Banabic D., Lazarescu L., Paraianu L., Ciobanu I., Nicodim I., Comsa, D.S., 2013, Development of a new procedure for the experimental determination of the Forming Limit Curves, Annales of CIRP, 62 (1), 255–258.
14. Marciniak Z., Duncan J.L., Hu S.J., 2002, Mechanics of Sheet Metal Forming, Butterworth-Heinemann, Oxford.

Sheet Material Characterization with the In-Plane Torsion Test: Cyclic Loading, Grooved Specimen and Twin Bridge Specimen

Heinrich Traphöner, Qing Yin, and A. Erman Tekkaya

1 Motivation

In today's industry the Finite-Element-Analysis (FEA) is essential for the design of parts and processes. Typical operations are the analysis of springback or failure. Due to the elastic component of the strain remaining after the forming of sheet metal components internal stresses occur, which ensure that the deformed components spring back elastically, whereby the final shape of the contour differs from the tools [1]. For the description of the plastic material behavior practically often isotropic hardening models are used as the determination of cyclic flow curves and characteristics necessitate high experimental effort. The correct prediction of springback cannot be achieved with simple isotropic hardening models, but only with isotropic-kinematic models for which the cyclic flow curves must be known [1]. For the determination of material parameters under cyclic load different tests can be used. These include tension-compression tests, cyclic shear tests or cyclic bending tests [2]. Different disadvantages are connected with most of the tests. For example, bending tests have an undefined stress state [2] and tension-compression tests have to be supported to prevent buckling under compression load [3] or the deformation zone has to be miniaturized [4].

The in-plane torsion test and its applications for sheet metal characterization will be presented. Beside the standard specimen that is used to measure multiple flow curves with a single specimen, two modifications of the test were introduced to measure anisotropic material behavior and to characterize damage under ideal shear load.

Heinrich Traphöner · Qing Yin · A. Erman Tekkaya ✉
TU Dortmund, Dortmund, Germany
e-mail: Erman.Tekkaya@iul.tu-dortmund.de

A. E. Tekkaya et al. (eds.), *60 Excellent Inventions in Metal Forming*,
DOI 10.1007/978-3-662-46312-3_3

2 Process Idea

The in-plane torsion test is a process for material characterization under simple shear load. It was first introduced by Marciniak and further developed by Tekkaya et al. [5, 6]. Figure 1a illustrates the principle of the test. A circular specimen is internally and externally clamped concentrically. By fixing the internal and rotating the external clamping device, an elastic and plastic deformation in the annular region between the clampings occur. Highest stress and strain appear at the inner clamp. In depency of the resulting torque M, the sheet thickness s and the radial distance r to the center of the radial motion the shear stress τ is calculated with

$$\tau = \frac{M}{2\pi s r^2}.$$

A decreasing stress gradient is formed between the clampings. In Fig. 1b a radial line is shown on the torsion specimen, which is distorted by rotation of the outer clamp. By tracking the deformation along the radial line, the shear strain can be determined. It is calculated with

$$\gamma = \tan(\eta) = r\frac{\mathrm{d}\vartheta}{\mathrm{d}r}.$$

With increasing radial distance the strain decreases and becomes purely elastic at a safe distance. Through continuous measurement of the deformation and shear stress at a constant radial distance, the flow curve of a material is determined. If the load direction is reversed, it is possible to also determine cyclic flow curves.

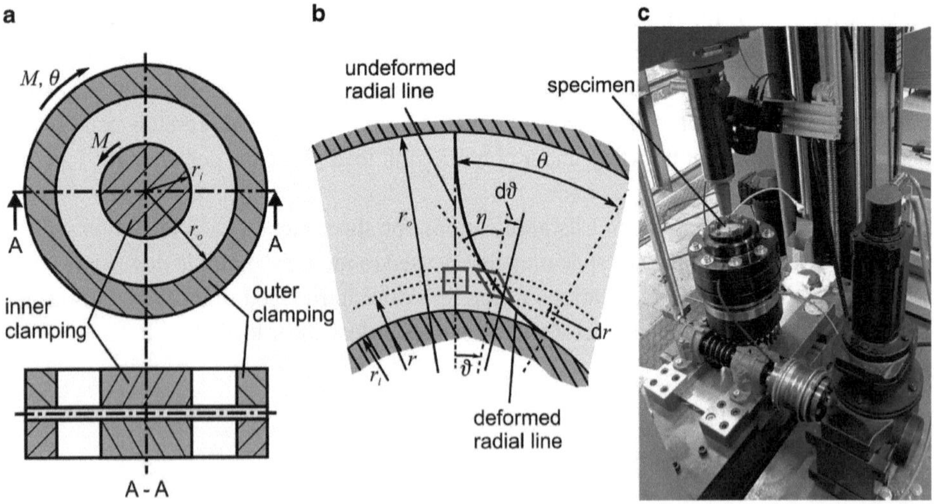

Fig. 1 **a** Process principle of the in-plane torsion test, **b** deformation of a radial line, **c** experimental setup [2]

The experimental test setup for the results of this work is described by Yin and uses the optical strain measurement for the determination of strain on the specimen [2]. Figure 1c shows the experimental setup. The entire construction is integrated in a universal testing device. A servo motor drives a worm gear to rotate the outer part of the torsion device, while rigid inner punches are mounted on the traverses of the testing machine. By moving down the traverse, the upper punch is used to clamp the specimens' center. Cameras are mounted at the upper traverse to continuously measure the specimens' displacement.

3 Results and Applications

Different specimen designs are used with the in-plane torsion test. Each specimen has its own application for the sheet metal characterization which will be presented below.

Determination of multiple shear curves with a full specimen

While the results of other shear tests are limited to a single cyclic flow curve for each specimen, an unlimited amount of curves can be determined with different pre-strains with the in-plane torsion test. Figure 2 illustrates this principle. By continuously evaluating the strain at different radial distances on a specimen, it is possible to determine cyclic flow curves with different pre-strains (Fig. 2b).

The determined cyclic flow curves are used to identify material parameters from isotropic-kinematic material models. It can be identified by analytical or inverse methods from the experimental curves. Due to the high number of experimental curves, dependencies of pre-strain can be easily investigated using fewer experiments. As an application to the use of multiple shear-curves the dependency of the Bauschinger-coefficient on pre-strain is showed in Fig. 2c for the aluminum alloy AA5182 and the DX54 steel. Each data point has been determined with one cyclic curve, but with only a few specimens. The higher the materials are pre-strained, the lower the Bauschinger-coefficients are for both materials.

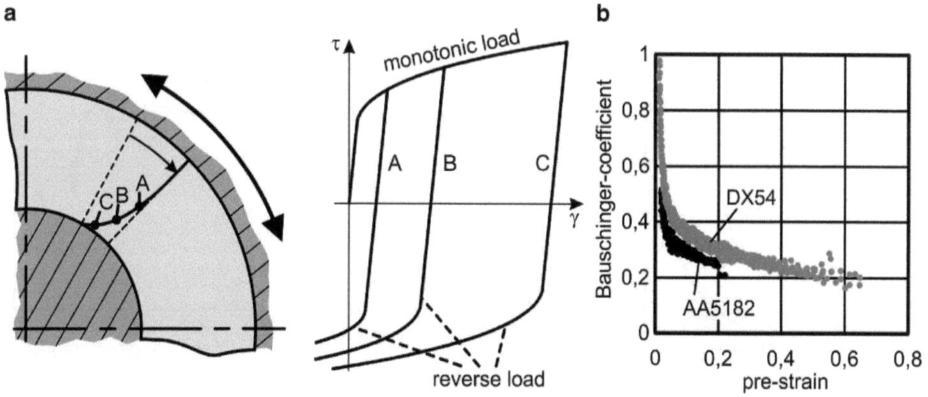

Fig. 2 **a** determing multiple shear curves [2], **b** Bauschinger-coefficient depending on pre-strain

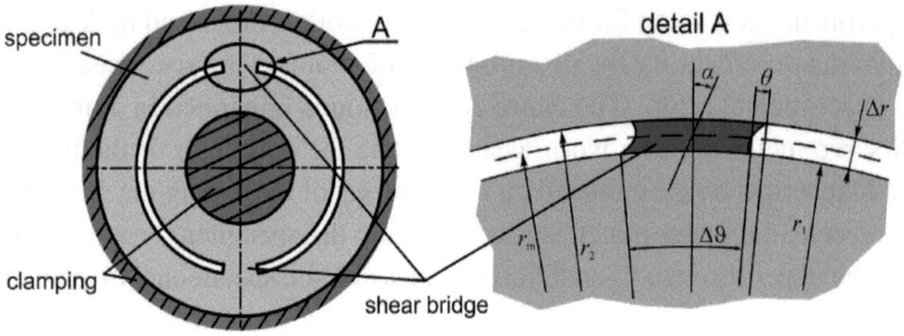

Fig. 3 Twin bridge shear specimen [2]

Determination of anisotropic material behavior with a twin bridge specimen

As a modification of the standard test, a twin bridge shear specimen has been developed to determine anisotropic material behavior [7]. The test specimen is shown in Fig. 3. Two shear bridges are arranged symmetrically by making partly tangential slots in the circular shape of the original test specimen. Hence the strain is localized in the bridges. By variing the position of the bridges an examination of anisotropic material behavior with defined direction is possible. Furthermore a reduction of torque permits the measurement of flow curves from high-strenght steel, since the danger of sliding through is reduced. Due to the symetric shape there is no danger of buckling and unwanted reaction moments, as are described for other cyclic tests.

Determination of damage under ideal shear load with a grooved specimen

An other introduced modification of the standard in-plane tortion test is a specimen with circumferential groove as shown in Fig. 4a [2]. A round circular groove has been cut out of one side of the specimen. Hence maximum shear strain located at the bottom of the groove. Crack initializion occur under ideal shear load without the influence of unwanted loads, as for example due to the clamping force. Failure due to shear load can be achieved reproducible as shown in Fig. 4c. A DP1000 steel is used to identify the point of fracture. One can see that the angles at fracture are similar within the three experimental curves.

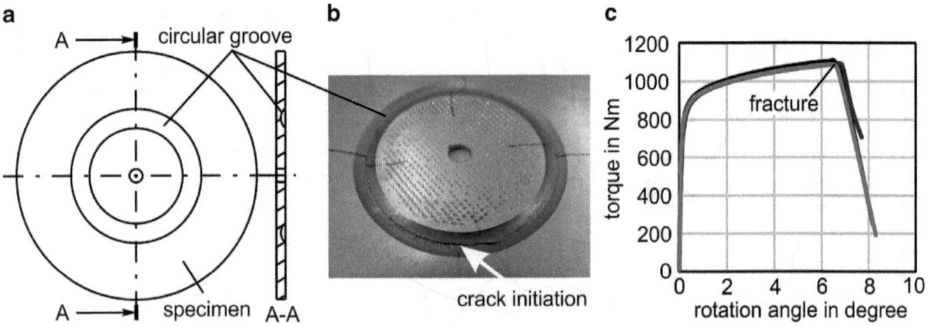

Fig. 4 **a** Specimen with round circular groove, **b** fracture of DP600 steel at the bottom of the groove **c** Measured torque and rotating angle to determine damage initiation of DP1000 steel [2]

The specimen with groove is also used to determine flow curves under shear load for high-strength steel since less clamping force is needed with less danger of slipping through.

4 Conclusion

The in-plane torsion test is simple in design and execution. It offers high potential for the identification of material parameters under cyclic ideal shear load and has significant advantages compared to other tests with cyclic load. The implementation of the reversal of load is possible without additional effort by changing the direction of rotation. High strain can be achieved in both directions of load. The optical strain measurement allows the determination of multiple flow curves with different pre-strain with a single test. Two modifications of the test specimen are introduced. A twin bridge specimen is used to examine anisotropic materials without unwanted additional stresses. Specimens with cylindrical groove are used to examine damage behavior under ideal shear load. Both are also applied to measure flow curves of high strength steels.

5 Acknowledgement

The authors gratefully acknowledge funding by the German Research Foundation (DFG) through Project PAK 250 and funding by AiF in the context of the program for support of Industrial Collective Research and Development (IGF) through the IGF project 17375 N – EFB-Nr. 06/112 of the European Research Association for Sheet Metal Working (EFB).

References

1. Vladimirov, I.N, Pietryga, M.P, Reese, S., 2009. Prediction of springback in sheet forming by a new finite strain model with nonlinear kinematic and isotropic hardening. Journal of Materials Processing Technology, vol. 209, 4062–4075.
2. Yin, Q., 2014. Verfestigungs- und Schädigungsverhalten von Blechwerkstoffen im ebenen Torsionsversuch. Shaker, Dortmund.
3. Yoshida, F., Uemori, T., Fujiwara, K., 2002. Elastic–plastic behavior of steel sheets under in-plane cyclic tension–compression at large strain. International Journal of Plasticity, vol. 18, 633–659.
4. Staud, D., Merklein, M., Borsutzki, M., Geisler, S., 2009. Zug-Druck-Versuche an Miniaturproben zur Erfassung von Parametern für kinematische Verfestigungsmodelle. Tagungsband Werkstoffprüfung, Stahleisen GmbH Düsseldorf, 211–218.
5. Marciniak, Z., 1961. Influence of the sign change of the load on the strain hardening curve of a copper test piece subject to torsion. Archiwum Mechaniki Stosowanej, vol. 13, 743–752.
6. Tekkaya, A.E, Pöhlandt, K., Lange, K., 1982. Determining stress-strain curves of sheet metal in the plane torsion test. CIRP Annals-Manufacturing Technology, vol. 31, 171–174.
7. Brosius, A., Yin, Q., Güner, A., Tekkaya, A.E, 2011. A New Shear Test for Sheet Metal Characterization. steel research int., vol. 82, 323–328.

Friction Analysis in Bulk Metal Forming

Laurent Dubar, André Dubois, and Mirentxu Dubar

1 Motivation

Friction is a key point in the setting up of a metal forming process. Friction may affect the plastic flow of the material, induce untimely tool wear, lead to workpiece surface defects, etc. The easiest and cheapest way to control friction is to use an efficient lubricant. Nonetheless, due to the great variety of forming processes, a lot of families of lubricants are commercialized: liquid or solid, environmentally friendly or not, resistant to high temperature, etc. Making a choice is then very difficult and uncertain: a good lubricant for a process can be ruinous for another. Testing lubricants directly on the industrial process could be a solution if it was not so expensive and time consuming. So the choice of a lubricant is often based on laboratory test results. But testing lubricants in a lab lead to uncertain results as contact conditions may be far from the industrial ones [1]. Performing friction tests where the lubricant behaves the same way as it will on the process is still a real challenge. As an example, the ring compression test was commonly used to test lubricants. This test is easy to manage and does not require specific apparatus. But it cannot dissociate the effect of contact pressure and plastic strain on friction and it leads to small sliding distance and small surface enlargement. The test can then provide good results for a lubricant which will be unable to behave properly in a real industrial process.

2 Methodology

Friction depends on so many physical, chemical and mechanical parameters that it is not possible to simulate all of them on a tribological test. Consequently, it is not possible to

Laurent Dubar ✉ · André Dubois · Mirentxu Dubar
University of Valenciennes and Hainaut Cambresis, Valenciennes, France
e-mail: laurent.dubar@univ-valenciennes.fr

A. E. Tekkaya et al. (eds.), *60 Excellent Inventions in Metal Forming*,
DOI 10.1007/978-3-662-46312-3_4

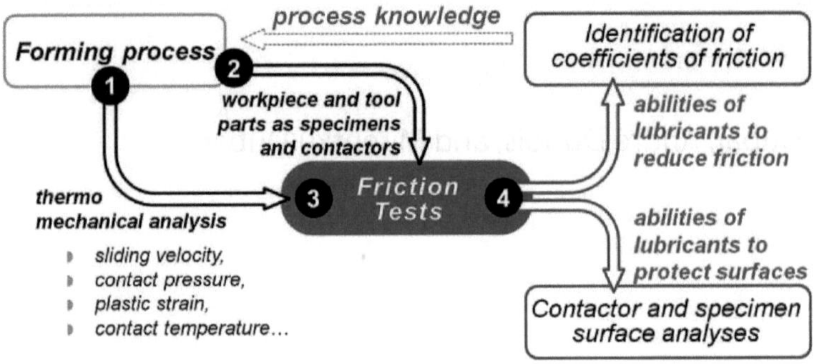

Fig. 1 Methodology to simulate industrial conditions of contact in a laboratory friction test

analyze the effect of all of these parameters on the efficiency of a lubricant. The idea is then to dissociate the mechanical parameters from the physical-chemical parameters by the use of real industrial surfaces for specimens and contactors. This lead to a testing methodology which operates in four steps (Fig. 1). First, a mechanical analysis of the process is performed to identify the level of key parameters to be simulated on the friction test. Depending if the forming process is cold or hot, key parameters are sliding velocities, tool and workpiece temperatures, contact pressure, strain rate, plastic strain and surface enlargement [2–4]. Second, specimens and contactors are machined from industrial workpieces and tools. By acting this way, the materials used on the bench test, their roughness, chemical reactivity, oxidation, hardness, etc., are the same as on the process. Third, the friction bench test loading is adjusted in order to simulate the key mechanical parameters identify at step 1, and tests are performed. As a consequence, the friction test must be designed in order to allow independent adjustments in terms of temperature, sliding velocity, strain and stress, and not just in normal force like most pin-on-disk tribometers. Fourth, test results are analyzed to deliver quantitative and qualitative information able to evaluate the efficiency of lubricants to reduce friction and to preserve the integrity of the contacting surfaces.

3 Test Benches and Applications

Four different test benches were designed to study cold and hot forging, cold and hot rolling (Fig. 2). All of them are based on the same principle:

- The contactor is fixed on a load cell specifically design for the bench test (Fig. 3). The active surface of contactor (the surface that will come in contact with specimens) is the original tool surface (Fig. 4). A hole is machined beneath active surface to place a heating cartridge. Contactor surface can be heated up to 300°C. The specimen is fixed on the moving part of a tensile testing machine. Specimens can be heated by induction up to 1200°C.

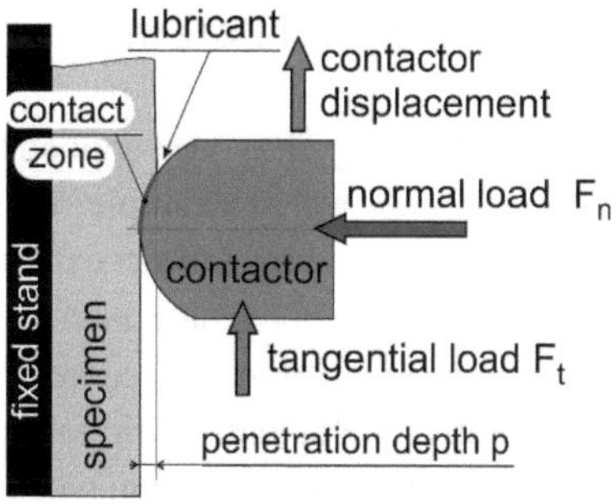

Fig. 2 Schematic view of the friction test

Fig. 3 Contactor stand with integrated load cell

Fig. 4 A real tool and a contactor machined from it

- Lubricants are applied either on contactors (hot forging/rolling) or on specimens (cold forging/rolling). They are sprayed or applied with a brush. The quantity of lubricant is controlled before testing (mass, thickness).
- The contactor moves towards the specimen, slightly penetrates its surface and then slides against it with a constant velocity and a constant penetration depth. The penetration depth is chosen in order to create a given plastic strain in the vicinity of the specimen contact surface. During the tests, normal and tangential forces, displacement and temperature are acquired. The contact surface is around 10 mm^2.
- After the tests, coefficients of friction related to various friction laws are identified from normal and tangential loads, contactor shape and penetration depth [5]. Contactor and tool surfaces are analyzed by SEM-EDS to detect residues of lubricant, occurrence of wear, galling, scratches or any other surface defect that may affect the industrial process efficiency. Finite element computations of the friction tests are performed to access to local data such as effective plastic strain, bulk yield stress, strain energy, damage [6], roughness flattening [7,8]. The accuracy of these computation is increased by the use of reliable stress-strain curves identified in the vicinity of the surface [9, 10].

4 Results and Potential

The methodology has been applied to study the lubrication of various cold and hot forming processes. In a first example, two consecutive forming sequences are studied: a wire drawing followed by a direct extrusion. Tests are performed on specimens that are coated in an industrial phosphate bath with an age varying from 2 to 30 days. Results show that the ageing of the surface treatment baths has no significant influence on the average Coulomb's coefficient of friction for the wire drawing operation (Fig. 5). However, the scatter observed is significant when the baths are new or at the end of cycle (30 days). This result is explained by a poor growth rate of the phosphate grains during these periods [11]. The coefficient of friction for the direct extrusion operation increases with ageing of the phosphate bath but the scatter observed is independent of the ageing. Indeed, surface analyzes shows that the drawing sequence prepared the coated wire by spreading out the coating, leading to more uniform distribution of phosphate grains [5].

The influence of the number of produced parts on friction is proposed in a second example. The friction test is adjusted to simulate the cold forming of the hexagonal head of a screw (Fig. 4). Two set of experiments were performed. The first set involved contactors machined from PVD TiN coating. Contactors of the second set were from CVD TiN coatings. The CVD TiN coated tool shown better results in terms of friction and lifetime (Fig. 6). Significant adhesion was observed followed by chipping of the transferred steel much later than for the PVD coating. This phenomenon leads to large fluctuation of friction forces which resulted in coefficients of friction varying from almost zero (lost of contact) to 0.6. In this case of forming process, the CVD TiN coating was the most appro-

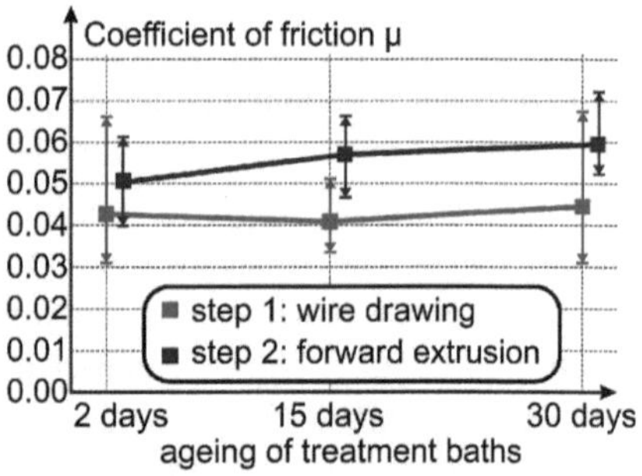

Fig. 5 Evolution of the coefficient of friction as a function of the ageing of the soaping phosphate line from [5]

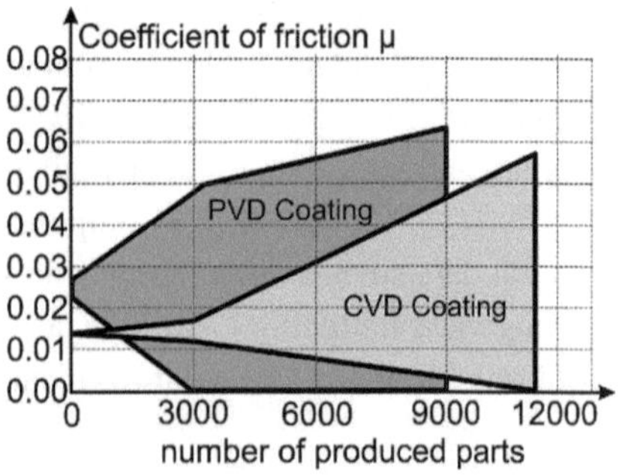

Fig. 6 Compared evolution of the friction coefficients between the two coatings from [12]

Table 1 Evolution of coefficient of friction and critical sliding length for several configurations of lubrication by dispersions of graphite in water [3]

	Lubricant film thickness			Lubricant particles size		
	10 µm	30 µm	40 µm	small	medium	large
Coefficient of friction µ	$0.46^{\pm 0.04}$	$0.28^{\pm 0.03}$	$0.16^{\pm 0.01}$	$0.20^{\pm 0.03}$	$0.28^{\pm 0.03}$	$0.36^{\pm 0.02}$
Critical length Lc	$0.83^{\pm 0.04}$	$1^{\pm 0.0}$	$1^{\pm 0.0}$	$1^{\pm 0.0}$	$1^{\pm 0.0}$	$0.9^{\pm 0.04}$

priate to delay friction scattering. This result was in good accordance with the obtained industrial tool lives [12].

A third example is related to the chemical composition of lubricants used for hot forging applications. Tests were performed in order to simulate the contact conditions encountered near the flash zone of the die, where tool wear is maximal and lubrication is the most critical. Tests were performed on steel specimens heated to 1100°C. Lubricants were 5 % of graphite in water. Two sets of experiments were run. The first one involved a medium particle size graphite and lubricant film thickness increasing from 10 to 40 µm. The second set involved small, medium and large particle size graphite and a lubricant film thickness equal to 30 µm. Main results were the coefficients of friction and the normalized sliding lengths of the contactors before any scratch appears along specimen surface (Lc). Critical sliding length Lc equal to 1 means that no scratch is observed: the lubricant perfectly protects specimen surface during the whole test. Critical sliding length Lc equal to 0 means scratches occur as soon as the contactor slides against the specimen. Results highlight that the film thickness of the lubricant layer has to be greater than or equal to 30 µm to prevent any scratch (Table 1). For thinner layer, the graphite is quickly "consume" during the sliding contact leading to the occurrence of scratches and an increase of the coefficient of friction. Particle sizes of graphite play a role in the structure of the graphite layer. Large particle sizes lead to heterogeneous graphite layers [13]. Parts of contactor surface are not protected, direct metal-to-metal contacts occur and scratches appear on the specimen surface, leading to an increase of the friction forces.

5 Conclusion

A methodology has been proposed to study in a reliable way the tribology of metal forming processes on laboratory friction tests. The methodology relies on the use of industrial workpieces and industrial tools as specimens and contactors. Specific friction bench tests have been designed so that they are adjustable in sliding velocities, contact pressure, plastic strain and temperature. Friction tests provide information such as coefficient of friction and sliding length before occurrence of scratches. These results characterize the ability of a lubricant to reduce friction and to protect tool and workpiece surfaces. The methodology has been applied with success to increase the knowledge of the lubrication of cold and hot forming operation.

6 Acknowledgements

The authors gratefully acknowledge for their support: the European Community, the Ministry of Higher Education and Research, and the National Center for Scientific Research, the International Campus on Safety and Intermodality in Transportation (CISIT), the French Nord-Pas-de-Calais Region and the Regional Delegation for Research and Technology.

References

1. P Groche, C Müller, J Stahlmann, S Zang, Mechanical conditions in bulk metal forming tribometers – Part one, Tribology International, Volume 62, 2013, Pages 223–231
2. L Dubar, A Dubois, M Dubar, Friction and wear phenomena in cold metal forming: an integrated approach, Proceedings of the Institution of Mechanical Engineers, Part B: Journal of Engineering Manufacture, volume 220, 2006, pp 1–10
3. A Dubois, M Dubar, L Dubar, Warm and Hot Upsetting Sliding Test: Tribology of Metal Processes at High Temperature, Procedia Engineering, Volume 81, 2014, 1964–1969
4. A Dubois, E Luc, M Dubar, L Dubar, C Thibaut, JM Damasse, Initiation of Sticking during Hot Rolling of Stainless Steel Plate, Procedia Engineering 81, 2014, 1958–1963
5. A Dubois, L Lazzarotto, L Dubar, J Oudin, A multi-step lubricant evaluation strategy for wire drawing–extrusion–cold heading sequence, Wear 249 (10), 951–961
6. M Dubar, S Huart, R Deltombe, A Dubois, L Dubar, Cold rolling of galvanized strips: A new approach to the evaluation of zinc fines formation, Journal of materials processing technology 177 (1), 2006, 505–508
7. A Belotserkovets, A Dubois, M Dubar, L Dubar, R Deltombe, H Vandekinderen, JM Damasse, 2D asperity deformation of stainless steel strip in cold rolling, International Journal of Material Forming, volume 1, 2008, 351–354
8. L Dubar, C Hubert, P Christiansen, N Bay, A Dubois, Analysis of fluid lubrication mechanisms in metal forming at mesoscopic scale, CIRP Annals-Manufacturing Technology 61 (1), 2012, 271–274
9. L Dubar, B Hauw, A Dubois, J Oudin, Numerical identification of coating bulk behaviour in metal forming, Computers & Structures, volume 79, 2001, 1027–1037
10. A Verleene, L Dubar, A Dubois, M Dubar, J Oudin, Hardening behaviour law versus rigid perfectly plastic law: application to a cold forging tool steel, International Journal of Plasticity, volume 8, 2002, 997–1011
11. L Lazzarotto, A Dubois, L Dubar, A Verleene, J Oudin, Testing of zinc phosphate/stearate coating according to true conditions of contact in wire drawing, Surface engineering, Volume 14, 1998, 335–338
12. M Dubar, A Dubois, L Dubar, Wear analysis of tools in cold forging: PVD versus CVD TiN coatings, Wear 259 (7), 2005, 1109–1116
13. E Daouben, A Dubois, M Dubar, L Dubar, R Deltombe, NGT Dinh, L Lazzarotto, Effects of lubricant and lubrication parameters on friction during hot steel forging, International Journal of Material Forming 1 (1), 2008, 1223–1226

Flow Stress Measurement in Upsetting Test with Grooved Platens

Kozo Osakada

1 Motivation

For estimating the load and the die pressure in metal forming, the flow stress of the deforming material under the working condition is the basis of calculation but it is difficult to measure the flow stress under the working condition, especially at large strains. Kudo et al. [1] determined accurate flow stress curves of various metals up to strains of about 4.0 by reshaping the deformed compression specimens by machining and renewing lubricant as shown in Fig. 1. This method, however, takes a long time and simple measuring are desired in the industry.

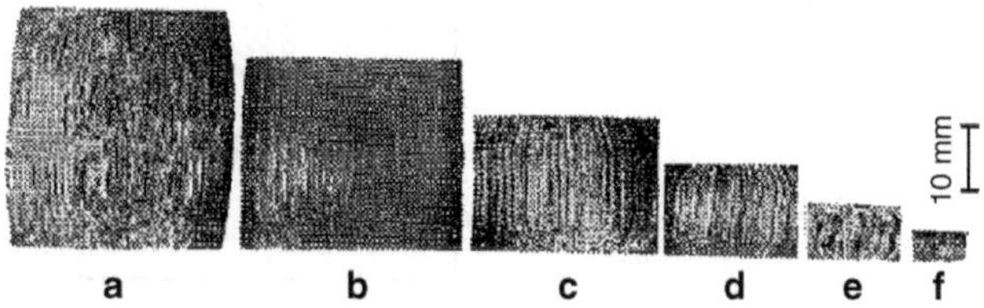

Fig. 1 Uniform compression test with repeated reshaping and lubrication [1]

Kozo Osakada ✉
Osaka University, Toyonaka, Osaka, Japan
e-mail: k.osakada@hm.h555.net

A. E. Tekkaya et al. (eds.), *60 Excellent Inventions in Metal Forming*,
DOI 10.1007/978-3-662-46312-3_5

2 Idea of Measuring Method

Kudo et al. [2] proposed a cold upsettablity test, in which a cylindrical specimen of D : H = 1 : 1.5 is compressed between the grooved platens shown in Fig. 2 by avoiding slipping completely irrespective of the frictional condition, and the critical reduction in height when a crack is first observed on the side surface is determined. This test is standardized by the Forging Committee of JSTP (Japan Society for Technology of Plasticity) and widely used to evaluate the cold forging steels in Japan.

Figure 3 illustrates the distributions of equivalent strain ε and strain rate $\dot{\varepsilon}$ in the specimen calculated by the rigid-plastic FEM at a reduction in height $e = 70\ \%$. Although the distributions are non-uniform, this test is simple to carry out and provides consistent result of load-stroke without being affected by friction. It is possible to calculate the average equivalent strain from this figure, and the average flow stress could be determined by dividing the measured average nominal working pressure by the constraining factor obtained with FEM simulation. Osakada et al. [3] proposed to use the curve of average flow stress and average equivalent strain of this test as an approximated flow stress curve.

By measuring the load P and the height h of specimen during upsetting, the average flow stress Y and the average equivalent strain ε in the specimen are determined by the following equations.

$$Y = \frac{1}{f} \cdot \frac{P}{A_0} \tag{5.1}$$

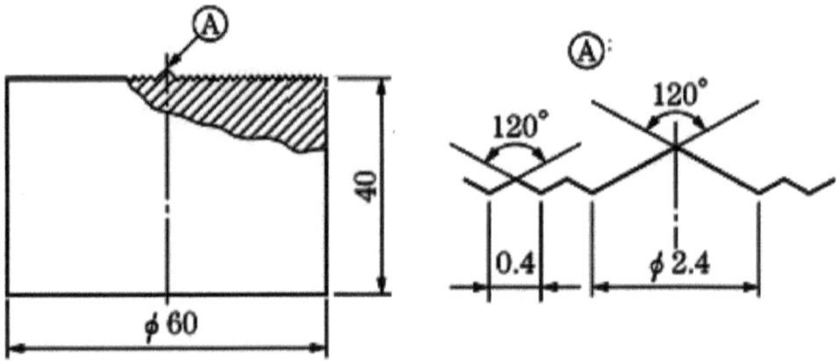

Fig. 2 Grooved platen for upsetting test (Source: Forging Commitee of JSTP)

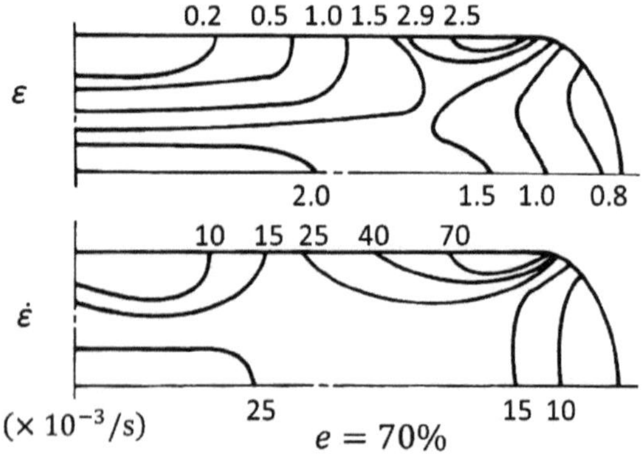

Fig. 3 Distributions of equivalent strain and strain rate in upsetting test specimen

Table 1 Average equivalent strain and constraining factor f in upsetting test

Reduction in height e (%)	Equivalent strain ε	f	Reduction in height e (%)	Equivalent strain ε	f
10	0.13	1.15	65	1.24	3.54
20	0.27	1.29	67.5	1.35	4.10
30	0.42	1.46	70	1.40	4.47
40	0.62	1.69	72.5	1.53	5.25
50	0.82	2.07	75	1.69	6.40
55	0.94	2.39	77.5	1.90	7.96
60	1.08	2.77	80	2.23	10.10

$$\varepsilon = F(e) = F\left(\frac{h_0 - h}{h_0}\right) \tag{5.2}$$

where h_0 and A_0 are the initial cross-sectional area of the specimen respectively, and P/A_0 is the nominal pressure. The nominal constraining factor f and the average equivalent strain ε are functions of reduction in height and they are determined by the use of the rigid-plastic FEM as given in Table 1.

3 Results at Room Temperature

The average flow stress–average equivalent curves obtained by the present method is compared with the flow curves of the uniform compression test with repeated reshaping and lubrication for various metals at room temperature as shown in Fig. 4. The two methods provide close results up to strains of about 2.0. Although cracks are observed in the upsetting test on the side surface of 2014Al and 2017Al specimens after certain reductions in height, the measured flow curves are not affected seriously as long as the cracks are confined to the part of bulged surface. It is confirmed that the error of the measured flow stress is less than ±5 %.

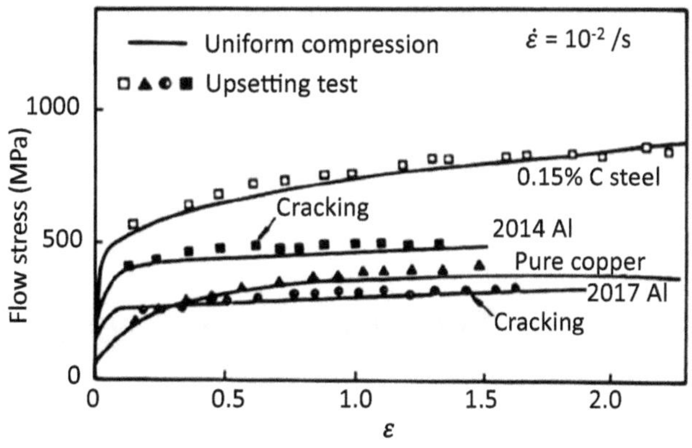

Fig. 4 Comparison of average flow stress curves obtained by upsetting test and flow stress curves in uniform compressing test with repeated reshaping and lubrication

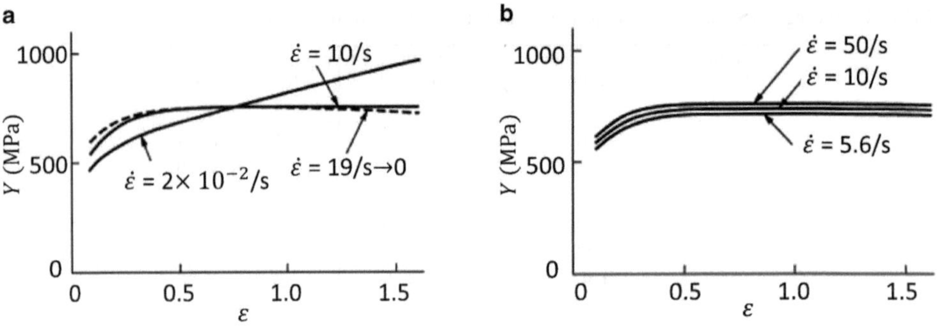

Fig. 5 Flow curves under different strain rates of 0.45 %C steel after shperoidizing annealing **a** Static testing machine and mechanical press, **b** Mechanical presses with different speeds

As the present method assumes that the average values of flow stress and strain in a non-uniformly deforming specimen are identical to those of the uniformly deforming specimen, this method provides a reasonable result only when the flow stress curve is smooth. Fortunately the flow stress curves do not have sudden changes under the cold forging conditions, and thus good coincident was obtained.

This measuring method of flow stress was recognized by the forging committee of JSTP as the standard procedure for testing cold forging steels together with determining ductility of the material. Kobe Steel Ltd. made a data acquisition system on the basis of this method [4,5] for customer service and some steel companies of manufacturing bar and wire for cold forging established similar systems.

By using this method, a cooperative work was carried out in JSTP to determine the flow stress and ductility of steels under practical forming condition [5,6]. Under the leadership of Prof. Y. Tozawa, 24 institutions including 12 from steel making and automotive industries measured the flow stress curves of 20 practical steels with various testing machines.

Figure 5a shows the flow curves of a spheroidized 0.45 %C steel at various strain rates measured in the cooperative work. Although work hardening was observed under the low strain rate of 2×10^{-2}/s up to a strain of 1.5, the flow stress under the press speed of 10/s became almost constant when the strains exceeds about 0.5 because work hardening and softening due to adiabatic temperature rise balanced. The change of strain rate from 19/s to zero during compression on a mechanical press did not result in significant difference. Figure 5b shows the flow stress curves determined with mechanical presses of different speeds. It was confirmed that almost all the tested steels exhibited saturation of work hardening at strains greater than 0.5 under the press working conditions.

4 Result in Warm Forging Temperatures

The proposed method can be carried out at elevated temperatures up to about 700°C, as long as the grooved tools are not deformed plastically. The flow stress curves of annealed 0.21 %C steel are measured on a mechanical press, and the flow stresses at strains of 0.5

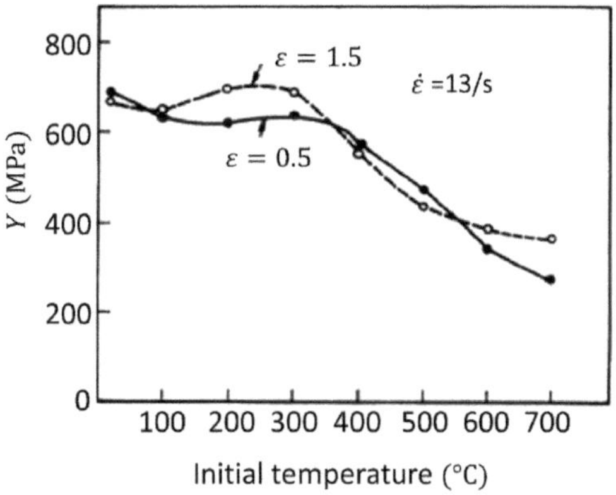

Fig. 6 Effect of initial temperature on flow stress of annealed 0.21%C steel at strains of 0.5 and 1.5

and 1.5 are plotted against the initial temperature in Fig. 6. It is observed that the peak of the flow stress due to blue brittleness shifts to lower initial temperature of specimen as the strain increases from 0.5 to 1.5. This phenomenon is related to adiabatic heating by plastic deformation, i. e. billet temperature rises as plastic strain increases.

5 Conclusion

The proposed method of determining flow stress curve is simple to carry out under actual forming conditions and has an accuracy of within ±5 %, and thus it is suitable for determining the flow stress curve under actual metal forming conditions, i. e. at large strains and high strain rate. Presently, this method is commonly used in Japan as a standard test of steels for cold forging.

References

1. Kudo, H., Sato, K., Sawano, I., 1965, Flow stress test and cold heading pressure of carbon steels. J. Japan Soc. Tech. Plasticity, 6 (56), 491–511. (in Japanese)
2. Kudo, H., Sato K., Aoi, K., 1968, A cold upsettability test, Annals of CIRP, 28(2), 309–318.
3. Osakada, K., Kawasaki, T, Mori, K., 1981, A method of determining flow stress under working conditions, Annals of CIRP, 30(1), 135–138.
4. Shiozaki, T., Kawasaki, K., 1986, Data sheets of flow stress and upsettability for carbon steel wire rods, J. Japan Soc. Tech. Plasticity, 27 (304), 568–572, (in Japanese).
5. Kawasaki, K., Tsuda, O, 1989, Automatic data processing system of flow stress during cold and warm forging, J. Japan Soc. Tech. Plasticity, 30 (343), 1127–1130, (in Japanese).
6. Tozawa, Y., 1989, Cooperative work on flow stress and ductility of steels for cold forging under forming conditions, J. Japan Soc. Tech. Plasticity, 30 (343), 1127–1130, (in Japanese).
7. Tozawa, Y., 1989, Flow stress and ductility of steels for cold forging under forming conditions [data], J. Japan Soc. Tech. Plasticity, 30 (343), 1131–1135, (in Japanese).

Equipment for Off-line Testing of Sheet Tribo-systems

Ermanno Ceron and Niels Bay

1 Motivation

Many sheet metal forming industries dealing with tribologically difficult materials e. g. stainless steel, aluminum and titanium alloys are searching for new, environmentally benign tribo-systems [1]. This is due to increasing demands as regards environmental issues dictated by legislation in Europe [2], Japan [3]. Industry is, however, reluctant to carry out production tests of new tribo-systems without any prior knowledge about the possible performance, since these tests are costly due to production stops, which may take even longer than anticipated, if the new system turns out to have poor performance and the tools therefore have to be dismounted and repolished.

This problem may be overcome by introducing off-line laboratory testing of the tribo-systems. In this way a larger number of potential new tribo-systems may be tested off-line and the most promising ones may be selected for production tests. Conventional sheet tribo-tests are, however, inadequate to emulate possible production problems e. g. galling, since pick-up of workpiece material on the tool surface is often a slowly developing process. A major Danish pump manufacturer with large experience in sheet metal forming of stainless steel components states that production tests should run successfully in at least 1500 strokes to judge the performance of a new tribo-system [4]. This was the reason for developing a new, universal sheet tribo-tester, which could carry out consecutive tests from coil at controlled speed and idle time between tests in order to allow cooling of the tool between strokes similar to production conditions [5]. The present paper is based on the two publications [6, 7].

Ermanno Ceron · Niels Bay ✉
Technical University of Denmark, Lyngby, Denmark
e-mail: nbay@mek.dtu.dk

A. E. Tekkaya et al. (eds.), *60 Excellent Inventions in Metal Forming*,
DOI 10.1007/978-3-662-46312-3_6

2 Description of Tribo-Tester

The tribo-tester can carry out all the main sheet tribo-tests, e. g. Plane Strip Test (PST), Bending Under Tension test (BUT), Draw Bead Test (DBT) and Strip Reduction Test (SRT) [8]. Figure 1 shows the machine, which has three individual axes that are all hydraulically driven and controlled by means of proportional electromechanical valves, which allows fast and accurate speed and load control. Axis 1 is the main drawing axis. In BUT testing the back tension is delivered by axis 2, whereas PST, DBT and SRT are carried out using axis 3 for the required test load. The sheet material is fed from coil, which may be more than 1000 m in length in case of 1 mm strip thickness implying that thousands of repetitions are possible. Lubrication is applied by two felt rolls. Control of the three axes is performed by a PLC, which is programmed for each test via a PC with a custom made LabView program. The range of operational parameters is: Sliding length: 0–250 mm, sliding speed: 0–150 mm/s, cycle time: 0–95spm, load on axis 1 and 2: 0–50 kN; load on axis 3: 0–10 kN. Figure 2 shows a picture of the machine.

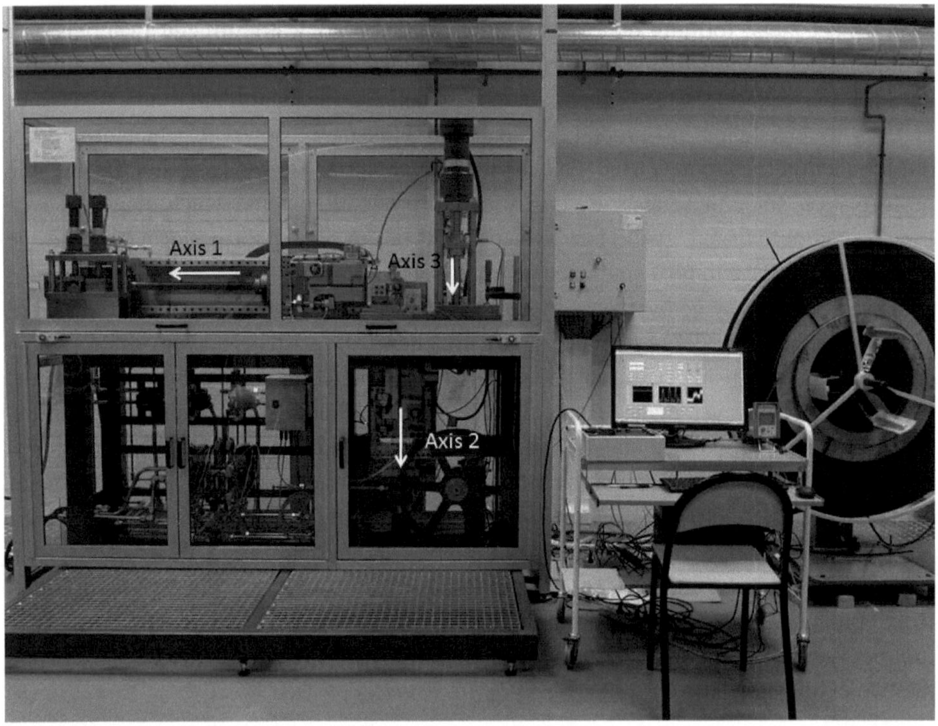

Fig. 1 New sheet tribo-tester [6, 7] (Copyright of photograph: Elsevier, Trans Tech Publ 2014)

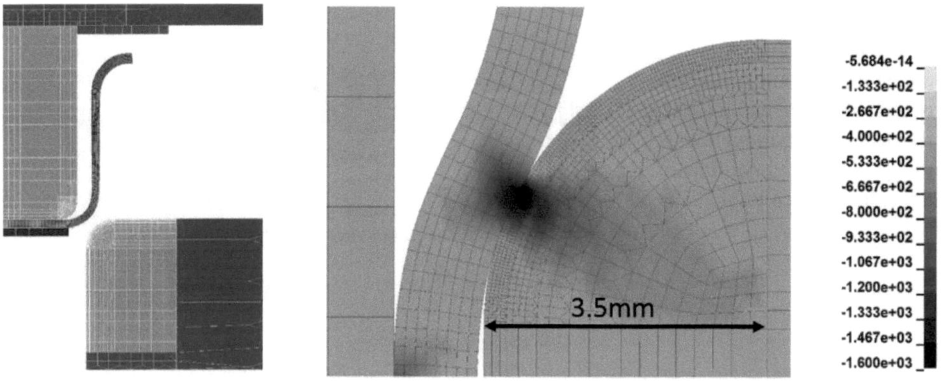

Fig. 2 Calculation of tool/workpiece normal pressure in the 2nd redrawing operation of the production tool [6, 7] (Copyright of illustrations: Elsevier, Trans Tech Publ 2014)

3 Test Design

The new test equipment has been applied in off-line testing of tribo-systems for a tribologically severe deep drawing production in a progressive tool including two redrawing operations after the initial deep drawing. FEA of the second redrawing operations showed the tool/workpiece interface pressure to reach 1600 MPa, which was obtained in the BUT test by designing the pin tool with a confined arc of contact, see Fig. 2 showing the calculated normal pressures in the production tool, whereas Fig. 3 shows the calculation for the BUT tool [6].

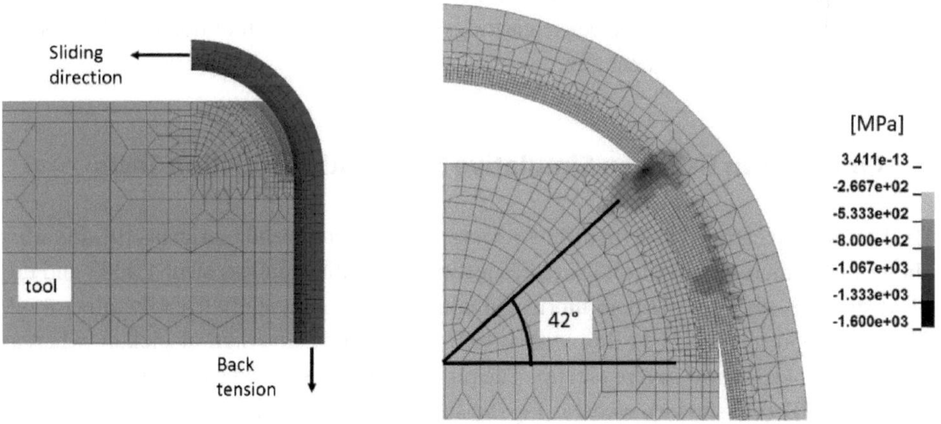

Fig. 3 Calculation of tool/workpiece normal pressure in the BUT tool with confined arc of contact [6, 7] (Copyright of illustrations: Elsevier, Trans Tech Publ 2014)

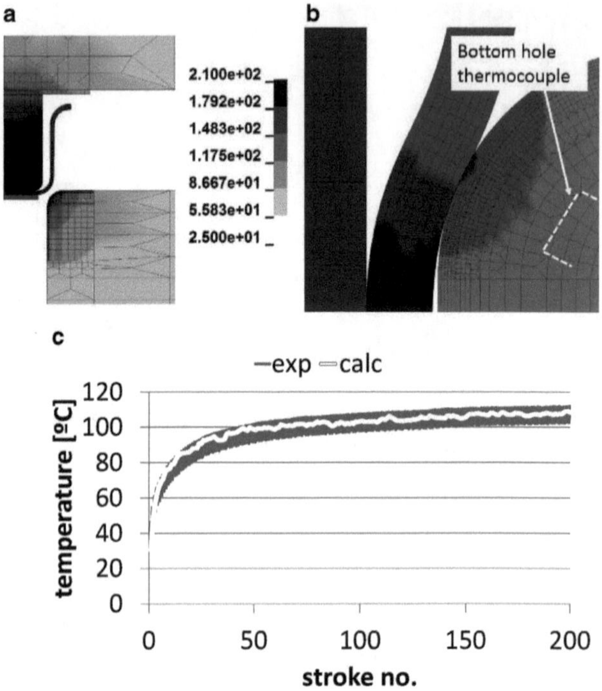

Fig. 4 **a, b** Temperature distribution in the 2nd redrawing operation, **c** Comparison between calculated and measured temperature at bottom hole in tool [7] (Copyright of illustrations: Trans Tech Publ 2014)

FEA analysis [9] showed the temperature in the tool/workpiece interface of the production tool in the second redraw to reach a maximum of 210°C and in the workpiece to reach 110°C before the second redraw, Fig. 4a and b, [2]. Figure 4c shows a comparison between measured and calculated temperature in the bottom hole 2 mm below the tool/workpiece interface in Fig. 4b. In order to obtain these high temperatures in the BUT test the tool was heated with electric cartridges mounted in the tool holder, whereas the strip was preheated with a heat gun.

4 Comparison Between Simulative Tests and Production Tests

BUT tests were carried out with different workpiece materials and lubricants [9]. Figure 5 shows SEM micrographs of tool surfaces testing austenitic stainless steel 1.4307 and duplex, stainless steel 1.4162 both lubricated with an environmentally friendly mineral oil. Figure 5a represents BUT testing with 1.4307, which shows no sign of pickup, whereas Fig. 5b for 1.4162 show severe pickup. The production tool showed similarly no pickup in case of 1.4307 (not shown), whereas Fig. 5c shows severe pickup for the case of 1.4162.

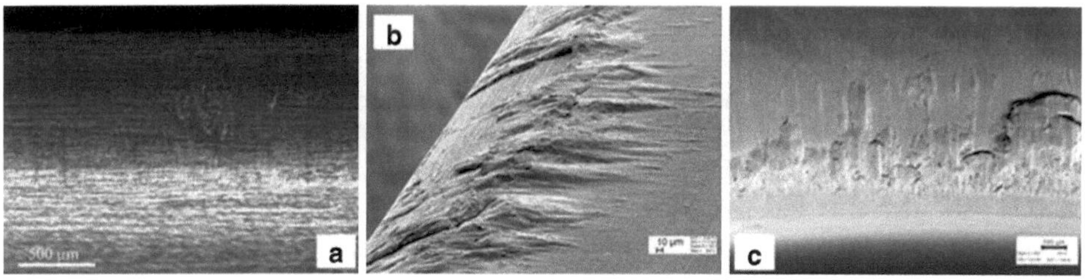

Fig. 5 SEM pictures of BUT tool surfaces run with **a** 1.4307, **b** 1.4162 and of production tool surface run with 1.4162 [9] (Copyright of photographs: Trans Tech Publ 2014)

5 Conclusion

New equipment for off-line testing of sheet tribo-systems has been developed and applied to an industrial case. The equipment allows repetitive testing under controlled conditions in order to emulate the slow buildup of pickup in case of lubricant film breakdown. Numerical analysis of the maximum die pressure and tool/workpiece interface temperature has formed the basis for designing the simulative test in order to emulate the tribological conditions in production. The good laboratory test results obtained with EN 1.4301 lubricated with an environmentally benign lubricant have led to the introduction of this lubricant in production instead of the chlorinated paraffin oil and to running without a previously applied tool coating and with an increase of production speed from 40 SPM to 90 spm.

6 Acknowledgement

The authors would like to acknowledge Grundfos A/S, IPU, Outokumpu Stainless AB, Outokumpu Stainless Research Foundation, SSAB EMEA AB and Uddeholms AB for their economic support.

References

1. Bay, N., Azushima, A., Groche, P., Ishibashi, I., Merklein, M., Morishita, H., Nakamura, T., Schmid, S., Yoshida, M., 2010. Environmentally Benign Tribo-systems for Metal Forming, CIRP Ann – Manuf Techn, 59 (2), 760–780.
2. European Parliament, Council, 2006, REACH, EC Regulation No 1907/2006 of the European Parliament and of the Council.
3. Ministry of Environment, Government of Japan, http://www.env.go.jp/en/laws/policy/basic/index.html, (accessed Nov. 2014).
4. Madsen, E., 2010 Personal communication.

5. Ceron, E., Bay, N., 2012, Testing and prediction of limits of lubrication in sheet metal forming. Proceedings of International Deep Drawing Research Group Conference, Mumbai, 251–257.
6. Ceron, E., Bay, N., 2013, A methodology for off-line evaluation of new environmentally friendly tribo-systems for sheet metal forming. CIRP Ann Manuf Techn, 62 (1), 231–234.
7. Bay, N., Ceron, E., 2014, Off-Line Testing of Tribo-Systems for Sheet Metal Forming Production. Adv Mater Res, 966–967, 3–20.
8. Bay, N., Olsson, D.D., Andreasen, J.L., 2008, Lubricant Test Methods for Sheet Metal Forming. Tribol Int 41 (9), 844–853.
9. Ceron, E., Martins, P.A.F., Bay, N., 2014. Thermal analysis of Bending Under Tension test. Procedia Eng. 81, 1805–1810.
10. Ceron, E., Olsson, M., Bay, N., 2014, Lubricant film breakdown and material pick-up in sheet forming of advanced high strength steels and stainless steels when using environmental friendly lubricants. Adv Mater Res, 966–967, 219–227.

Part II
Modelling

Anisotropic Yield Functions

Frédéric Barlat and Hyuk Jong Bong

1 Motivation

With advances in modern technology, the increasing complexity of a forming process and the greater accuracy required for the product make the parameter optimization a very difficult task. For instance, bipolar plates for fuel cell applications can be formed from extra-thin (100 μm) ferritic stainless steel sheets using the servo-press technology. However, the unlimited choice and complexity of slide motions available to successfully produce such a thin product render the process optimization extremely challenging and beyond the possibility of an experimental trial-and-error approach. Thus, in order to determine the optimum conditions for a sheet metal forming process, finite element (FE) analyses are conducted. Of course, experimental adjustments are needed but the model provides a detailed understanding of the process and narrows significantly its parameter window. All the aspects of the FE modeling must be carefully addressed to predict the best possible conditions. In this article, the plastic behavior aspect, particularly anisotropy, is discussed. Ideally, a multi-scale constitutive description would be desired because the process can be tailored to the material microstructure. However, the computation time would be too large with this approach for practical applications in industry. For this reason, continuum constitutive models are still very powerful.

Frédéric Barlat ✉ · Hyuk Jong Bong
Pohang University of Science and Technology, Pohang, Korea
e-mail: f.barlat@postech.ac.kr

A. E. Tekkaya et al. (eds.), *60 Excellent Inventions in Metal Forming*,
DOI 10.1007/978-3-662-46312-3_7

2 Flow Theory of Plasticity

For multiaxial loading, the stress is defined by a tensor $\boldsymbol{\sigma}$ or by its deviator **s** for pressure-independent plastic behavior. Therefore, the yield condition may be written in the form

$$\Phi = \bar{\sigma}(\mathbf{s}) - \sigma_r(\bar{\varepsilon}) = 0 \tag{7.1}$$

where $\bar{\sigma}(\mathbf{s})$ is the yield function, which defines the shape of the yield surface, and $\bar{\varepsilon}$ is the effective strain, which defines its size through the reference flow curve σ_r. $\bar{\sigma}(\mathbf{s})$ is mathematically a homogeneous function of first degree with respect to the stress components. This property allows the yield function to be considered as an effective stress and the loading state to be computed very simply. The relationship in Eq. 7.1 contains two of the three elements of a plasticity theory, namely, the yield condition and strain hardening. The third element, the flow rule, assumes that the plastic strain increment derive from the yield function. The so-called associated flow rule translates mathematically as

$$d\varepsilon_{pq} = d\bar{\varepsilon}\frac{\partial\bar{\sigma}}{\partial\sigma_{pq}} \tag{7.2}$$

for a homogenous yield function of first degree. Theoretical considerations on crystal plasticity indicate that the yield function is convex and that the associated flow rule is very relevant for metals. The effective strain $\bar{\varepsilon}$ is a monotonically increasing state variable, which describes the amount of accumulated plasticity in the material. For isotropic hardening, the shape of the yield surface is preserved during plastic deformation while its size only varies. In this case, a consistent effective strain increment is obtained from the plastic work equivalence $\bar{\sigma}(\mathbf{s})\,d\bar{\varepsilon} = \sigma_{pq}d\varepsilon_{pq}$.

3 Isotropic Yield Conditions

For an isotropic material, the plastic behavior is independent of the loading direction and should be described with stress tensor invariants only. This includes the three principal stresses, which are particularly convenient to verify the convexity of a yield function. Classical isotropic yield conditions are those of Tresca and von Mises. However, any yield function written in terms of invariants is acceptable such as that proposed by Hershey [1]

$$\bar{\sigma} = 2^{-1/a}\left\{|s_1 - s_2|^a + |s_2 - s_3|^a + |s_3 - s_1|^a\right\}^{1/a} = \sigma_r \tag{7.3}$$

which reduces to von Mises for $a=2$ or 4 and to Tresca for $a=1$ or in the limit case $a \to \infty$. This convex yield function was developed as a good approximation of self-consistent crystal plasticity calculations. Hosford [2] showed that, with exponents of 6 and 8, this equation leads to an almost perfect representation of the yield surfaces for isotropic BCC and FCC materials, respectively, as computed with crystal plasticity.

4 Anisotropic Yield Conditions

For plastic anisotropy, the above discussion is valid except that it is necessary to choose a material reference frame in which the plasticity equations are expressed. Moreover, the stress invariants must be, somehow, modified in a consistent manner. Although many theories of plastic anisotropy are available, the approach based on linear transformation of the stress tensor (or deviator) is considered here. This approach allows any isotropic yield function expressed with the principal stresses as invariants to be extended to anisotropy while preserving its property of convexity and homogeneity. For incompressible materials, a linear transformation is performed on the stress deviator, **s**, and the components of the transformed tensor, $\tilde{\mathbf{s}}$, are

$$\tilde{s}_{ij} = C_{ijkl} s_{kl}. \tag{7.4}$$

C is a fourth order tensor, which contains the anisotropy coefficients and accounts for the macroscopic symmetries of the material. In the present theory, an isotropic yield function $\bar{\sigma}$ can be generalized to anisotropy by replacing the principal deviatoric stress components by the principal values $\tilde{s}_1$, $\tilde{s}_2$ and $\tilde{s}_3$ of the transformed stress deviator $\tilde{\mathbf{s}}$. A complete extension of Eq. 7.3 for an orthotropic material using a linear transformation was proposed in [3]. It is also possible to use two (or more) linear transformations as long as the yield function is isotropic with respect to all the principal values of the transformed tensors. $\tilde{\mathbf{s}}'$ and $\tilde{\mathbf{s}}''$ are calculated in the same way as with one transformation. Applied to plane stress, the anisotropic yield function called Yld2000-2d was introduced by Barlat et al. [4]

$$\bar{\sigma}(\mathbf{s}) = 2^{-1/a}\left\{\left|\tilde{s}_1' - \tilde{s}_2'\right|^a + \left|2\tilde{s}_2'' + \tilde{s}_1''\right|^a + \left|2\tilde{s}_1'' + \tilde{s}_2''\right|^a\right\}^{1/a} = \sigma_r \tag{7.5}$$

where the matrix representation of the transformation tensor can be written as follows

$$\left[\tilde{\mathbf{s}}'\right] = \left[\mathbf{C}'\right][\mathbf{s}] = \begin{bmatrix} \alpha_1 & 0 & 0 \\ 0 & \alpha_2 & 0 \\ 0 & 0 & \alpha_7 \end{bmatrix}\begin{bmatrix} s_{xx} \\ s_{yy} \\ s_{xy} \end{bmatrix}, \quad \left[\mathbf{C}''\right] = \frac{1}{3}\begin{bmatrix} 4\alpha_5 - \alpha_3 & 2\alpha_6 - 2\alpha_4 & 0 \\ 2\alpha_3 - 2\alpha_5 & 4\alpha_4 - \alpha_6 & 0 \\ 0 & 0 & 3\alpha_8 \end{bmatrix}. \tag{7.6}$$

Eight independent anisotropy coefficients α_k are available in the formulation. All are equal to 1 in the isotropic case for which Hershey's yield condition is recovered. Note that Yld2000-2d is found in the literature under different forms [5]. An earlier yield function proposed in 1989, the so-called Yld89 [6]

$$\bar{\sigma}(\mathbf{s}) = 2^{-1/a}\left\{\bar{c}\left|\tilde{\sigma}_1 - \tilde{\sigma}_2\right|^a + \bar{a}\left|\tilde{\sigma}_1\right|^a + \bar{a}\left|\tilde{\sigma}_2\right|^a\right\}^{1/a} = \sigma_r \tag{7.7}$$

where

$$\tilde{\sigma}_k = \frac{\sigma_{xx} + \bar{h}\sigma_{yy}}{2} \pm \sqrt{\left(\frac{\sigma_{xx} - \bar{h}\sigma_{yy}}{2}\right)^2 + \left(\bar{p}\sigma_{xy}\right)^2} \tag{7.8}$$

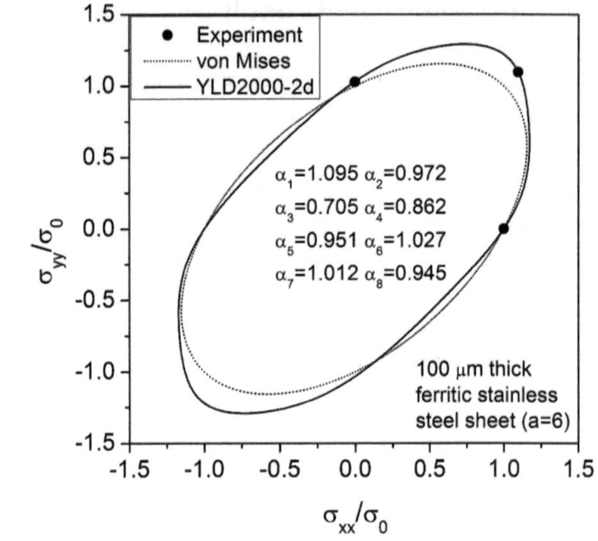

Fig. 1 Normalized yield loci calculated with von Mises and Yld2000-2d yield functions for ferritic stainless steel sheet sample

turns out to be a particular case of Yld2000-2d when the following relationships are used

$$\alpha_1 = \left(2 - \bar{h}\right)(\bar{c})^{1/a},\ \alpha_2 = \left(2\bar{h} - 1\right)(\bar{c})^{1/a},\ \alpha_3 = \alpha_4 = \bar{h}\,(\bar{a})^{1/a},\alpha_5 = \alpha_6 = (\bar{a})^{1/a},$$
$$\alpha_7 = \bar{p}\,(\bar{c})^{1/a},\ \alpha_8 = \bar{p}\,(\bar{a})^{1/a}. \tag{7.9}$$

For a general stress state, Eq. 7.3 was extended as a yield function called Yld2004-18p [7] in which, out of the 18 available coefficients, only 16 are independent [8]. For

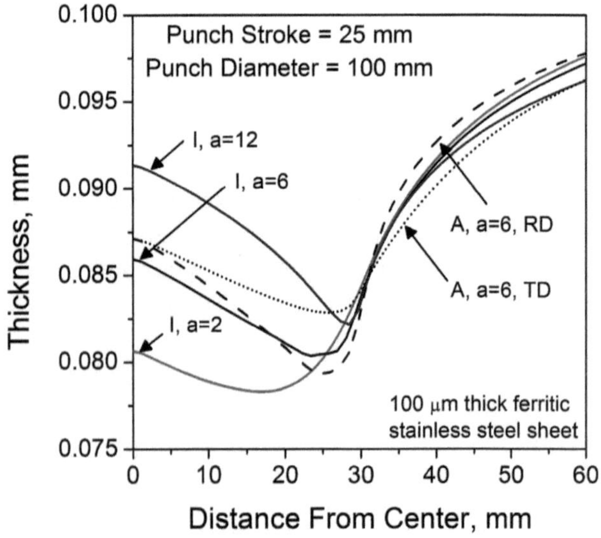

Fig. 2 FE simulated thickness distribution using different yield functions (*I* isotropic; *A* anisotropic) in hemi-spherical punch stretching

the sake of conciseness, the present authors cited mostly their own work but many other excellent contributions, as reviewed in Reference [9], can be found in the literature.

As an example, Fig. 1 represents the two-dimensional yield locus for a ferritic stainless steel sheet sample. The corresponding constitutive relationships were used in FE analyses of the Nakazima spherical punch-stretching test. Figure 2 represents the thickness distribution of the specimen as a function of the distance from the pole of the specimen. This figure illustrates the drastic influence of plastic anisotropy as well as the yield function exponent or, in other words, of the material crystal structure, on the predicted thickness distribution.

5 Other Yield Conditions

Although HCP materials exhibit a plastic behavior, which is independent of the mean stress, their tension and compression yield stresses are different. This so-called strength-differential effect has to do with their deformation mechanisms, slip and twinning. Cazacu et al. [10] proposed an isotropic yield criterion for these materials, which is expressed as a function of the principal deviatoric stresses

$$\bar{\sigma}(\mathbf{s}) = \xi^{-1/a}\left\{\left||s_1| - ks_1\right|^a + \left||s_2| - ks_2\right|^a + \left||s_3| - ks_3\right|^a\right\}^{1/a} = \sigma_r \tag{7.10}$$

ξ is a coefficient and k controls the tension-compression asymmetry. This yield function is convex, homogeneous and was extended to plastic anisotropy using the linear transformation approach [10].

When load becomes suddenly non-proportional during plastic deformation, the materials response usually differs from that imposed by isotropic hardening. A typical example is the Bauschinger effect, which occurs during reverse loading. Traditionally, this behavior is modeled using the kinematic hardening assumption, i. e., a translation of the yield surface in stress space. In a recent approach [11], the Bauschinger effect and other anisotropic hardening phenomena are captured exclusively by a distortion of the yield surface as expressed by the following yield condition

$$\Phi(\mathbf{s}) = \left[\bar{\sigma}\left(\mathbf{s}\right)^q + f_1^q \left|\hat{\mathbf{h}} : \mathbf{s} - \left|\hat{\mathbf{h}} : \mathbf{s}\right|\right|^q + f_2^q \left|\hat{\mathbf{h}} : \mathbf{s} + \left|\hat{\mathbf{h}} : \mathbf{s}\right|\right|^q\right]^{\frac{1}{q}} - \sigma_r = 0 \tag{7.11}$$

The so-called microstructure deviator $\hat{\mathbf{h}}$ characterizes the material deformation history. $\bar{\sigma}\left(\mathbf{s}\right)$ represents any isotropic or anisotropic homogeneous yield function of first degree as discussed in the previous sections. A number of state variables were introduced in this constitutive equation, including f_1, f_2 and $\hat{\mathbf{h}}$, but the foundation of the present theory of plasticity for anisotropic materials as well as the homogeneity and convexity of the yield function are preserved.

6 Conclusion

This brief summary describes the principles guiding the development of a plasticity theory for anisotropic metals. The derivation of the constitutive equations is consistent with considerations and simulation results of lower scale crystal plasticity. Consequently, microstructural information is included, albeit, in an approximate manner, in this approach. The corresponding anisotropic yield functions are simple enough to be conveniently used for the optimization of sheet metal forming processes in industry.

7 Acknowledgment

Parts of this work were supported by POSCO to whom the authors are very grateful.

References

1. Hershey, A.V., 1954. The plasticity of an isotropic aggregate of anisotropic face-centered cubic crystals, ASME J. Appl. Mech. 21, 241–249.
2. Hosford, W.F., 1972. A generalized isotropic yield criterion. ASME J. Appl. Mech. Trans. 39, 607–609.
3. Barlat, F., Lege, D.J., Brem, J.C., 1991. A six-component yield function for anisotropic materials. Int. J. Plasticity 7, 693–712.
4. Barlat, F., Brem, J.C., Yoon, J.W., Chung, K., Dick, R.E., Lege, D.J., Pourboghrat, F., Choi, S.-H., Chu, E., 2003. Plane stress yield function for aluminum alloy sheets–Part I: Theory. Int. J. Plasticity 19, 1297–1319.
5. Barlat, F., Yoon, J.W., Cazacu, O., 2007. On linear transformations of stress tensors for the description of plastic anisotropy. Int. J. Plasticity 23, 876–896.
6. Barlat, F., Lian, J., 1989. Plastic behavior and stretchability of sheet metals. Part I: A yield function for orthotropic sheets under plane stress conditions. Int. J. Plasticity 5, 51–66.
7. Barlat, F., Aretz H., Yoon, J.W., Karabin, M.E., Brem J.C., Dick R.E., 2005. Linear transformation-based anisotropic yield functions. Int. J. Plasticity 21, 1009–1039.
8. Van den Boogaard, A.H., Havinga, J., Belin, A., Barlat, F., 2015. Parameter reduction for the Yld2004-18p yield criterion. Int. J. Material Forming, in press, doi:10.1007/s12289-015-1221-3.
9. Banabic, D., Barlat, F., Cazacu, O., Kuwabara, T., 2010. Advances in Anisotropy and Formability. Int. J. Material Forming 3, 165–189.
10. Cazacu, O., Plunkett, B., Barlat, F., 2006. Orthotropic yield criterion for hexagonal close packed metals. Int. J. Plasticity 22, 1171–1194.
11. Barlat, F., Vincze, G., Grácio, J.J., Lee, M.G., Rauch, E.F., Tomé, C., 2014. Enhancements of homogenous anisotropic hardening model and application to mild and dual-phase steels. Int. J. Plasticity 58, 201–218.

BBC2005 Yield Criterion Used in the Numerical Simulation of Sheet Metal Forming Processes

Dorel Banabic and Dan-Sorin Comsa

1 Motivation

Many researchers have been involved in the development of more accurate yield criteria during the last four decades [1]. The performances of such models are closely related to the flexibility of their mathematical formulation. In general, the flexibility is enhanced by including a larger number of material parameters in the yield criterion. As a consequence, the identification procedure needs more experimental data as input. The tests performed for obtaining this data have gradually extended their investigation area from uniaxial tension to biaxial tension, plane-strain tension and even pure shearing. The yield criteria involving a large number of material parameters are also more complex from the mathematical point of view. This characteristic represents in many cases a serious drawback, especially when the computational efficiency is sought by the users. When the plasticity of highly-anisotropic sheet metals must be described, the use of more complex models is unavoidable. In these situations, the quality of the simulation results cannot be ensured without having an accurate description of the yield surface. A reasonable balance between the accuracy, computational efficiency, identification costs, and mathematical complexity is achieved by the models that use seven or eight experimental values as input. Such yield criteria have been proposed by Barlat et al. [2], Cazacu and Barlat [3], Vegter [4]. The BBC family of yield criteria [5, 6] also belong to this class of material models. As compared with other formulations proposed during the last decades, the BBC yield criteria do not use linear transformations of the stress tensor. As a consequence, their computational efficiency is superior in the simulation of sheet metal forming processes. At present, BBC 2005 is one of the most widely used yield criteria in industrial applications due to the fact that it has been implemented as a standard material model in the FE code AutoForm 4.1.1 [6, 7].

Dorel Banabic ✉ · Dan-Sorin Comsa
Technical University of Cluj-Napoca, Cluj Napoca, Romania
e-mail: banabic@tcm.utcluj.ro

A. E. Tekkaya et al. (eds.), *60 Excellent Inventions in Metal Forming*,
DOI 10.1007/978-3-662-46312-3_8

2 Formulation of the BBC 2005 Yield Criterion

The BBC 2005 model assumes that metallic sheets behave as plastically orthotropic membranes under plane-stress conditions. This hypothesis allows describing the yield surface by means of the equation

$$\bar{\sigma}(\sigma_{11}, \sigma_{22}, \sigma_{12} = \sigma_{21}) = Y, \tag{8.1}$$

in which $\bar{\sigma}$ is the BBC 2005 equivalent stress (see below), Y is a yield parameter, and $\sigma_{\alpha\beta}$ ($\alpha, \beta = 1, 2$) are planar components of the stress tensor expressed in an orthonormal basis superimposed to the axes of plastic orthotropy: (1) rolling direction (RD), (2) transverse direction (TD), and (3) normal direction (ND). The other stress components are subjected to the restriction $\sigma_{i3} = \sigma_{3i} = 0$ ($i = 1, 2, 3$) arising from the plane-stress hypothesis. The BBC 2005 equivalent stress is defined by the following formula [7]:

$$\bar{\sigma} = \left[a(\Lambda + \Gamma)^{2k} + a(\Lambda - \Gamma)^{2k} + b(\Lambda + \Psi)^{2k} + b(\Lambda - \Psi)^{2k}\right]^{\frac{1}{2k}}, \tag{8.2}$$

where $k \in \mathbf{N}_{>0}$ and $a, b > 0$ are material parameters, while Γ, Λ and Ψ are functions depending on the planar components of the stress tensor:

$$\begin{aligned} \Gamma &= L\sigma_{11} + M\sigma_{22}, \quad \Lambda = \sqrt{(N\sigma_{11} - P\sigma_{22})^2 + \sigma_{12}\sigma_{21}}, \\ \Psi &= \sqrt{(Q\sigma_{11} - R\sigma_{22})^2 + \sigma_{12}\sigma_{21}}. \end{aligned} \tag{8.3}$$

The coefficients L, M, N, P, Q, and R involved in Eq. 8.3 are also material parameters. The conditions $k \in \mathbf{N}_{>0}$ and $a, b > 0$ ensure the convexity of the yield surface defined by Eqs. 8.1–8.3. The coefficients L, M, N, P, Q, and R are not subjected to any constraint from this point of view.

One may notice that nine material parameters are involved in the formulation of the BBC 2005 equivalent stress: k, a, b, L, M, N, P, Q, and R. The integer exponent k has a special status, due to the fact that its value is fixed from the very beginning in accordance with the crystallographic structure of the material: $k = 3$ for BCC materials and $k = 4$ for FCC materials.

The identification procedure calculates the other parameters (a, b, L, M, N, P, Q, and R) by forcing the constitutive equations to reproduce the following experimental data (see [5] for details):

- Uniaxial yield stresses associated to the directions defined by 0°, 45° and 90° angles measured from RD
- Coefficients of uniaxial plastic anisotropy associated to the directions defined by 0°, 45° and 90° angles measured from RD
- Balanced biaxial yield stress associated to RD and TD
- Coefficient of balanced biaxial plastic anisotropy associated to RD and TD

3 Results and Applications

The Numisheet 2005 Benchmark #1 [6, 8] is used to prove the predictive capabilities of the BBC 2005 yield criterion. The forming process under analysis involves forming,

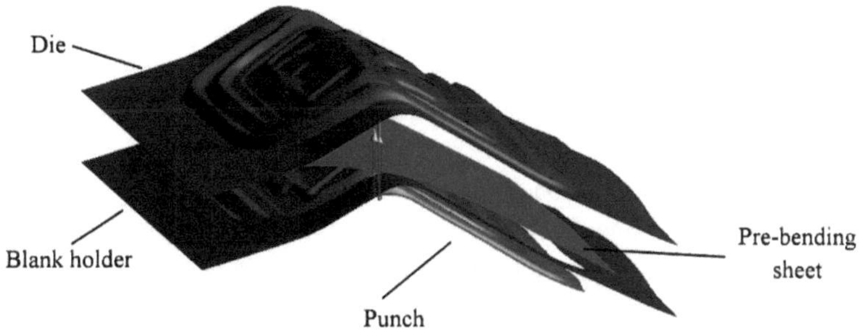

Fig. 1 Geometry of the blank and tools used in the deep-drawing process [6] (Copyright of picture: Springer Verlag 2010)

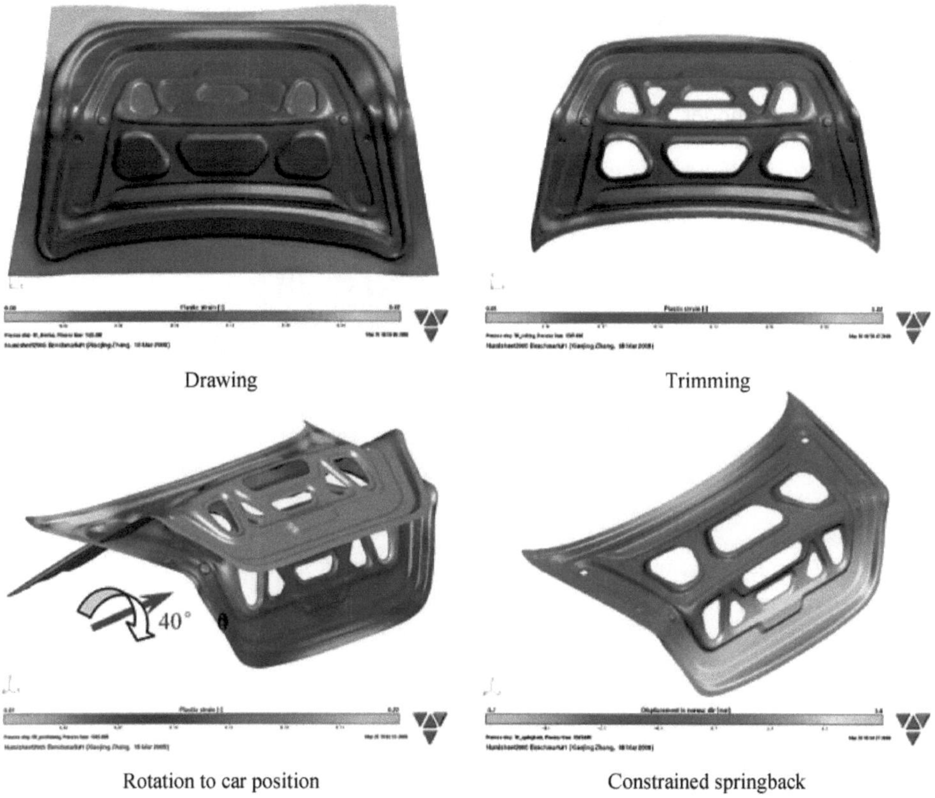

Fig. 2 Simulation of the deep-drawing, trimming, rigid body rotation in car position, and constrained springback [6] (Copyright of picture: Springer Verlag 2010)

trimming and constrained springback. The objective of the benchmark is to assess the springback prediction capability of various simulation programs.

The geometry of the blank and tools used in the deep-drawing process is illustrated in Fig. 1. One may notice that the positions of the die, punch and binder are specific to the forming operations performed on a single-action press. A thickness equalizer [8] is placed between the die and binder to ensure a constant gap. Therefore, a zero holding force is applied on the blank during the simulation of the deep-drawing process. The blank is cut from an AL 6111-T4P aluminum alloy sheet with the nominal thickness 0.9 mm (see [8] for a detailed presentation of the mechanical parameters used for calibrating the BBC 2005 yield criterion).

Figure 2 illustrates the simulation of the deep-drawing, trimming, rigid body rotation in the car position, and constrained springback. The results shown on the colored maps have been obtained with the AutoForm 4.1.1 finite-element program. Figure 3 and 4 give the comparison between simulation and experimental data referring to the springback amount at predefined measurement points. The statistical analysis shows that the standard deviation between the calculated springback and the experimental results is well-controlled under the value of 0.65 mm.

It is well-known that springback simulation is highly dependent on the capability of the yield criteria to describe the anisotropic response of metallic sheets. From this point of view, the results of the Numisheet 2005 Benchmark #1 prove that BBC 2005 is a very accurate material model.

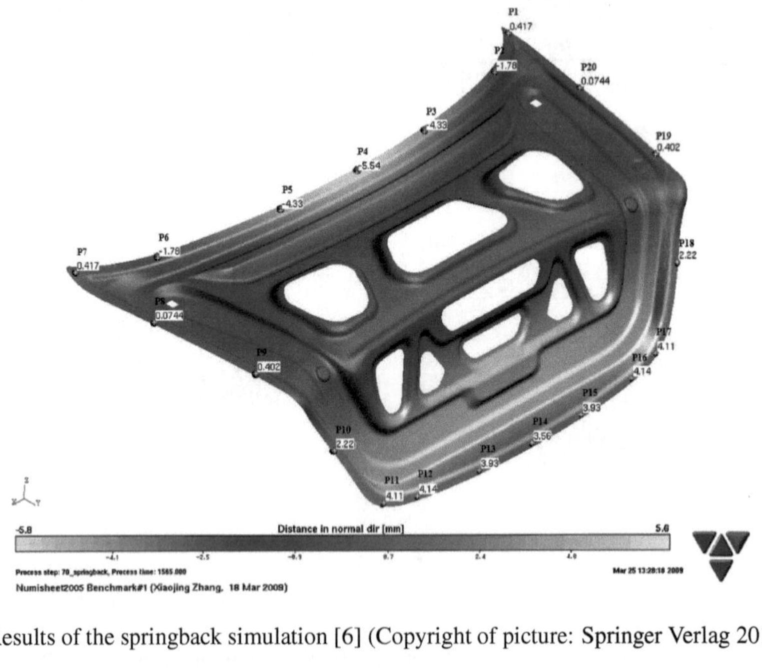

Fig. 3 Results of the springback simulation [6] (Copyright of picture: Springer Verlag 2010)

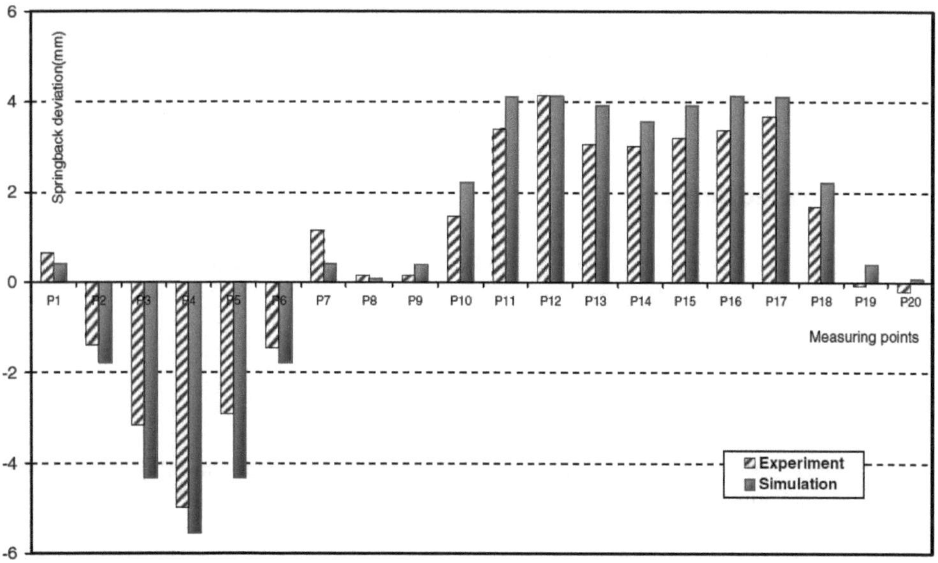

Fig. 4 Comparison between simulation and experimental data referring to the springback amount at predefined measurement points [6] (Copyright of picture: Springer Verlag 2010)

4 Potential

Numerous tests envisaging industrial forming processes have confirmed not only the predictive capabilities of the BBC 2005 yield criterion, but also its computational efficiency which is a consequence of the following features [6, 7]:

1. Simplicity of the mathematical formulation arising from the fact that BBC 2005 does not use linear transformations of the stress tensor
2. Flexibility and stability of the identification procedure
3. Possibility of being extended to more general formulations (for example BBC 2008) [9]
4. Integration in a hierarchical multi-scale (HMS) framework using the ALAMEL crystal plasticity model [10]

5 Conclusion

The results presented in this paper prove the ability of the BBC2005 yield criterion to provide an accurate description of the anisotropic behavior of cold-rolled sheet metals. The performances of the model have been evaluated using a benchmark test referring to the deep-drawing and springback of a decklid inner panel. A very good agreement with experimental data referring to the springback amount has been obtained. This fact together with the flexibility of the identification procedure recommends the use of the BBC2005 yield criterion in industrial applications.

6 Acknowledgement

The work has been kindly supported by the Romanian National Research Council (CNCS) under the grant number PCCE 100/2009.

References

1. Banabic D., Barlat F., Cazacu O., Kuwabara T., 2007, Anisotropy and formability, In: Advances in Material Forming-ESAFORM 10 Years on, (Editors: Chinesta F., Cueto, E.), Springer, Berlin, 143–173.
2. Barlat F., Brem J.C., Yoon J.W., Chung K., Dick R.E., Lege D.J., Pourboghrat F., Choi S.H., Chu E., 2003, Plane stress yield function for aluminum alloy sheets. Part I: Theory, International Journal of Plasticity, 19(12), 1297–1319.
3. Cazacu O., Barlat F., 2003, Application of representation theory to describe yielding of anisotropic aluminum alloys. International Journal of Engineering Science, 41(12), 1367–1385
4. Vegter H., van den Boogaard A.H., 2006, A plane stress yield function for anisotropic sheet material by interpolation of biaxial stress states. International Journal of Plasticity, 22(3), 557–580.
5. Banabic D., Aretz H., Comsa D.S., Paraianu L., 2005, An improved analytical description of orthotropy in metallic sheets, International Journal of Plasticity, 21(3), 493–512.
6. Banabic D., 2010, Sheet Metal Forming Processes: Constitutive Modelling and Numerical Simulation, Springer, Berlin.
7. Banabic D., Comsa D.S., Sester M., Selig M., Kubli W., Mattiasson K., Sigvant M., 2008, Influence of constitutive equations on the accuracy of prediction in sheet metal forming simulation, Proc.7th Int. Conf. and Workshop on Numerical Simulation of 3D Sheet Metal Forming Processes, Numisheet 2008, Interlaken, 37–42.
8. Smith L.M., Pourboghrat F., Yoon J.W., Stoughton T.B. (eds), 2005, Proc. 6th Int. Conf. and Workshop on Numerical Simulation of 3D Sheet Metal Forming Processes – Verification of Simulation with Experiment, Numisheet 2005, Detroit.
9. Comsa D.S., Banabic D., 2008, Plane-stress yield criterion for highly anisotropic sheet metals, Proc.7th Int. Conf. and Workshop on Numerical Simulation of 3D Sheet Metal Forming Processes, Numisheet 2008, Interlaken, 43–48.
10. Gawad J., Banabic D., Comsa D.S., Gologanu M., Van Bael A., Eyckens P., Van Houtte P., Roose D., 2014, Evolving texture-informed anisotropic yield criterion for sheet forming, , Proc. 9th Int. Conf. and Workshop on Numerical Simulation of 3D Sheet Metal Forming Processes, Numisheet 2014, Melbourne, 350–355.

The Impact of M-K Model on Development of Formability Assessment in Sheet Metal Forming Processes

Andrzej Kocańda

1 Motivation

Formability refers to the shaping of sheet metal and it describes the ease of shaping a metal through plastic deformation. Assessment of the formability of a metal usually includes measurement of the amount of deformation leading to failure in the form of fracture or local thinning. The industry has been hardly pushed in the last decades to increase the quality control and assurance in all areas of industrial production. Over 100 years ago, A.M. Erichsen invented the cupping test which was the first special test method to determine the level of quality of strip and sheet metal [1]. It involves pressing the spherically ended punch into the tested sheet metal until the crack appears throughout the thickness of sheet [2]. The primary use of this test allows the mutual comparison of different grades of sheet metal on the basis of the sample height (cupping index IE) in the appearance of cracks. Without this test the batch production of parts made of sheet metal would have been really difficult. In 1932 A.M. Erichsen introduced the deep drawing cup test. It is a method by which a circular blank is formed into a cup by means of a die and a punch. The biggest possible ratio between the blank and the punch diameters, obtained without a fracture of a formed cup, is the limiting drawing ratio indicating "formability of sheet metal". In the following years many assessment methods of sheet metal formability have been proposed and improved. Among them the forming limit diagram (FLD) concept originally developed by S.P. Keeler [3, 4] and G.M. Goodwin [5] proved to be very useful both in industry and research laboratories. Keller and Goodwin used experimental methods to measure critical strains, after local thinning had started, on various production and laboratory parts formed from cold-rolled steel sheet. Major and minor strain values were obtained by measurement of deformation of circle grids marked on the blank surface.

Andrzej Kocańda ✉
Warsaw University of Technology, Warsaw, Poland
e-mail: a.kocanda@wip.pw.edu.pl

A. E. Tekkaya et al. (eds.), *60 Excellent Inventions in Metal Forming*,
DOI 10.1007/978-3-662-46312-3_9

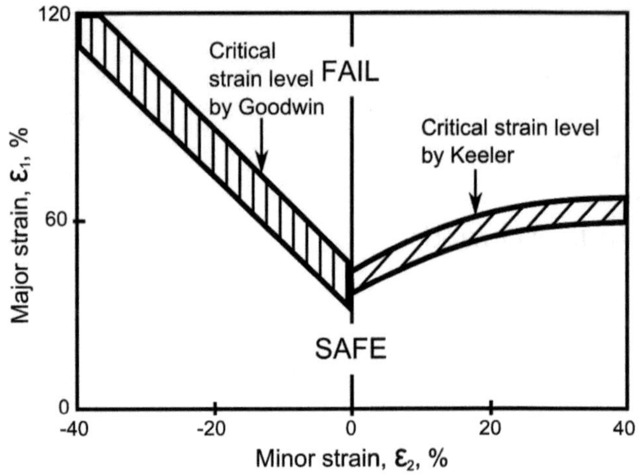

Fig. 1 General presentation of the forming limit diagram showing tension-compression strains (**a**) and tension-tension strains (**b**) – compilation of experimental results

Compilation of these strains resulted in the forming limit diagram shown in Fig. 1. There are bands, which represent scatter of critical strains dependent on different definitions of the critical strains and the accuracy of experimental methods. Production shops have considered visible localized thinning or splits/cracks unacceptable (roughly – upper limit of the band). On the other hand laboratories usually preferred the onset of local thinning as a critical moment. Anyway, it has been possible to show safe combinations of major and minor strains for sheet metal forming processes. Next, there have been used other experimental methods to find "formability of sheet metal", e. g. Marciniak biaxial stretching test, hydraulic bulging test, Marciniak in-plane sheet torsion test, Miyauchi shear test, Nakazima test, Olsen and Fukui tests.

In addition to the development of experimental methods, there have been many attempts made to develop theoretical models for calculation of the forming limit curve (FLC). First in 1952, Swift [6] and Hill [7] proposed models of so-called diffuse necking and localized necking respectively, which had been helpful to predict FLC. However, they assumed that sheet metals were homogeneous. Only after some years in 1965, Marciniak [8] introduced a model of plastic deformation of shells taking into account the geometric and microstructure non-homogeneities. Next in 1967, Marciniak and Kuczyński [9] developed a forming limit curve prediction model based on imperfections in sheet metal. This so-called M-K model has been the most commonly used for analytical prediction of FLC and has been the most crucial for further development of formability assessment in sheet metal forming processes. According to the researchers and scientists, the level of interest in this model could be well highlighted by huge amount of citations (many thousands) and organization of special sessions during the international conferences, e. g. IDDRG, NUMISHEET, NUMIFORM, ESAFORM [10].

2 M-K Model

M-K model was originally associated with the biaxial stretch forming processes, e. g. bulging or stretch drawing. Such processes lead to an increase in the sheet surface area at the expense of reduced thickness. Limit strains for biaxial stretching are larger than for uniaxial tension as has been shown experimentally, e. g. by Keeler, Fig. 1. So something delays local necking. Marciniak and Kuczyński indicated, that a completely new mechanism of the loss of stability might appear in these processes [8, 9]. This phenomenon does not depend on the loss of stability of the whole shell but starts from an initial non-homogeneity where strain begins to concentrate and a groove appears on the surface of sheet metal. Hence, it was assumed in the M-K model that fracture of sheet metal did not occur abruptly, but it was preceded by the loss of stability in a limited area.

Initial imperfections are related to the microstructure (inclusions, discontinuities, voids) and geometry (changes in thickness). They result from sheet metal manufacturing processes. Let us take into account a sheet metal sample with geometric imperfection in the form of groove B normal to the direction of the largest principal tensile stress σ_1 and having a thickness t_B, Fig. 2a. Region A outside the groove has the uniform thickness t_A. The geometric imperfection is described by the so-called coefficient of geometric non-homogeneity $f = t_B/t_A$. The ratio of principal stresses in region A remains constant during the loading process. Hence, there is a proportional deformation in this region. However, the stress ratio changes gradually in the region B when non-homogeneity develops. The equilibrium of "the force per unit width" perpendicular to the groove and transmitted through the regions A and B requires $\sigma_{1A} t_A = \sigma_{1B} t_B$ under condition of equal strains $\varepsilon_{2B} = \varepsilon_{2A}$. After reformulation of the equation we obtain $\sigma_{1A}/\sigma_{1B} = t_B/t_A = f$. It means that σ_{1B} is always larger than σ_{1A} because f is slightly less than unity. What more,

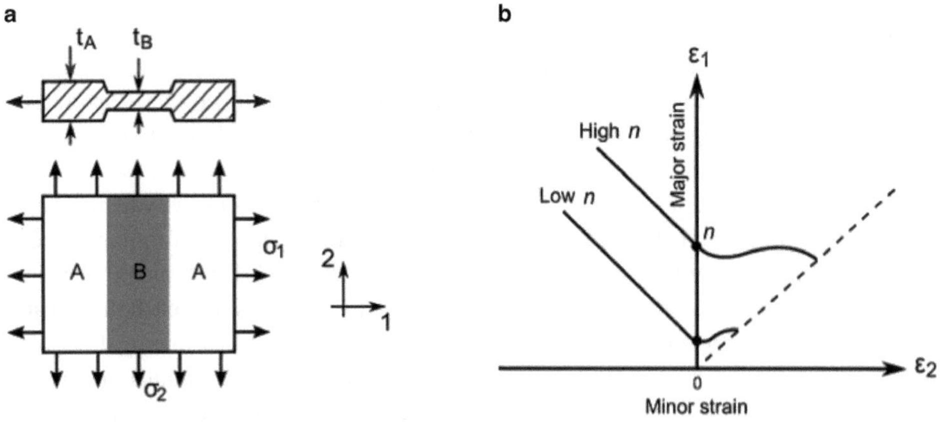

Fig. 2 Groove B in a region A of uniform strain (**a**) and the effect of strain hardening exponent n on the FLC (**b**)

the increase of the stress σ_{1B} in the groove leads to a different point on the yield surface as the depth of the groove increases. Finally plane strain is reached and the sheet tears along the groove [11].

When the force transmitted by the groove increases, consequently the region A will be also gradually deformed until the strain ε_{1A} reaches the limit value ε_{1A}^{*}. Theoretically, this phenomenon takes place when entire deformation is localized in groove B and yields to a complete loss of stability although in practice there have been used other approaches. This ε_{1A}^{*} strain together with ε_{2A}^{*} indicates a point belonging to the FLC.

3 Development of Formability Assessment Based on M-K Model

The physical meaning of the M-K model has been well received and easily understood. It could be used to predict the influence of process parameters and material properties on the limiting value of strain and easily linked with numerical simulation results for sheet metal forming processes. Based on the above-mentioned considerations, Sect. 9.2, Marciniak and Kuczyński proposed the equation determining the value of the limit strain outside the groove at a complete loss of stability. What more, it was possible to show analytically that the FLC is a material property dependent on strain hardening exponent n, normal anisotropy coefficient R, initial strain value ε_0, and coefficient of geometric non-homogeneity f. The higher values of n, R, ε_0 *and* f, the bigger limit strain will be obtained as shown exemplary in Fig. 2b. Decreasing n by cold working of the sheet metal will decrease the limit strain very rapidly in plane strain condition. On the other hand, the limit strain in equal biaxial tension will decrease less rapidly because of limitation in development of the neck. Strain hardening is usually the most important factor influencing formability in sheet metal forming processes [11].

Prediction of the FLC depends strongly on the quality of the constitutive models. Linking the phenomenological and polycrystalline models allows better understanding of the influence of temperature, strain rate and evolution of microstructure on the FLC. It also opens the possibility to analyze strain localization in high speed forming, superplastic forming, hydroforming or incremental sheet forming. An extended literature review on developments of M-K model could be found elsewhere [10].

The main drawback of M-K model of the FLC prediction is its sensitivity to non-linear strain paths. It means that the FLC is closely related with a particular forming process. Complex strain paths influence the position and shape of the FLC. This limitation inspired the development of the FEM computational models based on M-K approach in the early 1980's. Some imperfections had to be introduced. It has become possible by describing some variations in finite element thickness. As a possible result of the simulation, the first element with strain localization would be identified. However, the disadvantage of the M-K approach related to strong dependence of the limit strain on the thickness of assumed imperfection has not been solved by these computations in a satisfactory way. Moreover, the imperfections of similar size become less significant as the thickness of sheet metal increases and the FLC becomes higher in sheets with higher thickness. Hence, the nature, size and distribution of imperfections must be considered in a probabilistic manner. One of the solutions to decrease the sensitivity of the FLC to strain path was the concept of Forming Limit Stress Diagram (FLSD) introduced in 1982 by Arrieux [12]. The FLSD could be obtained by transformation of strains from the normal FLC to principal stresses curve. In spite of advantages of the FLSD, the concept has not been widely used because stresses are seen as less intuitive variables than strains and difficult to determine experimentally. What more, values of stresses calculated by means of FEM simulations show an oscillating behavior. This problem should be additionally analyzed and overcome.

Currently used FLC models are good enough to predict limit strains though they suffer from some limitations. Each model is specific for a given group of processes, strain path, microstructure, and process parameters. Future research works should be aimed at better understanding of phenomena related to necking and fracture, modeling of microstructure with imperfections, development of commercial codes enabling accurate calculations of forming limit curves for complex strain paths.

4 Conclusions

50 years have already passed since Marciniak's model of plastic biaxial stretching of shell was introduced with taking into account the geometric and microstructure non-homogeneities. Thus, Marciniak and Kuczyński suggested FLC prediction model based on imperfections in sheet metal. This so-called M-K model has become very important for the further development of formability assessment in sheet metal forming processes.

It was clearly shown by the M-K model that development of the neck in biaxial tension is considerably delayed by geometric constraints and the additional strain is greater than in the uniaxial tension. What more, analytically predicted FLC could be presented as a material property dependent on strain hardening exponent n, normal anisotropy coefficient R, initial strain value ε_0, coefficient of geometric non-homogeneity f, and other parameters. Thus M-K model proved to be very useful to investigate the individual effect of these parameters on the FLC. This physical meaning of the M-K model has been well received and easily understood. Still, it has a big potential for developments of formability assessment.

References

1. Erichsen, A.M., 1913, Process and apparatus for testing metal sheets and plates, GB Patent no. 91228571 (A), applied 1912-12-11
2. Erichsen cupping test: ISO standard 20482:2013
3. Keeler, S.P., 1961, Plastic instability and fracture in sheets stretched over rigid punches, PhD Thesis, Massachusetts Institute of Technology, Boston

4. Keeler, S.P., Backofen W.A., 1963, Plastic instability and fracture in sheets stretched over rigid punches, ASM Trans, 56, 25–48
5. Goodwin G.M, 1968. Application of strain analysis to sheet metal forming problems in the press shop, SAE Technical Paper 680093, 380–387
6. Swift H.W., 1952, Plastic instability under plane stress, Journal of the Mechanics and Physics of Solids, 1, 1–16
7. Hill, R., 1952, On discontinuous plastic states, with special reference to localized necking in thin sheets, Journal of the Mechanics and Physics of Solids, 1, 19–30
8. Marciniak, Z, 1965, Stability of plastic shells under tension with kinematic boundary condition, Archive of Applied Mechanics (Archiwum Mechaniki Stosowanej), 17, 577–592
9. Marciniak, Z., Kuczyński, K., 1967, Limit strains in the process of stretch-forming sheet metal, International Journal of Mechanical Sciences, 9, 609–620
10. Banabic, D., 2010, A review on recent developments of Marciniak-Kuczyński model, Computer Methods in Materials Science, 10, 225–237
11. Marciniak Z., Duncan J., 1992, Mechanics of sheet metal forming, Edward Arnold
12. Arrieux, R., Bedrin, C., Boivin, M., 1982, Determination of an intrinsic forming limit stress diagram for isotropic metal sheets, Proc. of the 12th IDDRG Congress, St. Margherita Ligure, 61–71

Cyclic Plasticity Model for Accurate Simulation of Springback of Sheet Metals

Fusahito Yoshida and Takeshi Uemori

1 Motivation

For lightweight constructions, high strength steel (HSS) sheets and aluminum sheets are increasingly used in many fields, especially in automotive industry. One of the serious problems in press forming of such metallic sheets is their very large springback. The prediction of springback is a key issue in the numerical simulation of sheet metal forming process. For the accurate simulation of forming and springback an appropriate constitutive model that properly describes material behavior is needed. In the springback analysis the description of elasto-plasticity stress-strain response in a reverse deformation is of vital importance. This is because, for example, in draw bending a material element is subjected to stretch bending and unbending when a sheet is drawn over a die-corner, it is a typical example of large-strain cyclic plasticity deformation. In such a case, a constitutive model of cyclic plasticity is essential for an accurate calculation of bending moment which is the driving force of springback. After a large plastic deformation in press forming, during the unloading process, springback takes place almost elastically, but in some cases re-yielding and the subsequent small-scale plastic deformation would occur. Specifically for HSS sheets, they have extremely high deformation resistance which induces a large driving force of springback, and furthermore, they exhibit very strong Bauschinger effect because of their heterogeneous micro structures, so that the early re-yielding and elastic-plastic deformation are more likely to occur during springback.

For modeling of large-strain cyclic plasticity of sheet metals, the following material behavior should be taken into account (see Fig. 1):

Fusahito Yoshida ✉
Hiroshima University, Higashi-Hiroshima, Japan
e-mail: fyoshida@hiroshima-u.ac.jp

Takeshi Uemori
Okayama University, Okayama, Japan

A. E. Tekkaya et al. (eds.), *60 Excellent Inventions in Metal Forming*,
DOI 10.1007/978-3-662-46312-3_10

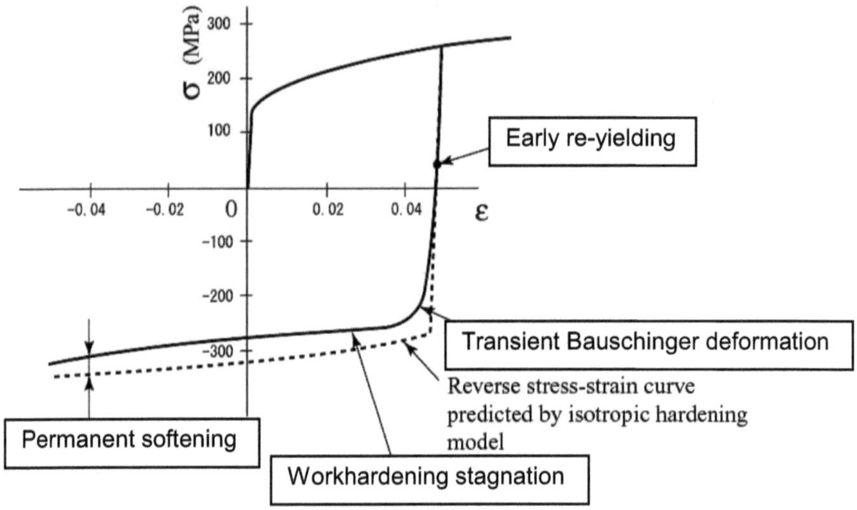

Fig. 1 Stress-strain response in a reverse deformation

- two stages of the Bauschinger effect: (i) *the transient Bauschinger deformation* characterized by early re-yielding and smooth elastic-plastic transition with a rapid change of workhardening rate; and (ii) *the permanent softening* observed in a region after the transient period;
- *the workhardening stagnation* appearing at a certain range of reverse deformation;
- strain-range and mean-strain dependency of cyclic hardening, e. g., the larger the cyclic strain range the larger the saturated stress amplitudes (refer to Yoshida et al. [1]).

To describe the above material behavior, the present authors proposed an advanced kinematic hardening model (so-called 'Yoshida-Uemori model', or 'Y-U model' for short, in 2002–2003, [1–3]). The Y-U model has been implemented into several FE commercial codes for sheet metal forming simulation, e. g., LS-DYNA (implemented in 2006), PAM-STAMP (in 2006), StamPack (in 2012) and ASU/P-form (in 2013), and nowadays it is widely used for springback simulation in press forming industry. The outline of the model, its performance in describing cyclic stress-strain responses and springback simulation applications are shown in the following Section.

2 Material Model: *Yoshida-Uemori Model*

The Y-U model is constructed within the framework of two-surface modeling, wherein the yield surface moves kinematically within a bounding surface, as schematically illustrated in Fig. 2a. Since the experimental stress-strain responses under reverse deformation show that the re-yielding starts at a very early stage of stress reversal (see [1]), only the kine-

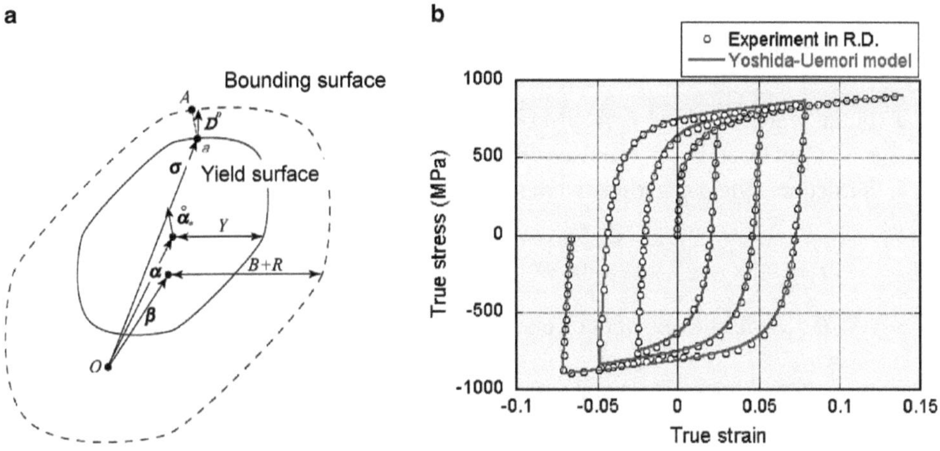

Fig. 2 The Yoshida-Uemori model. **a** schematic illustration of two-surface modeling, **b** example of simulation by the model for cyclic plasticity stress-strain response on a 780-MPa HSS sheet

matic hardening for the yield surface is assumed, while for the bounding surface mixed isotropic-kinematic hardening. In this model, workhardening characteristics are expressed by a newly proposed evolution equation of the backstress i. e., kinematic hardening law.

The global workhardening is expressed by the isotropic hardening of the bounding surface, and the permanent softening in a reverse deformation is described by its kinematic hardening. Furthermore, one of the features of the Y-U model is that it is able to describe the *workhardening stagnation* that appears in a reverse stress-strain curve for a certain range of reverse deformation (for details of modeling, refer to [2–3]).

In the present model, the size of yield surface is held constant. However, if we carefully observe the stress-strain response during unloading after plastic deformation, we find that the stress-strain curve is no longer linear but rather is slightly curved due to very early re-yielding and the Bauschinger effect. To describe this phenomenon, in the model, the equation of *plastic-strain dependent Young's modulus* is introduced.

The outline of the model is shown in the box "Outline of the Yoshida-Uemori model". Figure 2b shows an example showing the performance of the model in describing cyclic plasticity behavior on a 780-MPa HSS sheet, where we can see that the calculated stress-strain response agrees very well with the experimental result.

Outline of the Yoshida-Uemori model

- Initial yield criterion: $f_0 = \phi(\boldsymbol{\sigma}) - Y = 0$,
  $\boldsymbol{\sigma}$ = Cauchy stress, ϕ = anisotropic yield function, Y = initial yield stress
- Subsequent yield criterion and the associated flow rule:

$$f = \phi(\boldsymbol{\sigma} - \boldsymbol{\alpha}) - Y = 0, \; \boldsymbol{D}^p = \frac{\partial f}{\partial \boldsymbol{\sigma}} \dot{\lambda},$$

  $\boldsymbol{D}^p$ = plastic rate of deformation, $\boldsymbol{\alpha}$ = backstress, $\dot{\lambda}$ = scalar multiplier

- Bounding surface: $F = \phi(\boldsymbol{\sigma} - \boldsymbol{\beta}) - (B + R) = 0$,

  $\boldsymbol{\beta}$ = kinematic hardening stress, R = isotropic hardening stress,
  B = initial size of bounding surface

- Evolution of backstress (kinematic hardening law):

$$\boldsymbol{\alpha}_* = \boldsymbol{\alpha} - \boldsymbol{\beta}, \quad \dot{\boldsymbol{\alpha}}_* = C\left[\left(\frac{a}{Y}\right)(\boldsymbol{\sigma} - \boldsymbol{\alpha}) - \sqrt{\frac{a}{\bar{\alpha}_*}}\boldsymbol{\alpha}_*\right]\dot{p},$$

  $\dot{p}$ = effective plastic strain rate, $\bar{\alpha}_* = \phi(\boldsymbol{\alpha}_*)$, $a = B + R - Y$

- Degradation of unloading stress-strain slope (*plastic-strain dependent Young's modulus*):
  $E = E_o - (E_o - E_a)[1 - \exp(-\xi p)]$,
  E_o: Young's modulus of virgin material,
  E_a: unloading stress-strain slope at infinitely large strain

3 Applications of the Model to Springback Simulation

As an example, Fig. 3 shows two results of springback simulation for a hat-shaped draw-bending on 980-MPa HSS sheet, one by the isotropic hardening model (Fig. 3b) and the other by the Y-U model (Fig. 3c), together with the experimental result (Fig. 3a). In both the simulations LS-DYNA-based FE code JSTAMP (JSOL Ltd.) was employed. The Y-U model captures large side-wall curl realistically, but the isotropic hardening model completely fails to simulate it. From this comparison, it is clear that the use of an accurate model of the Bauschinger effect, including the plastic-strain-dependent Young's modulus, is essential for springback simulation. The second example is springback in S-rail forming (Numisheet '96) of 980-MPa HSS sheet. In the simulation, FE code PAM-STAMP 2G (ESI Co. Ltd.) with the Y-U model was used. As shown in Fig. 4, the simulation results

Fig. 3 Springback of hat-shaped draw bending of 980-MPa HSS: experimental result and simulations. **a** Experiment, **b** Isotropic hardening model, **c** Y-U model

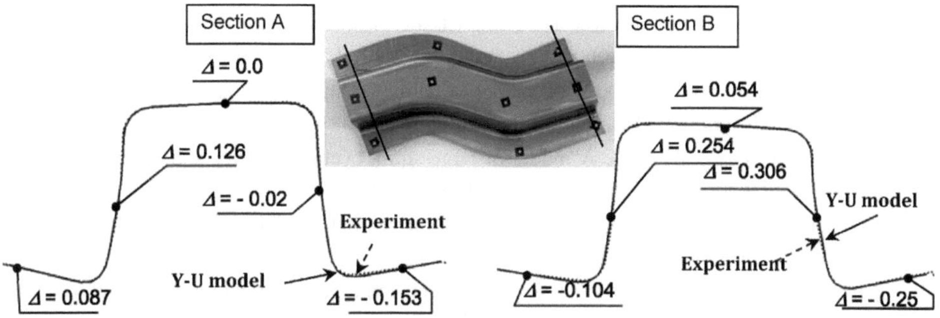

Fig. 4 Comparison of FE simulation (Y-U model) with experiment of springback of 980-MPa HSS S-rail. **Δ** = [Simulation] – [Experiment] in mm

agree very well with the experimental observations. Accuracy of the Y-U model and its high performance in springback simulation have been verified through several benchmark tests (e. g., see [4–6]).

4 Concluding Remarks

For springback prediction, its control and compensation, FE simulation taking account of materials' Bauschinger effect is essential. By using the Y-U model, accurate FE simulation of springback is made possible. The significant features of the Y-U model are summarized as follows:

1. It has a high capability of describing various cyclic plasticity characteristics, such as the Bauschinger effect, the workhardening stagnation, strain-range dependent cyclic workhardening, as well as the degradation of unloading stress-strain slope with increasing plastic strain.
2. It requires a limited number of material parameters (7 or 8 plasticity parameters and 3 elasticity parameters [including Young's modulus]). The scheme of material parameter identification and material tests for it have been clearly presented (refer to [2]).

3. To describe the anisotropy of sheets, it is able to use any type of anisotropic yield function.
4. It has obtained popularity in sheet metal forming industry since it has been already implemented into several FE commercial codes and is widely used for springback simulation.

References

1. Yoshida, F., Uemori, T., Fujiwara, K., 2002. Elastic-plastic behavior of steel sheets under in-plane cyclic tension-compression at large strain. Int. J. Plast. 18, 633–659.
2. Yoshida, F., Uemori, T., 2002. A model of large-strain cyclic plasticity describing the Bauschinger effect and workhardening stagnation. Int. J. Plast. 18, 661–686.
3. Yoshida, F., Uemori, T., 2003. A model of large-strain cyclic plasticity and its application to springback simulation. Int. J. Mech. Sci. 45, 1687–1702.
4. Eggertsen, P.-A., Mattiasson, K., 2010. On constitutive modeling of springback analysis, Int. J. Mech. Sci. 52, 804–818. Int. J. Mech. Sci. 52, 804–818.
5. Ghaei, A. Green, D. E., Taherizadeh, A., 2010. Semi-implicit numerical integration of Yoshida-Uemori two-surface plasticity model, Int. J. Mech. Sci. 52, 531–540.
6. The NUMISHEET 2011 Benchmark Study, BM4: Pre-strain effect of spring-back of 2-D draw bending, in Proceedings of the 8th International Conference and Workshop on Numerical Simulation of 3D Metal Forming Processes, Part C.

Fast Semi-analytical Approach for Deep Drawing Processes

Alexander Brosius and Tim Cwiekala

1 Introduction

For process planning of sheet metal forming processes the use of the Finite-Element-Method is currently state-of-the-art. Due to the significant draw-back of a time-consuming simulation for achieving a high accuracy (see Fig. 1), real-time modeling and simulation of production processes impossible for the early stage of process planning or optimization. In order to comprehensively study the process sensitivities during the process itself alternative methods are needed.

Oehler and Kaiser [2] as well as Doege and Rambke [3] have developed analytical approaches with limited accuracy in order to simplify the determination of deformation fields that are used for approximation of stress levels and forming energy needed.

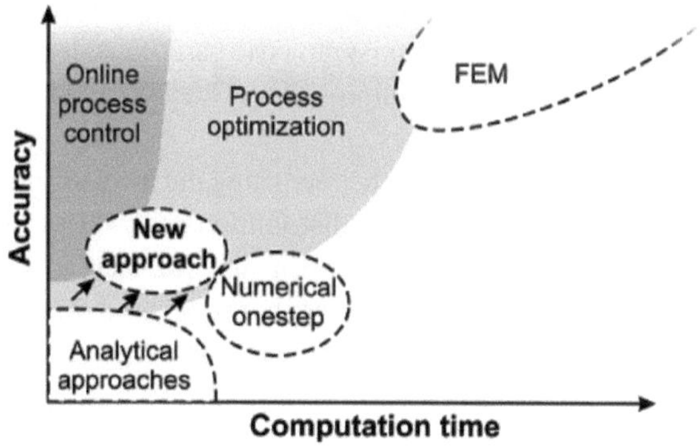

Fig. 1 Relationship between accuracy and computation time [1]

Alexander Brosius ✉ · Tim Cwiekala
TU Dresden, Dresden, Germany
e-mail: Alexander.Brosius@tu-dresden.de

A. E. Tekkaya et al. (eds.), *60 Excellent Inventions in Metal Forming*,
DOI 10.1007/978-3-662-46312-3_11

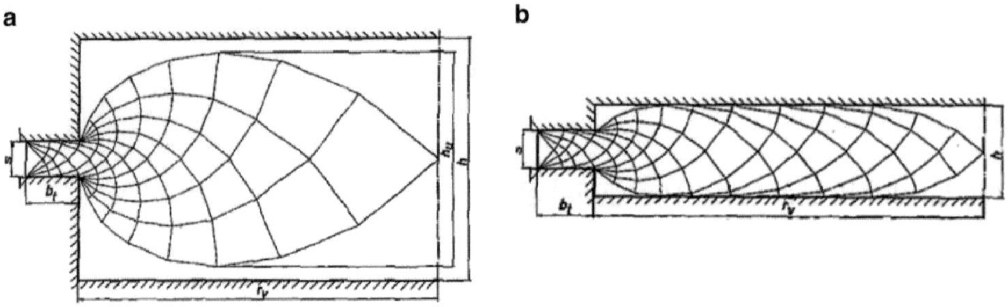

Fig. 2 Slip-line field for closed-die press forming [7]

The insufficient accuracy of these methods has its origin in the implemented simplifications for the analytical calculation. With the introduction of the slip-line field theory which is based on the work of Hill [4] and Hodge [5] the authors intend to reach a higher degree of accuracy while maintaining a fast semi-analytical approach that allows for real-time applications. Lange [6] has presented a very compact summary of the theory for application, demonstrating the great advantage of having relatively simple equations for calculating inhomogeneous strain fields. Originally, the slip-line theory was a graphical oriented procedure that has been transformed into a powerful fast simulation tool by the combination with conventional numerical algorithms like numerical integration and solvers for equation systems. An interesting example for the application of the theory was presented by Voelkner [7]. For a closed-die press forming process he determined the influence of die geometry and flash on the stress field, see Fig. 2.

The accuracy of the slip-line theory for calculating the stress distribution in a given part is reduced mainly by the necessary assumption of plain strain deformation. Nevertheless, several authors like Glöckl in [8] and Kim and Kobayashi [9] implemented algorithms for predicting the initial blank shape. Additionally, Siebel [10] implemented the theory for calculating the drawing force needed for axisymmetric parts considering ideal, friction and bending forces which was consecutively improved by Doege et al. [11] using the principal of virtual work for predicting punch forces.

Complementary new numerical approaches are using the inverse finite element method, the so called one-step method where strain distributions as well as blank geometry are predicted starting from a given final geometry in just one-step as shown by Lee [12]. Despite the inherent non-linearity of this approach, the method delivers fast results with a simulation time between seconds and minutes while working at an acceptable level of accuracy.

2 Theory of the new Approach

To increase the overall speed of a deep-drawing simulation, the usual application of the time-consuming finite-element method has to be avoided. Therefore, a geometry-based one-step mapping algorithm has been developed. It is based on a mapping algorithm, that estimates the blank geometry from a given sheet metal part. The calculated differences between the geometries is the basis to approximate the strain distributions in the deep-drawn part. Contrary to so called one-step solvers which use the inverse finite-element analysis [13, 14], the developed mapping algorithm uses a straight forward approach with only a small iteration loop. This results in a faster simulation for two-dimensional part like an axisymmetric cup. Because the algorithm for the three-dimensional part partly bases on the principle of the two-dimensional calculation, the idea will be explained only for the latter. The mapping algorithm contains several steps and starts with a first mapping by assuming volume constancy. This causes some errors compared to the exact simulation, but it serves as an initial guess for the above mentioned small iteration loop. One major assumption that has to be made for the geometric mapping is a constant sheet thickness for the whole part. To obtain the initial blank geometry, all elements are projected to a flat surface (see Fig. 3).

The calculation of the strains in each element is based on the geometric difference in the two geometries. Afterwards, this result is used to calculate a thickness distribution in the deep-drawn part with a corrected mapping and strain calculation. Therefore, the next step in the simulation process in order to have an enhanced accuracy of the result is the consideration of a varying sheet thickness for the sheet metal part. This distribution takes into account the two main influences on thickness change during the deep-drawing operation, namely tension-compression in the flange and tension-bending at the drawing radius. A third but minor influence is considered as well with the effect of the biaxial-

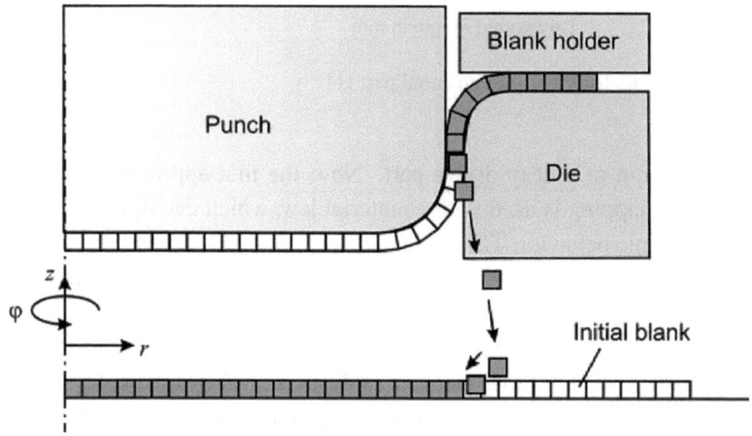

Fig. 3 Mapping strategy for two-dimensional part geometries [15]

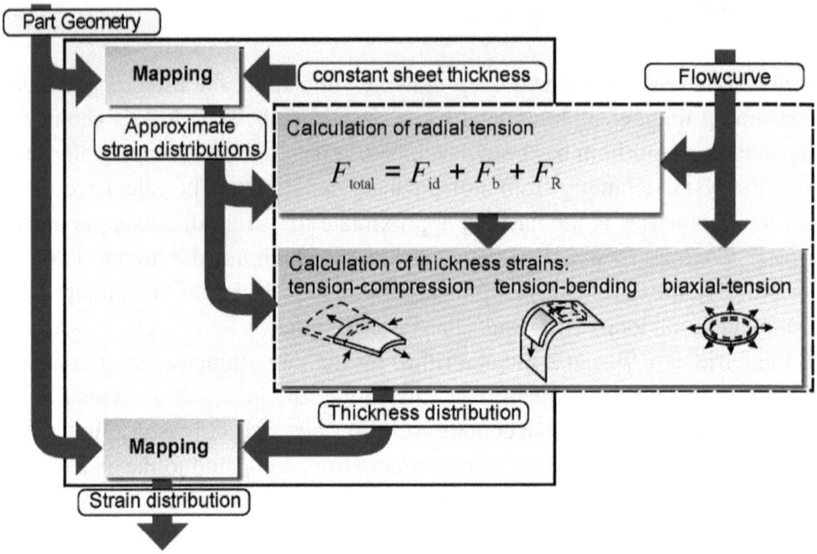

Fig. 4 Flow chart for strain calculation [1]

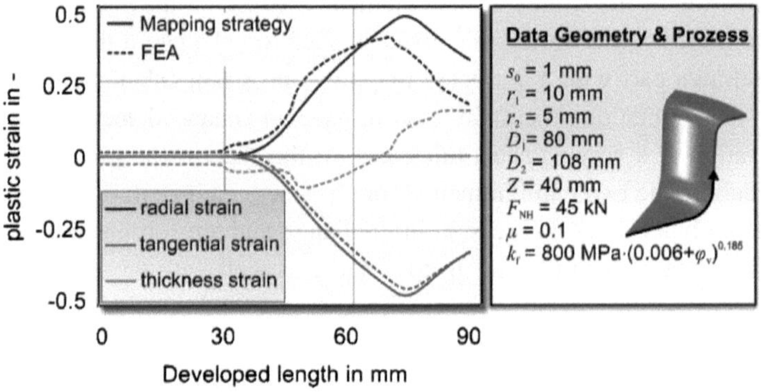

Fig. 5 Achieved results for a two-dimensional part [1]

tension in the bottom of a deep drawn part. Now, the first approximation of the strains by the geometric mapping is used with a material law, which can describe an isotropic as well as an anisotropic behavior. The results are the radial tension and the thickness strain, whereby the complete description of the analytical equations can be found in [16]. The flow chart is shown in Fig. 4.

Figure 5 shows the result of the developed mapping algorithms compared to a result from an incremental implicit finite-element analysis. The diagram displays the high accuracy of the approach for an axisymmetric part despite the fact that the simulation time for the semi-analytical approach is below one second.

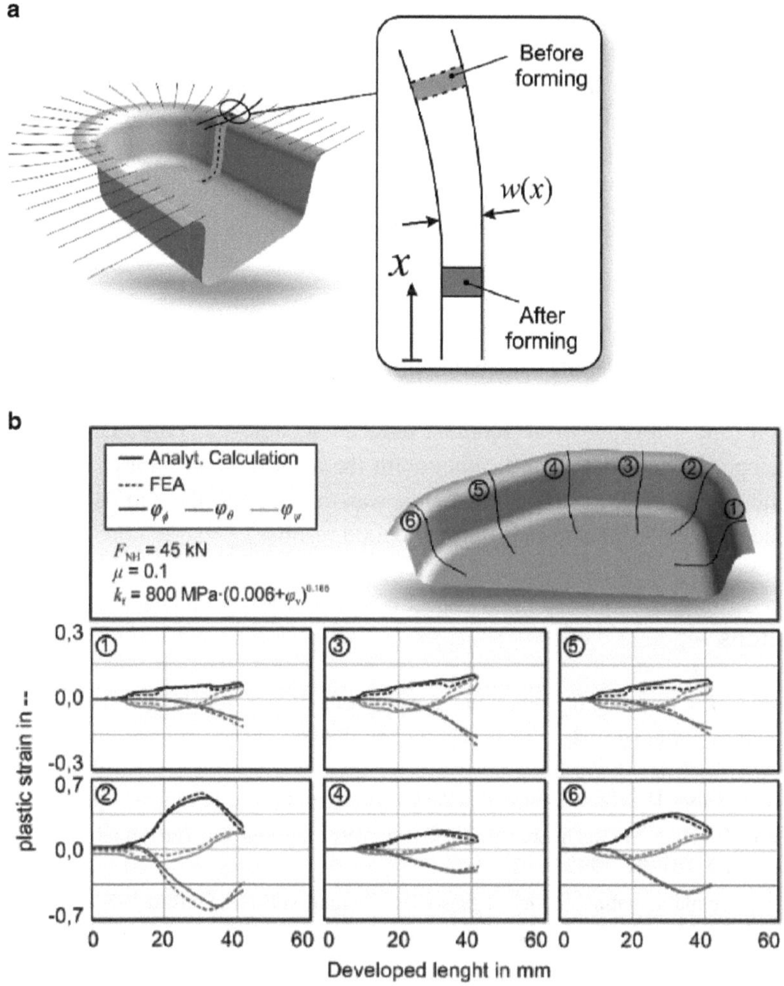

Fig. 6 **a** Flow channels [12], **b** Obtained Results [1]

In order to apply this time efficient and accurate approach for three-dimensional parts some changes have to be performed. The reason for these modifications is the missing possibility of the pure geometric mapping based on the volume constancy. Therefore, some radial flow channels are defined, where the major deformation takes place. The complete number of flow channels (compare Fig. 6a) has to be constructed by using the assumption that the first two principal strains are in the direction of the material movement and perpendicular to it. These flow fields enables the determination of the deformation behavior based on the semi-analytical approach described above for the two-dimensional parts. Therefore, the radial and tangential strains as well as the change of sheet thickness can be determined in a comfortable way.

An example of the capability is shown in Fig. 6b, which shows the obtained results from the semi-analytical approach compared to an incremental implicit finite-element analysis. All three strain-levels are in very good accordance to the complex finite-element simulation. While the simulation time needed for the semi-analytical approach is up to 1,000 times faster than the FEA.

3 Summary

The slip-line theory is a powerful method to determine the state as well as the strain distribution under some simplifying assumptions. The drawback can be seen in deviations compared to the finite-element solution, but the great advantage is the extremely short calculation time. In order to overcome a reduced accuracy, a semi-analytical approach was developed to combine the efficiency with the achievable accuracy. In combination with numerical algorithms, the applicability was extended to the simulation of realistic three-dimensional part geometries with very good results.

References

1. Cwiekala, T.: Entwicklung einer Simulationsmethode zur zeiteffizienten Berechnung von Tiefziehprozessen, Dr.-Ing. Dissertation TU Dortmund, Shaker-Verlag, ISBN 3844007016, 2011.
2. Oehler G, Kaiser F. Schnitt- Stanz- und Ziehwerkzeuge. Berlin: Springer-Verlag, 1975.
3. Doege E, Rambke M. Rechnergestützte Zuschnittsermittlung für Tiefziehteile, Blech, Rohre, Profile 42., p. 187–92, 1995.
4. Hill, R.: The mathematical theory of plasticity. Clarendon Press, Oxford 1950.
5. Prager, W.; Hodge, P.G.: Theorie idealplastischer Körper. Wien: Springer-Verlag, 1954.
6. Lange, K.: Umformtechnik – Band 3: Blechbearbeitung (2. Auflage). Springer-Verlag, Berlin Heidelberg, ISBN: 3540-50039-1, 1990.
7. Voelkner, W.: Zur theoretischen Spannungsermittlung beim Formpressen mit Grat, Wissenschaftliche Zeitschrift der TU Dresden 17, Heft 5, S. 1273–1281, 1968.
8. Glöckl, H.: Rechnerunterstützte Optimierung des Tiefziehens unregelmäßiger Blechteile, Ber. Nr. 68 Inst. f. Umformtechnik der Universität Stuttgart. Berlin, Heidelberg, New York, Tokyo: Springer-Verlag; 1983.
9. Kim, N.; Kobayashi, S.: Blank design in rectangular cup drawing by an approximate method. Int. J Mach Tool Des Res, 26(2):125–35, 1986.
10. Siebel, E. Die Formgebung im bildsamen Zustande, Düsseldorf, Verlag Stahleisen m.b.H., 1932.
11. Doege, E.; Behrens B. A.: Springub, B.: Improved analytical model for deep drawing processes of rotationally symmetric cups. Proc. NUMIFORM conference, 844–50, 2004.
12. Lee, CH.; Huh, H.: Blank design and strain estimates for sheet metal forming processes by a finite element inverse approach with initial guess of linear deformation. J. Mater. Process Technology, 82:145–55, 1998.

13. Guo, Y.Q.; Batoz, J.L.; et al.: Recent developments on the analysis and optimum design of sheet metal forming parts using a simplified inverse approach. Computers and structures 78, p. 133–148, 2000.

14. J. Lan, X. Dong, Z. Li: Inverse finite element approach and its application in sheet metal forming. J. Mater. Process. Technol. 170, pp. 624–631, 2005.

15. Cwiekala, T.; Brosius, A.; Tekkaya, A.E.; Grydin, O.; Schaper, M.; Bach, F.-W.; Barthel, C.; Svendsen, B.: Efficient Modelling and Simulation of Process Chains in Sheet Metal Forming and Processing. Steel research int. 79, No. 10, pp. 731–737, 2008.

16. Z. Marciniak, J.L. Duncan, S.J. Hu: Mechanics of Sheet Metal Forming, Butterworth-Heinemann, ISBN: 0 7506 5300 0, 2002.

Part III

Sheet Metal Forming

Vaporizing Foil Actuator: a Tool for Creating High-Pressure Impulses for Metalworking

Anupam Vivek, Geoffrey A. Taber, Jason R. Johnson, and Glenn S. Daehn

1 Introduction

Impulse based metalworking is interesting due to several advantages it offers over a corresponding quasi-static process. Increased forming limits, reduced springback, low cost tooling and reduced wrinkling are some of the documented advantages of impact forming [1, 2]. Shearing at high speeds has been shown to reduce sliver formation and provide increased dimensional tolerance [3]. Impact welding is a solid state process that allows the joining of dissimilar metals with little to no heat affected zone [4].

Explosives [5], magnetic pulse [1], and underwater electric discharge [6] are a few of the commonly applied tools for impulse metalworking. Using explosives runs into issues such as the inability to scale the process to thin sheets or small weld zones because explosives do not detonate reproducibly below a certain size known as their critical diameter. Furthermore, increasingly stringent safety regulations inhibit their use. The most significant drawback of using magnetic pulse for metalworking is the limited longevity of actuators above a certain magnetic pressure and operating cycle frequency [7]. Besides, magnetic pulse based actuators are efficient only with workpieces that are electrically conductive. Conductive driver sheets can be used to launch materials of lesser conductivity, but that adds to process time and cost. Underwater discharge for electrohydraulic metalworking has been found to pose problems such as inefficiency and pressure heterogeneity due to cavitation resulting from locally high pressures in the working media. The recently developed, vaporizing foil actuator (VFA) technique enables impulse based metalworking at size scales similar to magnetic pulse and electrohydraulic forming.

Anupam Vivek · Geoffrey A. Taber · Jason R. Johnson · Glenn S. Daehn ✉
Ohio State University, Columbus, USA
e-mail: daehn.1@osu.edu

A. E. Tekkaya et al. (eds.), *60 Excellent Inventions in Metal Forming*,
DOI 10.1007/978-3-662-46312-3_12

2 Process Idea

The process idea was developed on the basis of the exploding conductor phenomenon [8] that has been a subject of, mostly, phenomenological studies for a few decades. When a capacitor bank-driven, large current is passed through a thin conductor, the latter quickly sublimates. The conductor could be in the form of foil, wire, or wool. The rapid vaporization of the conductor creates a large pressure pulse, which is accompanied by light and sound energy. The pressure pulse, if directly properly, can propel workpieces to significantly high velocities, which can subsequently undergo metalworking operations [9]. The adaptation for metalworking purposes has been termed as VFA [10]. Aluminum has been found to be the material of choice for the foil, because, upon vaporization it reacts exothermically with nitrogen, oxygen, and carbon to increase the driving pressure even further. The thickness of the foil actuators is, typically, between 50 μm and 150 μm. Figure 1 shows the configurations for some of the applications of VFA. In each adaptation, in order to efficiently direct the pressure generated from foil vaporization toward the flyer, the foil is backed by a thick block of steel. The foil is electrically insulated from the steel fixture and sheet metal workpiece by polyester tape. The shape of the foil controls the distribution of the generated pressure: higher pressures are obtained above the narrowed section of the

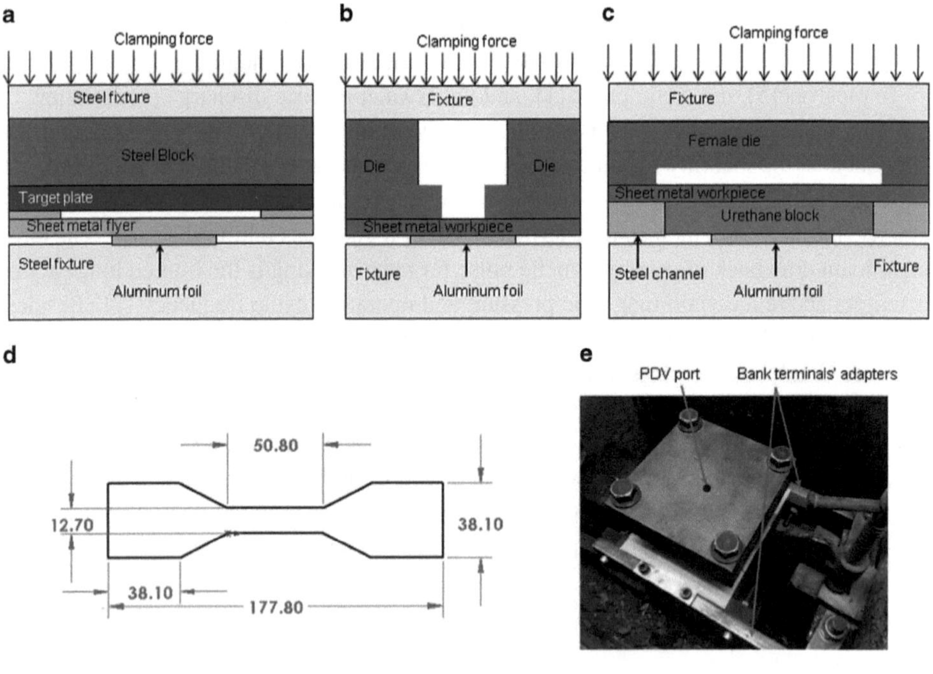

Fig. 1 Schematics for set up for **a** Impact welding, **b** shearing, **c** forming and embossing, **d** An example of the typical shape of a foil actuator (dimensions are in mm), **e** actual representation of the vaporizing foil actuator assembly

foil. For impact welding and shearing, the sheet metal is placed directly against the foil actuator. For forming and embossing an elastomer layer could be placed in between the foil and the workpiece in order to spread the pressure uniformly over a larger area than the active, narrow region of the foil. Measurements of flyer velocity, current, and discharge voltage are conducted to ensure efficiency of the processes.

3 Results and Applications

Figure 2 shows the temporal evolution of the velocity of a 0.5 mm thick copper sheet launched with 8 kJ of input energy into a 0.0762 mm thick aluminum foil actuator. This figure also illustrates the variation of current and voltage with time, indicating that a maximum current of 127 kA was reached within 10 μs of discharge and the foil burst occurred shortly after. The foil burst corresponds with the local maxima depicted in the voltage curve. The copper sheet rapidly accelerates to 550 m/s within 5 μs of foil burst. As plotted in Fig. 2b, this corresponds to a travel distance of mere 1.6 mm. Furthermore, once the foil vaporizes its conductivity decreases rapidly and the electrical circuit is effectively opened. Therefore, the reversal of the direction of current is inhibited, which is good for the capacitor bank's circuitry. A basic analytical model, based on input energy and thicknesses of

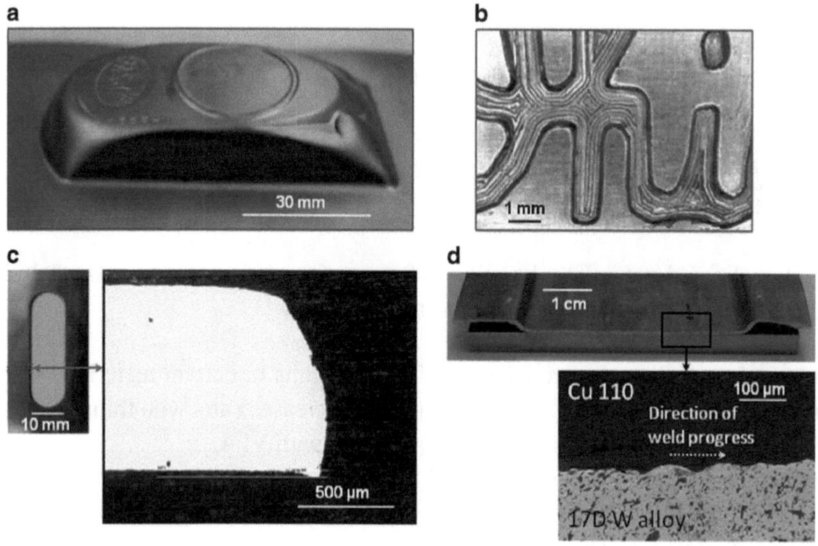

Fig. 2 Some salient applications of VFA. **a** A 0.5 mm thick CP-Ti sheet formed into a cell phone case die, **b** A 0.5 mm thick AA 2024-T3 sheet embossed into a die with machining marks, **c** A 0.8 mm thick high strength ferritic steel sheared with very little sliver formation, **d** impact weld between a 0.5 mm thick copper sheet and 3 mm thick tungsten alloy plate

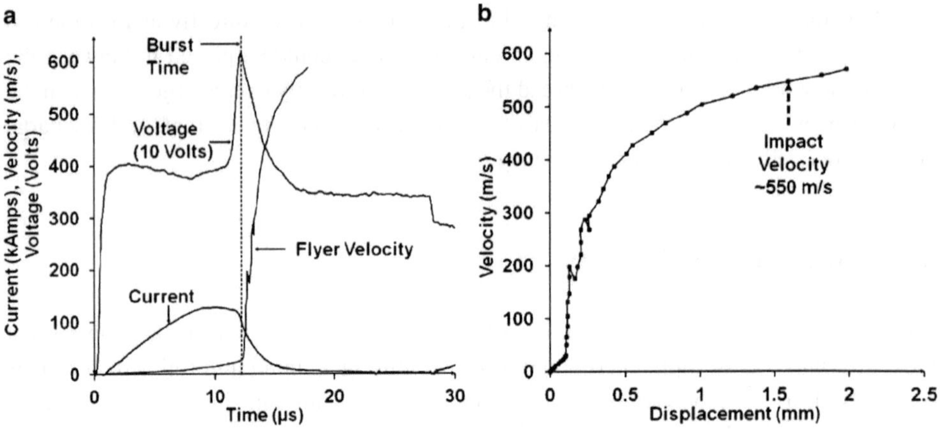

Fig. 3 Results of the diagnostics run during the VFA experiment. **a** Temporal evolution of current, voltage and velocity with 8 kJ input energy into a 0.0762 mm thick aluminum foil, **b** Velocity of the flyer plate plotted against distance travelled indicating impact with target plate at 550 m/s

foil and workpiece has been created for predicting flyer sheet velocity, and it agrees well with the experimental data [11].

The aforementioned adaptations of VFA have been successfully used for impact welding [10], forming and embossing [12], and shearing of sheet metals [9]. Figure 3 shows some results of these processes. Other processes such as springback removal, tube joining [13], and dynamic powder compaction [14] have also been implemented, but not shown here.

4 Potential

The VFA tool could potentially be utilized for a variety of impulse based metalworking tasks at an industrial level. Some of those are listed below:

1. Improve formability of materials: The forming limits of certain materials, when deformed at high strain rates, have been found to increase. Parts with thinner gauge and tighter bend radius could potentially be produced with VFA.
2. Embossing of sharp features: With the high pressures that VFA can create, embossing or coining of sharp features on automotive body panels could be implemented.
3. Springback removal: The ability to calibrate aerospace and automotive components without undergoing heat treatment cycles is a unique advantage of VFA. Elastomer pads that compliment the desired final shape of the part would need to be used as pressure transfer mediums.
4. Sheared edges with reduced burr and rollover: Flat sheared edges with low burr height could be obtained with VFA-assisted shearing of high strength materials.

5. Solid-state welding of dissimilar metals: Vehicle lightweighting calls for use of multi-material structures, which are difficult to assemble by traditional, fusion-based welding technique. VFA has been successfully used for welding together material systems, such as Al-Fe, Al-Mg, that are relevant to the automotive industry.

5 Conclusion

VFA is a versatile tool for impulse based metalworking operations. It operates at a size scale that is smaller than explosives and similar to magnetic pulse and electrohydraulic-based techniques. A concerted effort toward its industrial adaptation needs to be made. Scalability, reproducibility, predictability, and automation of the process should be addressed in future work. Furthermore, the byproducts of the process, such as formed gases and noise should be assessed carefully to ensure good health and safety in workplace.

6 Acknowledgement

The work reported here has resulted from several projects. The authors are particularly grateful for the financial support provided by US Department of Energy (DOE) under award number (DE-EE0006451). Additional funding provided by Alcoa Foundation's Advancing Sustainability in Research is deeply appreciated. Early work occurred during the period when financial support was provided by Sandia National Laboratories, to whom the authors are also very thankful. This paper builds on the work done in collaboration with many colleagues, including Erman Tekkaya , Christian Weddeling, Ryan Brune, Marlon Hahn, Steven Hansen, and Bert Liu. Their contributions are sincerely acknowledged.

References

1. Psyk, V., Risch, D., Kinsey, B. L., Tekkaya, A. E., Kleiner, M., 2011, Electromagnetic forming – A review. Journal of Materials Processing Technology, 211(5), 787–829.
2. Daehn, G. S., 2006, High-velocity metal forming. Metalworking: Sheet Forming, ASM Handbook Volume 14 B, 405–418.
3. Breitling, J., 1998, The challenges and benefits of high-speed-blanking. Dr.-Ing. Dissertation. Universität Stuttgart, ISBN 3-88355-266-6
4. Aizawa, T., Kashani, M., Okagawa, K., 2007, Application of magnetic pulse welding for aluminium alloys and SPCC steel sheet joints. Welding Journal 86, 119 s–124 s.
5. Blazynski, T. Z. (1983). *Explosive welding, forming, and compaction*. Elsevier Applied Science. New York, NY.
6. Felts, R. Application of the electrohydraulic effect to metal forming. Creative Manufacturing Seminars. Advanced High Energy Rate Forming, 1961–62. Paper SP62–08, 10p.

7. Golovashchenko, S., 2007, Material formability and coil design in electromagnetic forming. Journal of Materials Engineering and Performance 16 (3), 314–320.
8. Chace, W. and Moore, H., 1959-63, *Exploding Wires*, 4 vols. Plenum Press.
9. Vivek, A., Taber, G. A., Johnson, J. R., Woodward, S. T., and Daehn, G. S., 2013, "Electrically Driven Plasma via Vaporization of Metallic Conductors: A Tool for Impulse Metal Working", *Journal of Materials Processing Technology, vol 213, issue 8, pp 1311–1326.*
10. Vivek, A., Hansen, S.H., Liu, B. C., Daehn, G. S., 2013, "Vaporizing Foil Actuator: A Tool for Collision Welding", *Journal of Materials Processing Technology, vol 213, issue 12, pp 2304–231*
11. Vivek, A., Hansen, S. R., & Daehn, G. S., 2014, High strain rate metalworking with vaporizing foil actuator: Control of flyer velocity by varying input energy and foil thickness. *Review of Scientific Instruments*, *85*(7), 075101.
12. Vivek, A., Brune, R. C., Hansen, S. R., Daehn, G. S., 2014, Vaporizing foil actuator used for impulse forming and embossing of titanium and aluminum alloys. Journal of Materials Processing Technology, 214(4), 865–875.
13. Vivek, A., Weddeling, C., Hahn, M., Daehn, G. S., & Tekkaya, A. E. (2014). Electrically Driven Plasma via Vaporization of Metallic Conductors: A Novel Tool for Joining Tubular Workpieces. *Procedia CIRP*, *18*, 62–67.
14. Vivek, A., DeFouw, J. D., Daehn, G. S., 2014, Dynamic Compaction of Titanium Powder by Vaporizing Foil Actuator Assisted Shearing. Powder Technology 254, pp. 181–186.

Hybrid Deep Drawing Tools for High Strength Steels

Thomas Mennecart, Jörg Kolbe, and Matthias Kleiner

1 Motivation

The fact that the automotive industry has to satisfy more and more customer requirements concerning the performance and the design of cars leads to an increase of derivatives. In Fig. 1 the change of number of derivatives of the Daimler AG are shown for the years 1995 up to 2008 [1].

Although more and more parts of cars can be used in several derivatives, some parts are just produced in low batch sizes up to 10,000 parts. Especially in design or in prototyping, the batch sizes are small and the use of expensive tools made out of steels is not necessary.

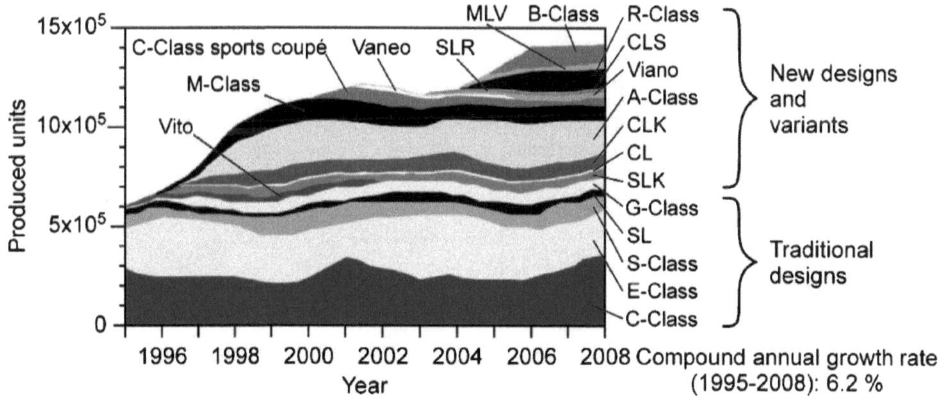

Fig. 1 Increased number of derivatives for the Daimler AG [1]

Thomas Mennecart ✉ · Jörg Kolbe · Matthias Kleiner
TU Dortmund, Dortmund, Germany
e-mail: Thomas.Mennecart@iul.tu-dortmund.de

A. E. Tekkaya et al. (eds.), *60 Excellent Inventions in Metal Forming*,
DOI 10.1007/978-3-662-46312-3_13

Several investigations exist which deal with alternative tool materials. Since the 1990's Polyurethane is used for the production of tools in prototyping processes [2]. In [3], Pinto et al. investigated the use of polyurethane and polyurethane densified wood tools for the deep drawing of DC04 steel.

For varying the flow of the material loads can be applied locally and temporally. As shown by Tommerup and Endelt in [4], loads of approximately 80 MPa are needed for a deformation of steel tools for the control of the material flow. Here, there is a benefit of the lower Young's Modulus of polymer, which enables the use of much lower forces to reach a tool deformation. With this method, the accuracy of the formed high strength steel parts can be adjusted and the tool can be used for different material types with different springback behaviours after unloading.

2 Tool Design

The production process of the hybrid tool and the main results which show the possibility of substituting steel tools are described in this section. The innovative hybrid tool is designed for the forming of batch sizes up to 10,000 parts. Figure 2 describes the process chain of the production of an hybrid tool.

The process window for the indirect application of an WC-FeCSiMn coating with a content of 50 % cast tungsten carbide (WSC) on a steel substrate by an electric arc wire spraying process has to be followed strictly to provide a removal of the coating from the substrate. Hereby, contrary to conventional coating processes, the spraying distance has to be increased up to $d_{spray} = 200\,mm$. The amperage of $I = 200\,A$ and the thermal spray velocity of $v_{spray} = 400\,mm/s$ are helpful for a homogeneous coating with a maxi-

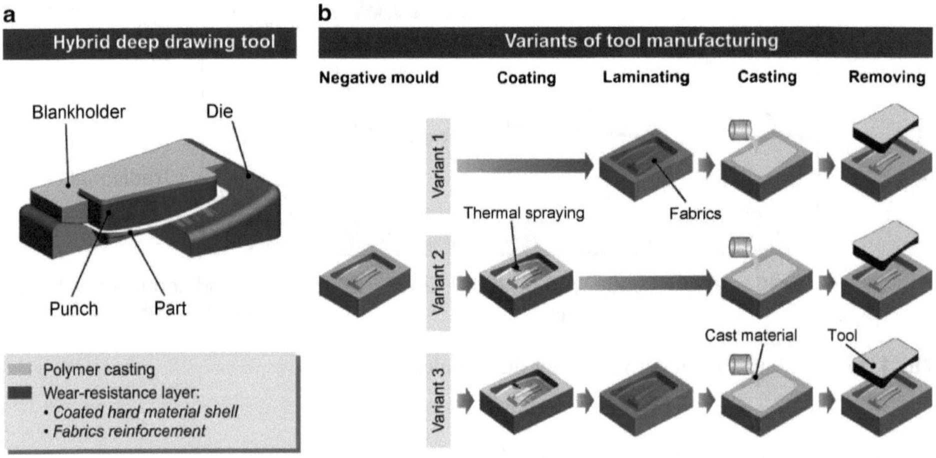

Fig. 2 Production processes of hybrid tools [5]

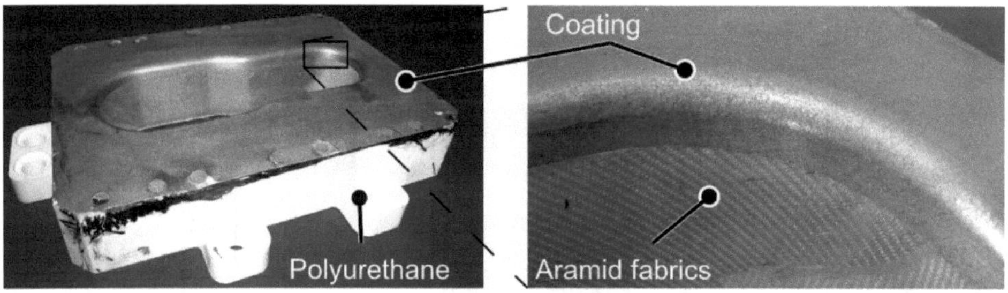

Fig. 3 Hybrid tool with coating and FRP reinforcement [6]

mum thickness of $t_{coat} = 0.35$ mm. The next step can be varied depending on the loads in forming process: If high loads are expected in the deep drawing process it is necessary to increase the stiffness of the tool to avoid deformations. In this case, the polymer can be strengthened by layers of fibre-reinforced plastics (FRP), as shown in Fig. 3, which are applied in different fibre direction on the coating with the use of a vacuum chamber. For lower forces in the process, this kind of reinforcement can be renounced and the castable polymer can be filled in directly after the coating process. After the polymer has curred, the hybrid tool can be taken out of the mould. Different polyurethanes are used with a range of shore hardness between Shore $D = 70 - 82$. A benefit of these deformable tools is the capability of varying the stiffness locally and temporally in an easy way with the use of passive elements (springs) or active elements (actuators) to adjust the geometry accuracy of the formed elements.

3 Results and Applications

For a successful use in forming of high strength steels for low batch sizes, the tools have to enable a high geometrical accuracy and have to withstand high forces without the appearance of wear on the tool surface. Therefore, different investigations were carried out, to show the possibility for a substitution of steel tools by hybrid tools. The main investigations were:

- Lifetime behaviour of hybrid tools
- Elastic behaviour of hybrid tools
- Adjustment of stiffness

For the better understanding of the process an analytical approach for the deformation of the tool has been developed and the numerical verification has been carried out through finite-element simulations [7]. It could be observed, that with a layer-thickness of FRP of 2 mm the deformation of the tool was minimised. With an application of a thickness

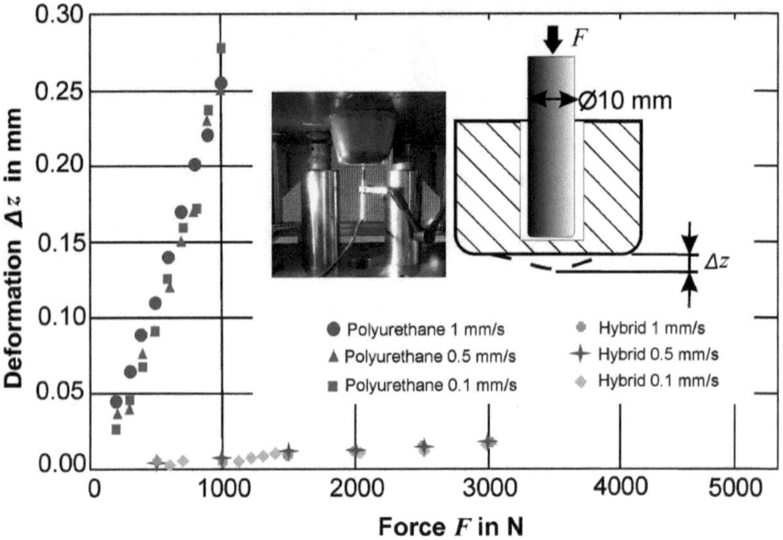

Fig. 4 Design of active and passive elements for material flow control

of 4 mm the minimisation of further deformation is not significant. Further investigations were dealing with the lifetime of the hybrid tools. Here, the hybrid tool with the FRP-reinforcement, a polymer tool made out of polyurethane with a 2 mm thick surface coating of epoxy and a conventional steel (1.2379) with a hardness of 60 HRC were compared concerning their wear behaviour when forming 1,000 parts of high strength steel HC380LA. In Fig. 4 the results of the deformation can be seen. After 200 deformed parts, the polymer tool deformed too much, so that the geometry accuracy of the formed part was not reached. The hybrid tool and the steel tool showed a reachable lifetime of 1,000 parts with just a small difference of the deformation in the die entering radius [7]. In case of different material properties caused by batch fluctuations or to equalize deformations like shown in figure but also to adjust the material flow of the blank it is possible to provide the tool with stiffer elements, which can act in an active (actuator) or passive (springs) way. For the layout of these elements experiments were carried out to measure the deformation of the tools when applying forces with different velocities. It can be shown, that the deformations for the hybrid tools is not affected by the velocity and achieve 0.022 mm with an applied force of 3 kN. For an elastic polymer tool without reinforcement and coating the deformation reached values of nearly 0.28 mm when a force of 1 kN was applied.

In forming experiments, where the actuators were placed on one side of the open channel, the geometry accuracy could be improved with the use of the stiffer elements in case of active and passive elements when forming DP600 high strength steel [8]. For the actuator a temporally variable load was applied to reduce the springback. The results show the capability to increase the geometry accuracy of the parts.

4 Potential

With the use of hybrid tools in deep drawing processes, different benefits can be used, so that steel tools can be substituted for low batch sizes of parts of high strength steels. The benefits listed below could be observed:

1. Construction of the tools is cheaper and easier compared to steel tools, in the case that low batch sizes will be formed with these tools.
2. Elastic deformation of the hybrid tool during the forming process can lead to a more homogeneous contact pressure, which minimise load peaks on the sheet.
3. Stiffness variation: With the use of passive and active elements in the hybrid tool, the accuracy of geometry could be increased.
4. Stiffness variation: Equalisation of batch fluctuation or equalisation of tool deformation can be reached with the use of stiffer elements. For an in-situ compensation of tool deformation, a closed loop control is necessary.
5. The hybrid tools show no wear after 1,000 formed parts.
6. The deformation of the die radius can be minimised with the use of FRP-layers.

5 Conclusion

For deep drawing processes of high strength steels, the use of hybrid tools can lead to cost and time efficiency. First, the design and production of these tools can be cheaper compared to the use of cast iron or steel tools made conventionally. Also the fact that moulds can be used several times, helps saving costs in production process. Inaccuracies of the hybrid tools due to the production or the deformations caused in the forming process or difference in material properties due to different fluctuations of material properties can be compensated by using elements which are able to induce loads locally and temporally. These benefits lead to the conclusion, that the hybrid tool can substitute steel tools for low batch sizes up to 10,000 parts.

6 Acknowledgement

This tool has been investigated in the subproject C1 in the Sonderforschungsbereich 708, which is kindly supported by the German Research Foundation (DFG).

References

1. Dannenberg, J. (Hrsg.); Gehr, F. (Hrsg.); Hellingrath, B. (Hrsg.); Kleinhans, C. (Hrsg.): Future Automotive Industry Structure (FAST) 2015 – die neue Arbeitsteilung in der Automobilindustrie, Materialien zur Automobilindustrie. Bd. 32. Frankfurt: Verband der Automobilindustrie, 2004
2. Baur, E. (Hrsg.); Brinkmann, S. (Hrsg.); Osswald, T. (Hrsg.); Schmachtenberg, E. (Hrsg.): Saechtling-Kunststoff-Taschenbuch. München: Carl Hanser Verlag, 2007
3. Pinto, M.; Santosa, A. D.; Teixeira, P. ; Bolt, P. J: Study on the usability and robustness of polymer and wood materials for tooling in sheet metal forming. Journal of Material Processing Technology, 202, 2008, 47–53
4. Tommerup, S.; Endelt, B.: Experimental verification of a deep drawing tool system for adaptive blankholder pressure distribution. In: Journal of Materials Processing Technology, 212, 2012, S. 2529–2540
5. Witulski, J.; Trompeter, M.; Tekkaya, A.E. : Fiber-reinforced polymers for rapid tooling of deep drawing tools of a high wear resistance. Proceeding IDDRG 2009, USA, 761–772
6. Witulski, J.; Trompeter, M.; Tekkaya, A. E.; Kleiner, M.: High wear resistant deep drawing tools made out of coated polymers. In: CIRP Annals Manufacturing Technology, 60, 2011, S. 311–314
7. Kolbe, J.: Thermisch beschichtete, faserverstärkte Polymerwerkzeuge für die Umformung höherfester Blechwerkstoffe; Dissertation, TU Dortmund, 2012, ISBN 978-3-8440-1353-5
8. Mennecart, T.; ul Hassan, H.; Fruth, J.; Wagner, T.; Güner, A.; Ben Khalifa, N.; Tekkaya, A. E.: Reduction of Springback by Use of Drawing Tools with Locally and Temporally Varying Stiffness. Conference Proceedings: TTP Conference 2013, September 19–20, Graz/Austria, 309–320

High-Accuracy & High-Rigidity Forming Machines (UL Presses)

Takaaki Imura

1 UL Press Features

The UL Press Series was developed based on the design concept of achieving "a high-accuracy machine that was more accurate than the die". It was designed in order to achieve the ultimate in high rigidity and high accuracy, and our aim was to be able to maintain high accuracy not only under static no-load conditions but also maintain dynamic accuracy under actual metalforming loading conditions.

As a result, using a UL press enables the customer to improve product accuracy and also dramatically improve die life. And it can also be utilized for a wide range of press forming applications, such as standard sheet forming, forging applications with concentrated loads, and multi-stage forming.

2 The Technologies That Enable High Accuracy

UL presses deliver high accuracy primarily through the application of the following three technologies.

① Zero-Clearance Slide Guides (Spherical Shoes, 4-Surface Guiding)
This minimizes the lateral motion of the slide under high-precision dynamic loading conditions.

② A Centrally-Located 1-Point Design without Connecting Rods
This enables a highly rigid, compact design and reduces the effects of temperature.

③ A Highly Rigid Unitized Frame
The frame design also factors in lateral rigidity.

Takaaki Imura ✉
AIDA ENGINEERING, Kanagawa, Japan
e-mail: imurat@aida.co.jp

A. E. Tekkaya et al. (eds.), *60 Excellent Inventions in Metal Forming*,
DOI 10.1007/978-3-662-46312-3_14

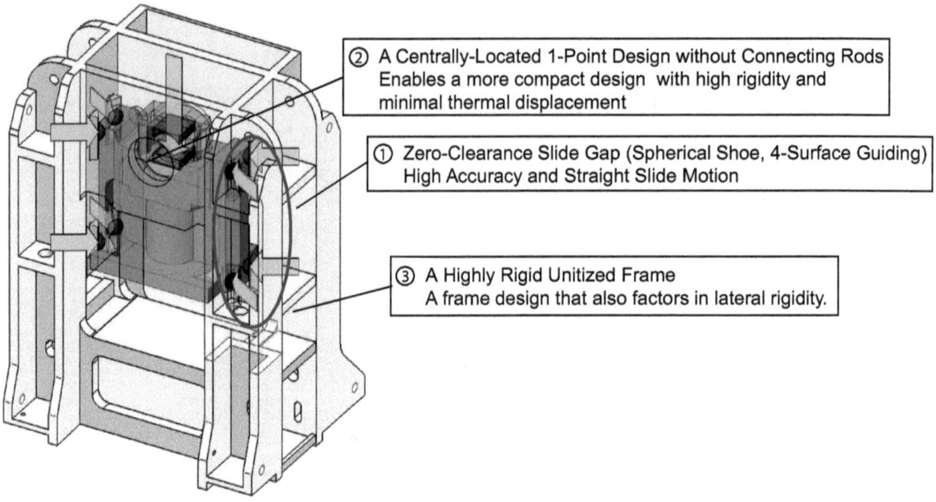

Fig. 1 The UL Press Design

Fig. 2 The Slide Guide Area Design

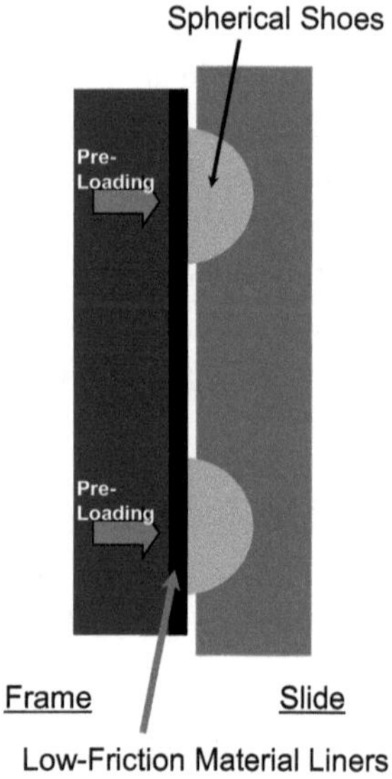

The design of a UL Press is shown in Fig. 1.

The usage of spherical shoes (at 4 locations each on upper and lower horizontal planes) on the back surface of the slide guide areas enables zero-clearance slide guiding. (Fig. 2)

Thanks to the spherical shoes, surface contact is always maintained regardless of the loading conditions. As a result, we have achieved static accuracy for perpendicularity between the bolster top and the vertical slide motion that is 10 μm or less.

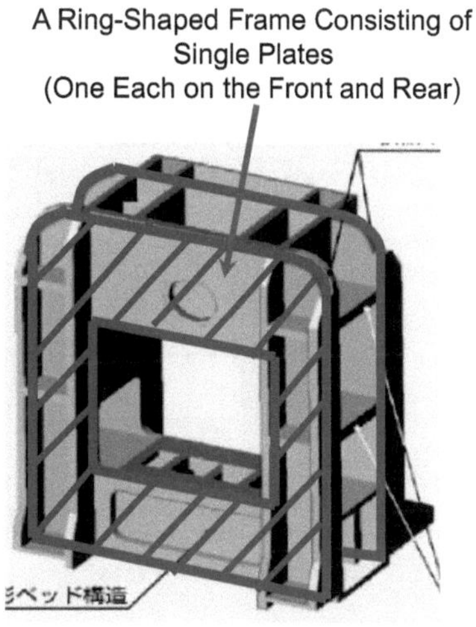

Fig. 3 The Frame Design

In order to assure not only elongation-related longitudinal rigidity but also lateral rigidity, we use a unibody frame design consisting of ring-shaped plates (except on some large UL presses) to withstand the forming loads. (Fig. 3)

The laterally-rigid frame suppresses lateral slide movement during forming.

When eccentric loads are applied during forming, there are five load-bearing points – ①: The slide drivetrain; ② & ③: The front and back lower slide guides; and ④ & ⑤: The front and back upper slide guides. Slide accuracy is maintained by these load-bearing areas and the lateral rigidity of the frame (Fig. 4).

The bed area is designed with load-bearing ribs between the front and rear ring plates, resulting in rigidity characteristics equivalent to a forging press. With its 1-point high-

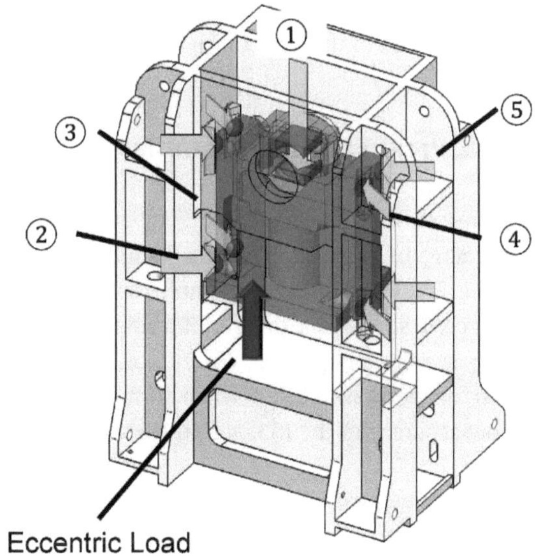

Fig. 4 Load-Bearing Areas When Eccentric Loads are Applied

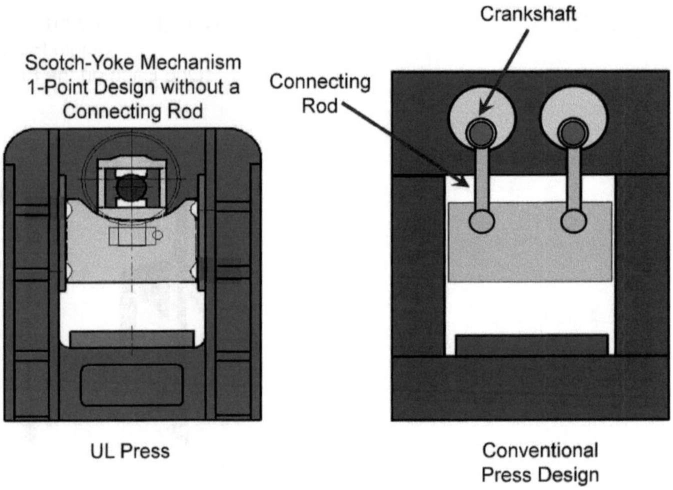

Fig. 5 Comparison between a UL Press and a Conventional Press

rigidity slide, it can also handle concentrated forming loads such as those commonly associated with forging applications.

The slide area is composed of a 1-point scotch-yoke mechanism that does not utilize connecting rods. Because the connecting rods found on conventional presses that connect the crankshaft to the slide are unnecessary, it enables a compact unibody frame, and the overall height is less than that of a standard press. (Fig. 5)

The lower overall height and the connecting rod-free design minimizes longitudinal elongation during forming.

The design of the slide and the frame (which factor in rigidity) along with the compact size of the press serve to minimize the thermal expansion that results from temperature variations in the press frame members, and also minimizes variation in the bottom dead center position of the slide during forming.

3 UL Press Forming Examples

UL presses with their high accuracy and high rigidity characteristics can be used for a wide range of press applications, including sheet metal forming, cold forging, and sheet forging. UL Press forming examples are shown below.

Figure 6 shows an example of a single-stage backward extrusion and piercing process for a part with a depth/I.D. (L/D) ratio of 6:1. The die is off-centered 300 mm from the center of the press. Even when forming under such eccentric loading conditions, the runout accuracy of the I.D. with respect to the O.D. is 0.15 mm.

Fig. 6 Piston Pin (Material: SCM415)

Figure 7 shows an example of high-accuracy shear forming of a gear out of 8 mm SPHC material. Even though the side clearance in this forming application of 10 μm is very small, the high-accuracy characteristics of a UL Press enable the forming of uniform fully-balanced shear surfaces with circumferential clearances.

Figure 8 shows a forming example of a CPU heat spreader. In order to meet the heat dissipation requirements, the amount of contact between this part and its mating part is

Fig. 7 Gear

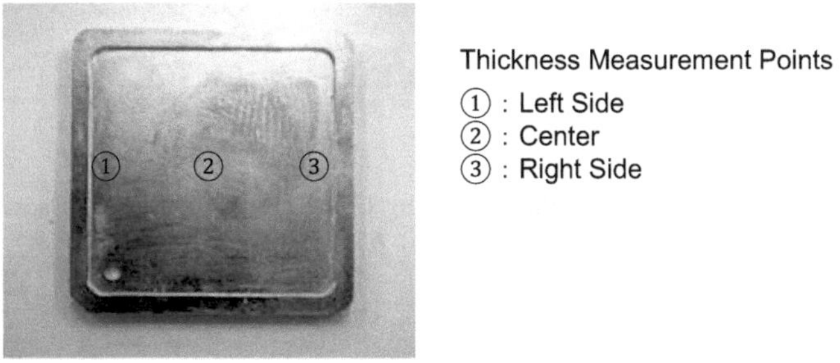

Fig. 8 Heat Spreader (Material: C1100)

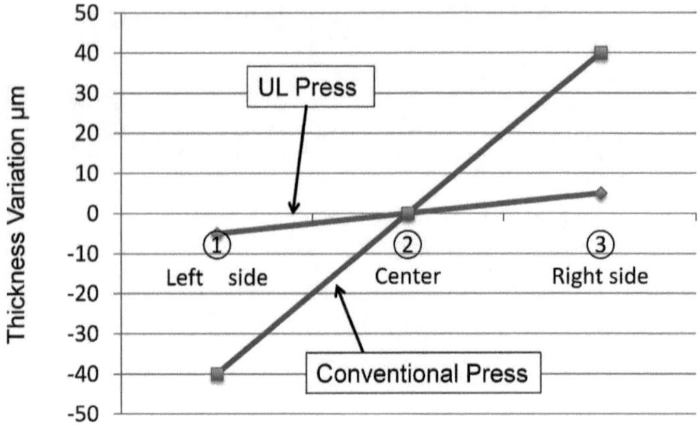

Fig. 9 Variations in Left-Right Squeeze-Formed Area Thicknesses When Using a UL Press Versus a Conventional Press

very critical. Accordingly, the flatness specification of the center inset area is 35 μm, and the thickness accuracy specification is 50 μm. The squeeze-forming of the square shape in the middle of the product is done at the center stage of the press, and the sizing forming of the square squeeze-formed area is done in the next stage, which is located to the right of the center stage. Because of this, differences in the left-right thicknesses in the square-shaped squeeze-formed area occur due to the effects of the load being applied on the right side compared to the press center stage. The right/left product thickness variation is influenced by the press rigidity value, and higher rigidity results in smaller thickness variations. Figure 9 shows product thickness measurement values. The benefits of the increased rigidity of a high-accuracy and high-rigidity press compared to a conventional press is plainly manifest.

4 Conclusion

Going forward, the high-accuracy, high-rigidity UL presses developed by AIDA will contribute to higher product accuracy and higher value-added content, and we expect that these presses will be leveraged to achieve even greater technological innovations.

References

1. Imura, T., 2014, High-Accuracy Shear Forming Using High-Accuracy Presses, Journal of The Japan Society for Technology of Plasticity, Vol. 55, No. 638, 175–179.

Short-Cycle-Stretch-Forming (SCS)

Mathias Liewald, Philipp Schmid, Matthias Schneider, and Apostolos Papaioanu

1 Introduction and Motivation

Body shells in automotive industry are mainly produced by deep drawing. Depending on material properties and on material hardening behaviour of parts during production, different mechanical properties like denting resistance and part stiffness result. For flat parts like door or roof panels such properties are insufficient due to low work hardening during deep drawing phase, especially in areas with low strain contents. This properties can be significantly increases by applying a pre stretching operation prior deep drawing [1]. In order to produce body shell parts a combination of pre-stretching with a subsequent deep drawing operation was realised and field-tested [2]. This new technology combines pre-stretching and deep drawing of sheet metal parts and is known as Cyril-Bath-process. Because of the high investment costs and high process time, use of stretch forming technologies is not qualified for the production of car body panels today.

In order to benefit from stretch forming technologies but also to save investment costs, a new technology named SCS (Short-Cycle-Stretch-Forming) was developed at the Institute for Metal Forming Technology at University of Stuttgart. The SCS technology allows a cost efficient production of high quality parts with a simple tool design working without any blank holder. Recent tool concept for producing door panels shows enormous potential and the applicability of this technology to real press shop applications [3]. In following chapters most important point such as process development, tribology, maximum applicable prestrain, dent resistance and energy consumption are discussed.

Mathias Liewald ✉ · Philipp Schmid · Matthias Schneider · Apostolos Papaioanu
University of Stuttgart, Stuttgart, Germany
e-mail: Mathias.Liewald@ifu.uni-stuttgart.de

A. E. Tekkaya et al. (eds.), *60 Excellent Inventions in Metal Forming*,
DOI 10.1007/978-3-662-46312-3_15

2 Process Idea

The SCS process is based on a simple tool design including two opposed bead sets which interlock while the tool closes. During this procedure the fringe of the blank becomes bend and unbend alternately which leads to an increased tensile force and finally to a stretching of the specimen. In addition, this effect is enhanced by the friction of the blank, while it slides over the bead radii of the lower and upper bead sets. Figure 1 shows fundamental SCS tool setup.

In order to stretch the blank to achieve a predefined strain value, the process gives the possibility of adjusting several process parameters. It is possible to adjust the shoulder radii of the upper and lower bead sets in order to modify the stretch phase and the strain value. The blank dimension and the friction conditions during process are also factors which influence the process. The feasibility of the SCS technology was proofed by pre-stretching a regular used mild steel DC04 alloy as well as a standard aluminium alloy of AA6016. Mild steel material was pre-stretched to $\varphi \approx 0.09$ while the aluminium material was pre-stretched to an effective strain value of $\varphi \approx 0.04$. These investigations were realised with small blank strips with dimensions of 50 mm × 500 mm. In order to enhance the pre-stretching of sheet material using SCS technology further investigations concerning the number of lower and upper bead sets has been conducted.

Friction between the blank and the tool indeed reveal tremendous influence on the stretching of the blank strips. Thus, the surface roughness of the tools can be an indicator for the maximum achievable strain value caused by the friction force. Sandblasted tool elements are cheap and easily to perform but the wear is very high and the robustness of the process is inconstant. For the implementation of a series tool based on SCS technology, the use of coated tool surfaces makes much more sense.

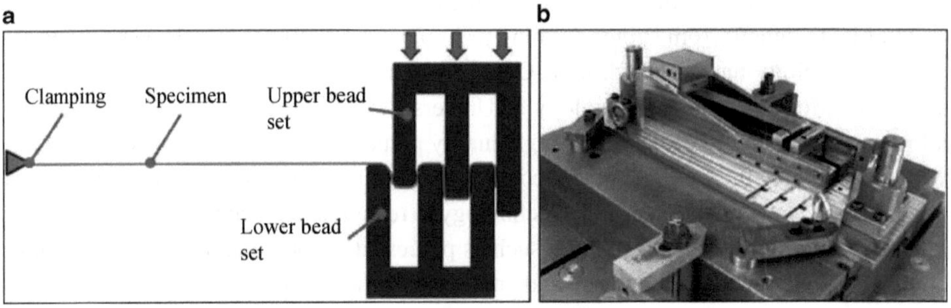

Fig. 1 **a** Fundamental SCS tool setup, **b** Prototype SCS tool

3 Results and Applications

Based on these studies a close-to-production tool was designed and realised according to the configuration in Fig. 2. At a first step bead sets with three upper and three lower bead elements were used for the SCS tool. In the first step of the process the fringe of the blank becomes threaded into the bead sets whereby the free region of the blank becomes laterally stretched. The deep drawing operation starts with the first contact of the punch with the pre-stretched blank within the same press-stroke. For this purpose a scaled door panel was designed for a mild steel alloy DC04 with a blank thickness of $s = 0.75$ mm. The tool allows for a pre-defined stretching of the blank: the maximum measured strain reached around 7 % [6].

In order to gain a higher restraining force in minor strain direction, additional bead elements were located in transversal direction. Due to this configuration wrinkles and cracks result in the corner region of the drawn part. In Fig. 3 the SCS pre-stretched and deep drawn part with plastic strain indicated is shown.

Static dent tests were performed using a dent test apparatus developed at IFU. With present denting device it is possible to perform both static and dynamic dent tests depending on the loading condition to be examined. Regarding the progression of plotted curve

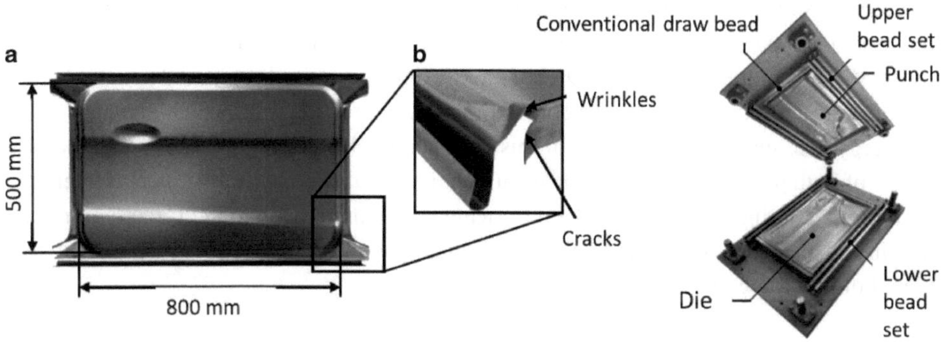

Fig. 2 SCS tool for producing front door panels

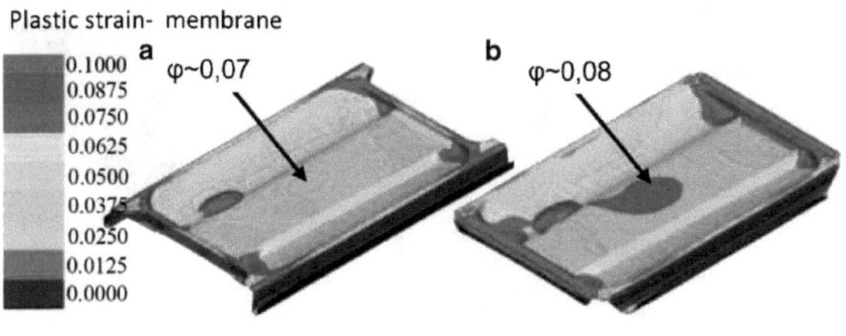

Fig. 3 Strain distribution before and after simulative optimisation

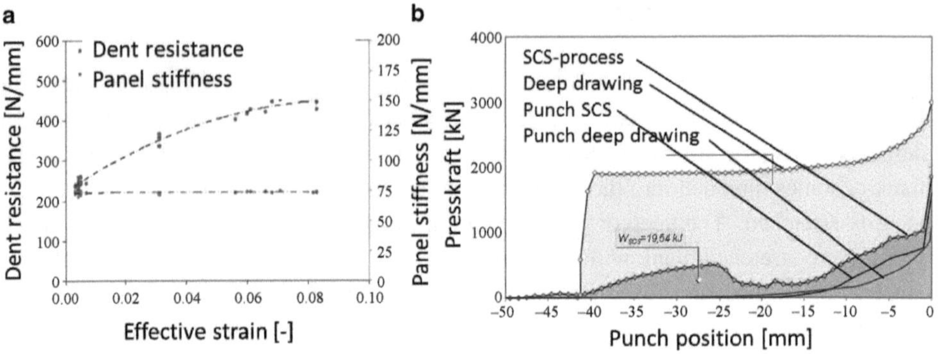

Fig. 4 Dent resistance distribution in dependency of effective strain in dent location **b** energy consumption of SCS in comparison to conventional deep drawing

in Fig. 4 dent resistance *Ds* reaches saturation at $\varphi eff \approx 0.07$. A further increase of effective strain in parts area being at risk against denting seems not to be advisable. Calculated dent resistance shows some scatter due to measuring inaccuracy but the increase of dent resistance can be clearly distinguished. Panel stiffness *Ss* is obviously not affected by the increase of effective strain at all. [7]

Comparison in Fig. 4b illustrates energy-consuming of a conventional deep-drawing process in comparison with SCS method. When forming thus mentioned front door panel using SCS technology amount of 19 kJ of energy only is required compared to manufacturing of same component by using a conventional deep drawing tool, which demands 84 kJ of energy, i. e. energy requirement in this case appear 4 times higher. The force requirements of both punches in Fig. 4 indicate almost identical curves, i. e. the increased energy requirement during conventional deep-drawing is, in comparison to the force required by the shaped elements during the SCS process, mainly explained by the process-related high blankholding force of FBH = 1,900 kN [8].

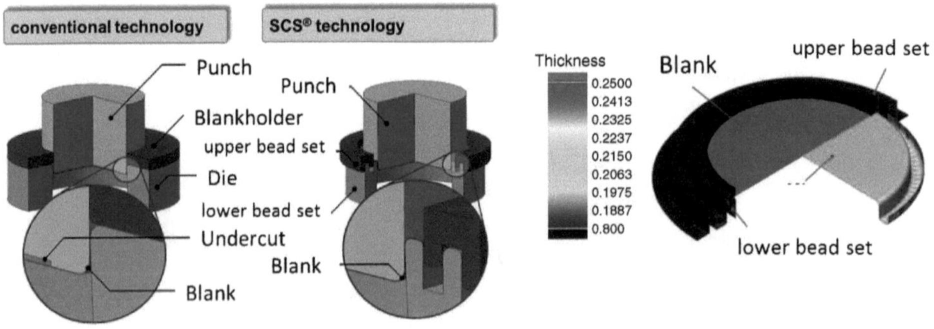

Fig. 5 Principle of SCS-Cupping

Current investigations deal with integration of SCS technology into first cupping processes used for beverage can production. Conventional cupping process normally consists of a simple rotational symmetric deep drawing, followed by multiple ironing processes. Bottom zone of part is not stretched by this technology, so initial sheet thickness remains until last process stage. In order to achieve material savings, SCS bead sets have been implemented into such cupping tool to reduce sheet thickness and to increase strength of material, see Fig 5. For thus given design of cup a thinning of bottom zone of component by approx. 5 % or more today can be reached. Main research issue is to avoid development of wrinkles during the early phase of deep drawing. First optimisations of SCS bead shape to modify tension, bending and friction simultaneously shows promising results [9].

4 Potential

The short-cycle-stretch-forming technology offers various potential for current and future forming processes:

- Replacement of blankholder and die cushion and thus reduction of investment costs.
- Reduction of tool size regarding space requirements.
- Pre-stretching of material with enhancement of dent resistance.
- Reduction of material thickness in bottom areas of part in terms of lightweight design.
- Reduction of required press force and energy and thus press size.

5 Conclusion

SCS technology provides multiple advantages over conventional forming methods and represents a very energy-efficient alternative in the field of the production of components made of sheet metal. Further investigations will deal with implementation of SCS technology to more complex part geometries and thus provide a energy efficient, low-cost tool concept for combined pre-straining and deep drawing processes.

6 Acknowledgement

This process has been investigated in several projects. The work has been kindly supported by Baden-Württemberg Stiftung Pro28, AiF ZIM project (KF2201106RU2) and AiF Fosta Projekt (IGF-Nr. 17422 N).

References

1. D. Vlahovic, M. Liewald, K. Siegert. Optimierung von Ziehteileigenschaften durch gezielte Vorverfestigung, International Conference "New Developments in Sheet Metal Forming", Stuttgart, Germany: MATInfo Verlag, 2006.
2. C. Bath. Streckziehen von Karosserieteilen, Werkstatt und Betrieb, 98. Jahrgang, Heft 3, 1965.
3. D. Vlahovic, M. Liewald. Neue Ansätze in der Streckziehtechnik, International Conference "New Developments in Sheet Metal Forming", Stuttgart, Germany, 2008.
4. Papaioanu, A.; Liewald, M.: Further development of the SCS Strech-Forming Technology with assistance of Forming Simulation. The 13th International ESAFORM Conference on Material Forming, Brescia, Italy, (2010)
5. Vlahivic, D: Neue technologische Ansätze zum kombinierten Recken und Tiefziehen von Außenhauptbeplankungen aus Feinblech, Dissertation, Universität Stuttgart, 2009
6. Papaioanu, A.; Liewald, M.: Influence of Process Parameter Variation on SCS Process. 4th ICMEN Conference, Thessaloniki, Greece, (2011)
7. Papaioanu, A.; Liewald, M.: Influence of Pre-Stretching on Dent Resistance of Stretch Formed Parts
8. Papaioanu, A.; Liewald, M.: SCS® – Energy-efficient procedure to produce automotive body shell components for car bodies, proceedings New Developments in Sheet Metal Forming, Fellbach, 2014, pp 185–195
9. Schneider, M.; Liewald, M.: Approach for implementation of short cycle stretch forming (SCS) to cupping processes, KEM 611-612 (2014), pp. 1096–1101

Sheet-Bulk Metal Forming

Daniel Gröbel, Thomas Schneider, and Marion Merklein

1 Motivation

One of the key priorities in future production processes will be the necessity to meet the demands that emerge through the ever increasing shortening of natural resources. This is particularly important in the automotive sector, both due to the strict emission standards and the reduction of fuel consumption requested by the customer. A promising answer to this challenge is the use of lightweight products, which can be achieved directly through a reduction of the amount of material used in the end product. However, the reduction of the part dimensions will typically result in less loading capacity, which cannot be tolerated in many instants. Hence, to encounter this problem, high strength materials are used to compensate this circumstance. Besides this, it is often possible to save material through functional integration in a single part, which will not reduce the load capacity. However, the disadvantage with this strategy is that the part geometry is increasing in complexity. Hence, it is oftentimes difficult to manufacture the part in question in conventional mass production processes, especially if high strength materials are to be used. As a result the process chain will lengthen and the production costs will increase substantially. An answer to this challenge is the introduction of the new process class SBMF, which core principle is the application of bulk forming operations on sheet metal. By means of this method, functional integrated parts of high quality and strength, like the examples in Fig. 1, can be mass produced in a cost efficient manner.

Currently, amongst others, SBMF is researched extensively in the Collaborative Research Center CRC/Transregio 73, funded by the German Research Foundation (DFG), at the Universities in Erlangen, Dortmund and Hannover.

Daniel Gröbel · Thomas Schneider · Marion Merklein ✉
Friedrich-Alexander-Universität Erlangen-Nürnberg, Erlangen, Germany
e-mail: Marion.Merklein@fau.de

A. E. Tekkaya et al. (eds.), *60 Excellent Inventions in Metal Forming*,
DOI 10.1007/978-3-662-46312-3_16

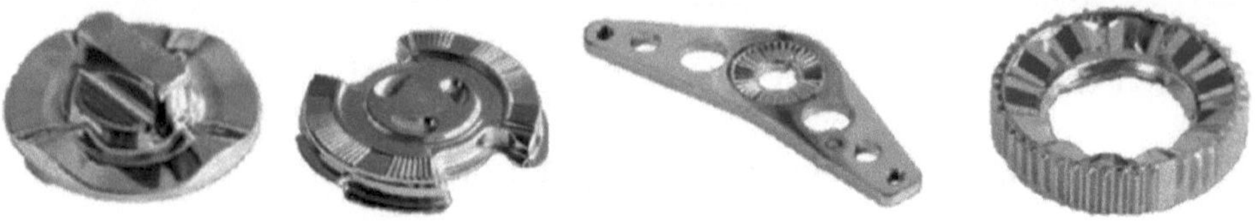

Fig. 1 Examples of SBMF parts with functional elements [1]

2 Process Definition and Classification

SBMF is defined as forming of sheets with intended three dimensional material flow as in bulk forming processes [2]. Such processes consist of several conventional processes that are combined and extended in their application. Bulk forming operations are applied to sheets or plates, whereby the initial sheet thickness of the semi-finished product varies between 1 and 5 mm. Most SBMF processes can also be done in combination with sheet forming operations. An integral characteristic of SBMF is the complex interaction between forming zones of high and low strains and locally varying 2D and 3D stress and strain states [2]. In many cases the intention is to form local functional elements out of the sheet plane with a dimension magnitude similar to the sheet thickness. The tools used for the forming can either possess a linear or rotational movement, which can be used to classify the processes in question according to Fig. 2. The resulting change of the sheet thickness depends on the tool kinematics and its position relative to the sheet, whereby the thickness itself can either be locally increased, decreased or remain equal [3].

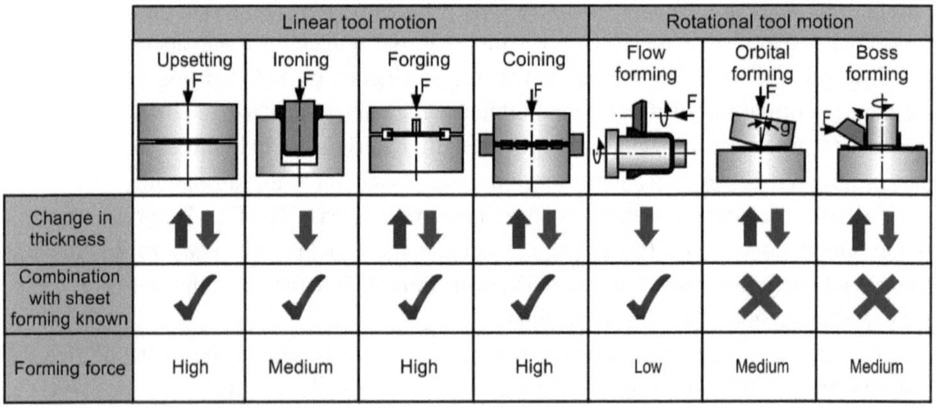

Fig. 2 Process classification and characteristics of common SBMF operations [3]

3 Fields of Application

Since many process combinations exist, that can be ranked as SBMF, the field of application is manifold. In the following the successful use of this process class will be demonstrated at the example of upsetting with deep drawing.

Upsetting is a sophisticated bulk forming operation, which is defined as the reduction of at least one workpiece dimension between two linearly moving tools. However, the application to sheets results in severe force requirements due to the large contact area between the punch and the workpiece. Hence, to encounter this challenge, upsetting operations are often combined with conventional sheet forming processes. In this manner, the forming forces can be applied within the sheet plane [4]. A common approach is the combination of deep drawing and upsetting. This enables a thickening of the cup wall by a reduction of the cup height. Besides circumferential thickening, also forming of variants by a geometrical control of the material flow is possible. In order to improve forming of the functional elements and to avoid unnecessary part weight, tailored blanks with adapted thickness characteristic are used. To manufacture those specimen flexible rolling and orbital forming, which can be both classified as incremental SBMF processes, can be utilized. Figure 3a illustrates the shape of the cup after deep drawing. By using a flexible rolled tailored blank out of a sheet with $t_0 = 2.0$ mm and a resulting maximum thickness of $t_{max} = 2.6$ mm, the preform of the gear shape after the first forming operation is improved to the same level of a blank with $t_0 = 2.6$ mm [5]. Figure 3b shows the final result after deep drawing and upsetting using a geometrically identical tailored blank made by orbital forming. The mold filling behavior of the gear teeth is improved, as buckling of the cup wall is inhibited by the revised preform after deep drawing due to the pre-distribution of material at the tailored blank. Furthermore, due to the strain hardened center of the specimen, folding at the fillet is reduced as well as an undesired material flow from the cup wall into the cup bottom [6].

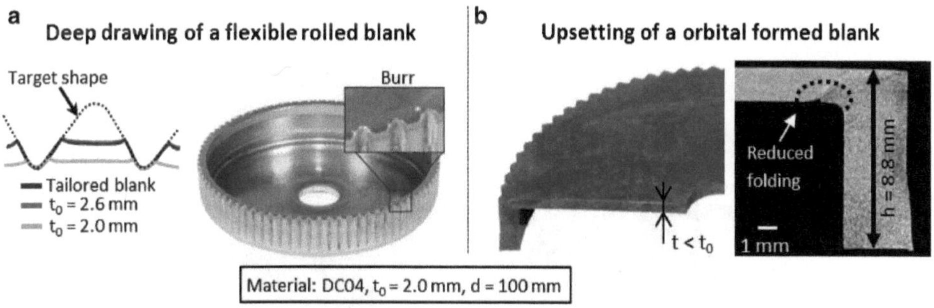

Fig. 3 Application of **a** deep drawn and **b** orbital formed tailored blanks in combined deep drawing and upsetting of geared components to improve the mold filling behavior and component properties

4 Special Characteristics and Considerations

SBMF processes disclose new challenges that have to be controlled if high quality parts are to be produced. Research in recent years has revealed that, in front of all material flow, friction considerations, tool load and numerical modelling are the most severe aspects in this regard. An important aspect that divides SBMF processes from other processes is that the area of the forming zone is low in comparison to the total blank volume. Hence, the material flow tends to be directed out of the high strain forming zone, as has been revealed by research on the forming of functional elements. As a consequence the intended forming of the elements is hampered [2]. In order to ensure proper forming despite this, the material flow has to be actively controlled. It is possible to meet this requirement by a local modification of the friction, as for example, by means of Tailored Surfaces on tool parts [7]. As an alternative, the blank itself can be modified as can been seen in Fig. 4. In this case, the forming of a tooth geometry on a blank has been improved considerably by sandblasting the sheet parts that surround the tooth cavity, whereby the area above the mold remained untreated. Due to the higher friction factor of the treated areas, the material flow in the sheet plane is hampered and, as a consequence, the flow is more pronounced in the direction of the tooth cavities. In this way the height of the functional element has been improved by 12 % under otherwise equal conditions.

Other aspects are the typically high local tool loads that are present in many SBMF Forming operations, especially in forging. They are a direct consequence of the large contact areas between the tool and the workpiece. In addition, due to oftentimes asymmetric positioning of functional elements, the loads are also asymmetric, which increases the risk of tool damage and reduces the tool life. Hence, innovative tool design, like non-circular reinforcements and high quality steels, like powder metallurgical steels, are necessary in many cases [2]. Since asymmetric forces have an additional negative impact on the preci-

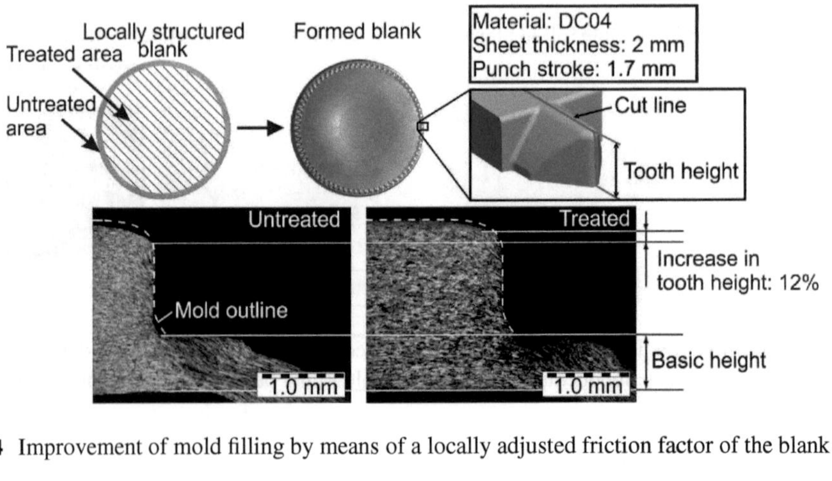

Fig. 4 Improvement of mold filling by means of a locally adjusted friction factor of the blank

sion of the part production, recent research also aims to reduce the tool dislocation during the forming operation. Exemplarily, the ram displacement of the forming machine has been successfully reduced by means of an electromagnetic system, as is evidenced by recent research at the University of Hannover [8]. Besides this, the modeling of the processes in conventional finite element software programs is of heightened difficulty. The typically small forming zone, which demands high numerical precision and the large remaining sheet areas with extensive contact areas are a challenge to current computational capacity. A powerful strategy to overcome this problem is the use of adaptive modeling, which enables the local modification of the numerical mesh. In this way it is possible to increase calculation precision while reducing the amount of calculation at the same time [2].

5 Potential

SBMF possesses a multitude of advantages compared to alternative processes that qualify it as an adequate production method for functional integrated parts:

1. High lightweight potential by means of functional integration and the use of high strength materials
2. Production of parts with geometrically complex elements possible
3. High cost efficiency as a consequence of a short process chain
4. Improved mechanical properties due to an unbroken fiber course and material hardening
5. Efficient production of high quality net-shape or near net-shape parts
6. Suitable for mass production

6 Conclusion

SBMF is an innovative process class that enables the manufacturing of complex functional integrated parts in a cost efficient manner. It is well suited to produce lightweight products of high strength and durability out of sheet metal and therefore meet the requirements of current and future production demands. The unique challenges associated with this process class arise from the necessity to control the complex material flow during the forming operation, the high tool loads, the complex tribological behavior, as well as the demanding simulation of the forming operation in conventional finite elements software programs. Thanks to recent research, many promising solutions could be suggested to solve the aforementioned difficulties.

7 Acknowledgement

This process class has been investigated in several projects within the scope of the collaborative research center CRC/Transregio 73. The work has been kindly supported by the German Research Foundation (DFG).

References

1. Feintool AG, 2012, Taumeltechnologie von Feintool – Hohe Qualität und Wirtschaftlichkeit beim Umformen. Fineblanking Technology.
2. Merklein, M., Allwood, J. M., Behrens, B.-A., Brosius, A., Hagenah, H., Kuzman, K., Mori, K., Tekkaya, E., Weckenmann, A., 2012, Bulk forming of sheet metal. CIRP Annals – 61 (2), 725–745.
3. Merklein, M., Hagenah, H., Schneider, T., 2013, Sheet-Bulk Metal Forming Processes – State of the Art and its Perspectives. TTP 2013 – Tools and Technologies for Processing Ultra High Strength Materials, Verlag der TU Graz, S. 197–204.
4. Maeda, A., Araki, K., 1996, Plate Gear, Japanese patent, 9-222158, (in Japanese).
5. Merklein, M., Plettke, R., Schneider, T., Opel, S., Vipavc, D, 2012, Manufacturing of sheet metal components with variants using process adapted semi-finished products. Key Eng. Mater. 504–506, 1023–1028.
6. Schneider, T., Vierzigmann, U., Merklein, M., 2014, Analysis of Varying Properties of Semi-finished Products in Sheet-bulk Metal Forming of Functional Components. NUMISHEET 2014: 9th Int. Conf. and Workshop on Numerical Simulation of 3D Sheet Metal Forming Processes, 930–933.
7. Biermann, D., Surmann, T., Odendahl, S., Steiner, M., Hense, R., Rausch, S., 2011, Creating Functional Surface Structures by Milling using Self-Excited Tool Vibrations. Proceedings ASPE 2011 Spring Topical Meeting – Structured and Freeform Surfaces, 55–58.
8. Gröne, M., Salfeld, V., Krimm, R., 2014, Design of an Electromagnetic System to Avoid Horizontal Ram Displacement. Advanced Materials Research 1018, 237–244.

Electromagnetically Assisted Sheet Metal Stamping and Deep Drawing

Glenn S. Daehn, Anupam Vivek, and Jianhui Shang

1 Introduction

Sheet metal forming by stamping and deep drawing is widely used in automotive, aerospace, and appliance manufacturing industries. In the vehicle industry, there is a constantly increasing need for reduction of structural weight in order to increase fuel efficiency. This is being accomplished by the use of materials such as high strength steel, and aluminum and magnesium alloys, which have a high strength to weight ratio [1]. However, these materials also tend to have lower formability. Impulse based processes, such as electrohydraulic [2] and electromagnetic [3] forming have been demonstrated to stretch many materials beyond their conventional forming limits. However, such increase in formability is not ubiquitous and one cannot rely on increased formability alone to form parts using a single step impulse based process. Another way is to control the strain distribution in the stamped part. Ideally, nowhere in the part should be excessive compressive strains causing wrinkling or excessive tensile strains that can cause tearing. Techniques like active draw bead control [4], active blank holder force control [5], lubrication [6] and blank optimization are used to control metal flow and distribute strains more uniformly in a stamped part. Here, a different approach, electromagnetically assisted sheet metal forming (EMAS), has been discussed.

Before delving into the process idea of EMAS, it is useful to describe electromagnetic forming (EMF), which is a well known process [7]. During EMF, the forming forces are generated from magnetic repulsion between a stationary primary coil and the workpiece. A high voltage discharge from a capacitor bank causes the flow of a transient current

Glenn S. Daehn ✉ · Anupam Vivek
Ohio State University, Columbus, USA
e-mail: daehn.1@osu.edu

Jianhui Shang
Edison Welding Institute, Columbus, USA

A. E. Tekkaya et al. (eds.), *60 Excellent Inventions in Metal Forming*,
DOI 10.1007/978-3-662-46312-3_17

through the primary coil which is placed proximally to the workpiece. The induced eddy currents in the workpiece are directionally opposite to the one in the primary coil. Therefore, a repulsive Lorentz force is created between the two current carrying conductors. By changing the initial charging voltage, the driving magnetic pressure pressure, which is typically a few hundred megapascals and lasts over a time period around 10 µs, can be controlled [8]. In EMAS, EMF augments a conventional stamping process This process was initially conceived and developed by Daehn and co-workers [9].

2 Process Idea

The main idea for EMAS is to apply multiple, small magnitude electromagnetic pulses in strategic locations of the sheet metal as the punch and the die of a conventional stamping set up move toward each other. This is accomplished by embedding EMF coils in certain regions of the punch, die, and blank holder. The desired shape is mainly formed by the conventional tools while the electromagnetic forces alter the strain distribution. Basically, the EMF coils are places in regions which undergo small strains in a corresponding conventional process. During stamping or deep-drawing, there are three main areas where different levels of strains are encountered [10]:

a. Flange: This is the region under the blank holder and undergoes small and often compressive strain. Wrinkling commonly occurs in this region. The amount of material drawn in from this region depends on the friction and the normal blank holder force.
b. Wall: This region transmits the forming forces to the flange region and undergoes largest amounts of strain.
c. Bottom: This is the area directly under the punch face and has relative small strain due to friction between punch and itself.

EMAS is intended to exploit the potential for further strains in the flange and bottom regions of the sheet metal part in order to encourage draw-in and therefore manufacture parts with higher depths than allowed by a conventional process. Figure 1 illustrates two adaptations of EMAS. In Fig. 1a [11] the EMF coils are positioned near the bottom of the formed part. First, the part is drawn to a large fraction of the possible depth while avoiding tearing in the wall or wrinkling in the flange region. Then a small EMF impulse is provided to stretch the face of the part and encourage draw-in. The punch is advanced into the formed cavity and then the EMF impulse is provided again. This progressive operation could be repeated until tearing happens somewhere in the part. Figure 1b shows the punch with the EMF coil embedded in it. In the second adaptation [12], illustrated in Fig. 1c, active control of draw bead is effected by an EM coil embedded in the blank holder which can increase the effective limiting draw ratio (LDR). A small region of the flange is electromagnetically formed into a cavity made in the blank holder. The punch is moved down and the material in the formed bubble is drawn in easily. Once the bubble

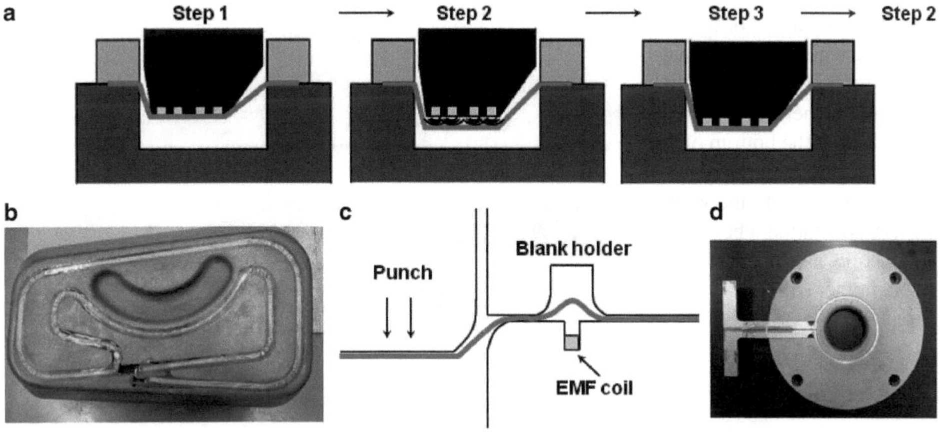

Fig. 1 Schematics for different adaptations of EMAS. **a** progressive bump forming, **b** a hybrid punch with the EMF coil embedded in it, **c** active draw bead control by EMF, **d** the embedded coil for draw bead control

is straightened the EMF process is repeated and followed by movement of the punch. The formed bubble makes the drawing process easier and better in three ways. First, it stretches the material which otherwise would not be strained. Second, it draws material from regions farther from the main die cavity. Third, it reduces the draw-in force by creating a geometry-discontinuity in the flange. When the punch draws the flange into the die cavity, the metal flowing out of the bubble can be considered to be undergoing free flow. The reduction in draw-in force helps in preventing rupture of the wall region, which has undergone the highest amount of stretching.

3 Results and Applications

Figure 2a shows the results of conventional stamping and EMAS of a 1 mm thick aluminum alloy 6111-T4 sheet. In this variant of EMAS, the EMF coil was embedded in the punch. The conventional stamping was done with a clamp load of 156 kN and an oil-based lubricant was used. The draw depth was found to be 44 mm. For the EMAS process, 22 EMF impulses each of 5.4 kJ were provided. The total depth before failure was found to be 63.5 mm with the EMAS process. Figure 2b shows the results of a deep draw process with a 1.5 mm thick aluminum alloy 2219-O sheet. The blank holder force was kept at 1334 N in both, conventional and EMF augmented methods. The conventional method resulted in a cup height of only 8.6 mm before failure, whereas, in case of EMAS a cup height of 30.9 mm was reached with the help of 25 EMF pulses. In this adaptation of EMAS, the EMF coil was embedded in the blank holder.

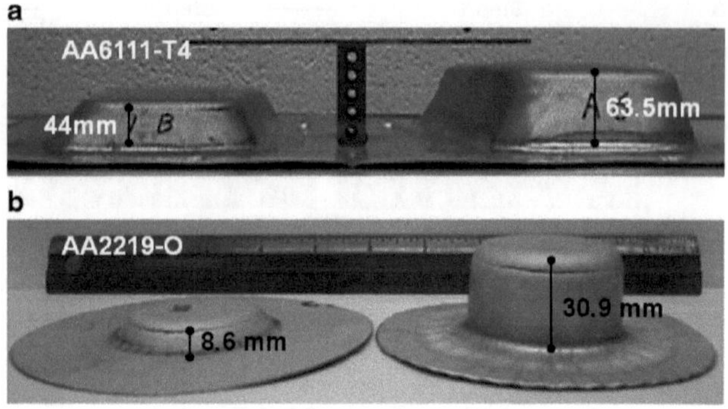

Fig. 2 Comparison of forming done with **a** conventional stamping (left) and EMAS adaptation where EMF coil embedded at the bottom of the punch (right), **b** conventional deep drawing (left) and with EMAS adaptation where the EMF coil was embedded in the blank holder (right)

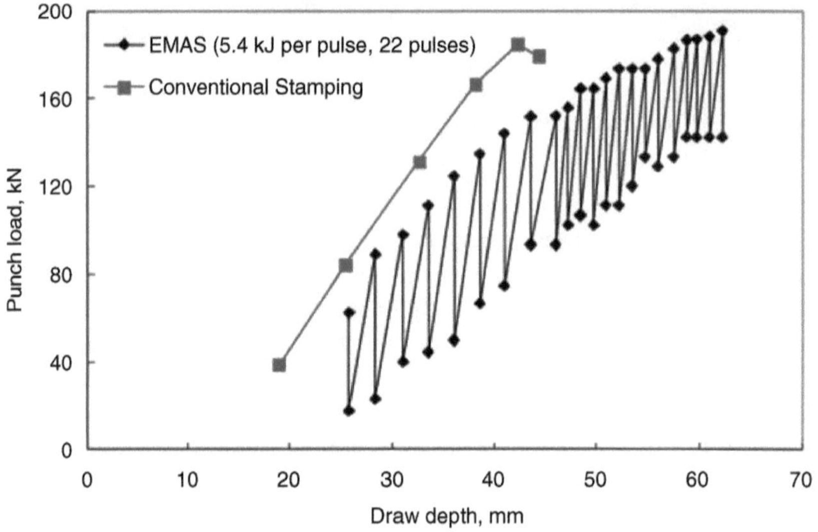

Fig. 3 Punch load versus draw depth for conventional stamping and EMAS [10]

Figure 3 shows the curved of punch load versus draw depth for the samples in Fig. 2a. In EMAS, after each EMF pulse there was a significant drop in punch load. The punch displacement in that region corresponds to filling of the bubble formed by the EMF of the bottom of the workpiece. When the punch starts directly stretching or drawing in the material the punch load increases. The cyclic unloading and loading of the punch continue until part failure happens. It can be noted that the maximum punch loads for EMAS as well as conventional stamping are approximately the same.

4 Potential

Due to its unique capability to effect uniform strain distributions and attain higher draw depths, EMAS could be a useful technique in the metal working industry. The system is easy to control and offers greater flexibility than conventional stamping and deep drawing. Here, EMF coils were embedded only in one type of area. It can be envisioned that multiple coils could be placed in strategic locations and fired simultaneously or successively to attain a desired result. Furthermore, the process shown here was much slower than ones in conventional stamping facilities. However, it should be noted that the actual EMF process occurs over a few 10 s of microseconds and the charging time of the capacitor bank can be cut down significantly by having a larger power supply. Alternatively, multiple capacitor banks could be utilized for firing the coils at different coils. Automation would also help reduce process cycle time. Recently, numerical modeling of the hybrid process has also been started [13], which would help with designing the process cycles optimally. This is also an area where robust simulation that includes both conventional sheet metal forming and electromagnetic forming would be quite useful as there are a large number of parameters that must be optimized and it is very difficult by trial and error.

5 Conclusions

EMAS has been demonstrated as a feasible technique for improving draw depth as compared to conventional stamping and deep drawing processes. This improvement is realized by distributing the strain more uniformly in the formed part and enabling larger amounts of draw in. High strain rate induced increased forming limits could have partially helped achieve the higher draw depths, even though they were not intentionally exploited in this body of work. There is significant scope for future work with regard to process modeling, automation, coil design, and pulsed power supply.

References

1. Kleiner, M., Geiger, M., & Klaus, A. (2003). Manufacturing of lightweight components by metal forming. *CIRP Annals-Manufacturing Technology, 52*(2), 521–542.
2. Balanethiram, V. S., & Daehn, G. S. (1994). Hyperplasticity: increased forming limits at high workpiece velocity. *Scripta Metallurgica et Materialia, 30*(4), 515–520.
3. Seth, M., Vohnout, V. J., & Daehn, G. S. (2005). Formability of steel sheet in high velocity impact. *Journal of Materials Processing Technology, 168*(3), 390–400.
4. Naceur, H., Guo, Y. Q., Batoz, J. L., & Knopf-Lenoir, C. (2001). Optimization of drawbead restraining forces and drawbead design in sheet metal forming process. *International Journal of Mechanical Sciences, 43*(10), 2407–2434.

5. Ahmetoglu, M. A., Altan, T., & Kinzel, G. L. (1992). Improvement of part quality in stamping by controlling blank-holder force and pressure. *Journal of materials processing technology*, *33*(1), 195–214.
6. Lovell, M., Higgs, C. F., Deshmukh, P., & Mobley, A. (2006). Increasing formability in sheet metal stamping operations using environmentally friendly lubricants. *Journal of materials processing technology*, *177*(1), 87–90.
7. Psyk, V., Risch, D., Kinsey, B. L., Tekkaya, A. E., & Kleiner, M. (2011). Electromagnetic forming – A review. *Journal of Materials Processing Technology*, *211*(5), 787–829.
8. Vivek, A., Kim, K. H., & Daehn, G. S. (2011). Simulation and instrumentation of electromagnetic compression of steel tubes. *Journal of Materials Processing Technology*, *211*(5), 840–850.
9. Daehn, G. S., & Vohnout, V. J. (2000). *U.S. Patent No. 6,050,121*. Washington, DC: U.S. Patent and Trademark Office.
10. Hu, J., Marciniak, Z., & Duncan, J. (Eds.). (2002). *Mechanics of sheet metal forming*. Oxford, England: Butterworth-Heinemann.
11. Shang, J., & Daehn, G. (2011). Electromagnetically assisted sheet metal stamping. *Journal of Materials Processing Technology*, *211*(5), 868–874.
12. Shang, J. (2006). *Electromagnetically assisted sheet metal stamping* (Doctoral dissertation, The Ohio State University).
13. Kiliclar, Y., Demir, O. K., Vladimirov, I. N., Kwiatkowski, L., Brosius, A., Reese, S., & Tekkaya, A. E. (2012). Combined simulation of quasi-static deep drawing and electromagnetic forming by means of a coupled damage–viscoplasticity model at finite strains. *ICHSF2012*, 325.

Dry Metal Forming – a Green Approach

Frank Vollertsen, Hendrik Flosky, and Thomas Seefeld

1 Vision and Definition of Dry Metal Forming

In production engineering, a sustainable use of raw materials is essential for saving resources, energy and production cost. In 1992 within the context of the United Nations Conference on Environment and Development (UNCED) the Agenda 21 was signed by 172 member states of the United Nations (UN). One part of Agenda 21 is the idea of using raw materials in a sustainable way in the focus of research and development. In this context the sub area production engineering has the obligation to enforce the suitable use of raw materials and other resources. The development of new green technologies is needed. To this end, resource efficiency and energy productivity are to be doubled until 2020. This includes the decrease of waste products and the abandonment of environmentally hazardous substances as well as an efficient utilization of resources.

The metal forming process is already known as one of the most material and energy efficient processes, but one disadvantage is the use of lubricants. These lubricants are used to reduce friction between work piece and forming tool, to reduce process forces and to protect semi-finished goods against corrosion. Moreover, transport of heat and metal swarf may play a role in fast forming processes [1]. In mostly all groups of forming processes lubricants are used and in some cases they are harmful to health and in most cases harmful to the environment. However, from both an economic as well as an ecological point of view, there exists a strong demand to avoid lubricants within metal forming processes. For following process steps after the metal forming steps, e. g. coating, joining and painting processes the lubricant has to be removed in an additional process step. This is carried out in post-treatment processes by usage of degreasing agents which are solvent-based (mostly toxic or irritating substances) and, thus, hazardous to both environment and health.

Frank Vollertsen ✉ · Hendrik Flosky · Thomas Seefeld
BIAS, Bremen, Germany
e-mail: vollertsen@bias.de

A. E. Tekkaya et al. (eds.), *60 Excellent Inventions in Metal Forming*,
DOI 10.1007/978-3-662-46312-3_18

Therefore, avoiding lubricants in metal forming would result in three significant advantages:

1. Reduction of process steps in production
2. Reduction of environmental impacts
3. Reduction of avoidable health burden

A definition of the term "dry metal forming" has been proposed recently that is based on the process chain and focusses the ecological and economical points of view, i. e. avoiding any cleaning necessity after forming [2]:

> Dry metal forming is a process where a work piece leaves the forming tool without the necessity of cleaning or drying before further production steps such as coating or joining processes.

2 Challenges and Benefits

To achieve the avoidance of lubricants and to reach the vision of a dry metal forming press plant, manifold challenges on the area of metal forming, especially regarding to the field of material science have to be faced. In most metal forming processes, lubricants have one principal function: to reduce friction within the forming process between work piece and forming tool. The absence of a lubricant results in an intensive mechanical interaction between work piece surface and forming tool surface. Without the use of lubricants during the metal forming processes the sheet metal slides over the forming tool's surface. Due to the adhesion tendency, this leads to adhesive wear between both objects. The consequence is a poor surface quality of the formed work piece, extensive tool wear, and, as worst case, the destruction of the tool [3]. This leads to large deviations from the given work piece geometry and to additional costs for tool repairs or tool changes.

Focused on the economic point of view, the cost-saving potential can be reached by dry processing amounts to about 2–17 % of work piece-specific production costs, depending on the selected production process [4]. The avoidance of lubricants in metal forming enables – besides a smaller environmental impact – an increase in productivity through the reduction of work steps and the rational use of resources. It can be assumed that the increased energy requirement through more complex tool manufacturing processes in dry metal forming is out of all proportion to the energy saving potential by elimination of production, transportation, application, removal and disposal of lubricants. Overall, an increase in sustainability compared to current process chains can be achieved. This leads to the assumption that dry metal forming exhibits a high potential as a green technology.

3 Approaches

Since the 1980's studies on dry metal forming without lubricant have shown that results highly depend on the material combinations. In [2], where a good overview of the state of research is given, it was pointed out that there are three different categories how to achieve a dry forming can be divided: ceramic tools, self-lubricating coating systems, and hard material coatings.

Investigations with uncoated forming tools under unlubricated conditions were made using different ceramic materials as tool material for a drawing die within a dry deep drawing process. Depending on the work piece material, different ceramics turned out to be promising. However, by providing a smaller limiting drawing ratio, compared to lubricated processes [5]. Another approach proved that ceramic materials ZrO_2-WC and Al_2O_3-TiC are suitable for dry metal forming processes. In this case, limiting drawing ratios close to the value of lubricated processes were achieved [6].

During storage and transport sheet metal is often protected by organic coatings [7]. Also in this case, DLC coatings proved to be advantageous as they had a low friction coefficient within a dry forming process with Zn-coated steels [8]. The friction coefficient was strongly influenced by the surface finish [9]. For different work piece materials (aluminum and steel), various a-C:H coatings proved to be suitable for dry metal forming [10].

Carbon based coatings (especially diamond-like coatings (DLC)) have shown to be a promising material in dry forming [11]. Within compression-spin testing, these coatings exhibited the lowest wear compared to layer systems based on BCN, WC/C-CrN, TiAlN, WC/C-TiAlN, TiC and TiN [12]. DLC layers were also rated to be advantageous in dry wire drawing. The operational lifespan of the layers depended significantly on the layer thickness [13].

4 Diamond Forming Die

For micro metal forming [14], materials like synthetic diamonds can be used as forming tool material because in the used geometry of some cubic millimeters the price is low. The advantages of diamonds are the low friction coefficient (~0.1) in combination with work piece materials as stainless steels, aluminum, copper etc. under dry conditions and of course the wear resistance. For micro deep drawing, a cubic diamond with the size of $3.5 \times 3.5 \times 2\,mm^3$ and a cylindrical hole of 1 mm was used as a micro deep drawing die. As shown in Fig. 1, the diamond drawing die (b) was pressed and glued into the tool holder (a), which is important for positioning the tool in the highly dynamic micro forming machine developed at BIAS [15].

For deep drawing of micro cups a blank holder is required, which was in this case made from the cold work steel X155CrVMo121 (hardened to 62 HRC). For comparing the influence of the lubricant micro cups were formed with and without lubricant between diamond and work piece, see Fig. 2. The lubricant HBO 947/11 was used in both cases in

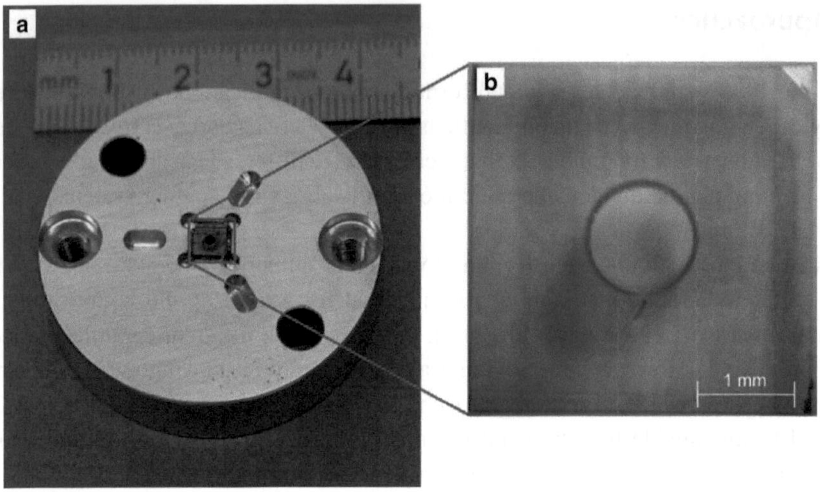

Fig. 1 Diamond insert in the tool holder

between work piece and blank holder in order to avoid any influence by dry friction not including the diamond deep drawing die.

On the left side a micro cup produced with lubricant shows wrinkles, while the right picture shows a dry formed micro cup without wrinkles. The dry formed micro cups has the same shape without wrinkles and bottom tears as a micro cup produced with a

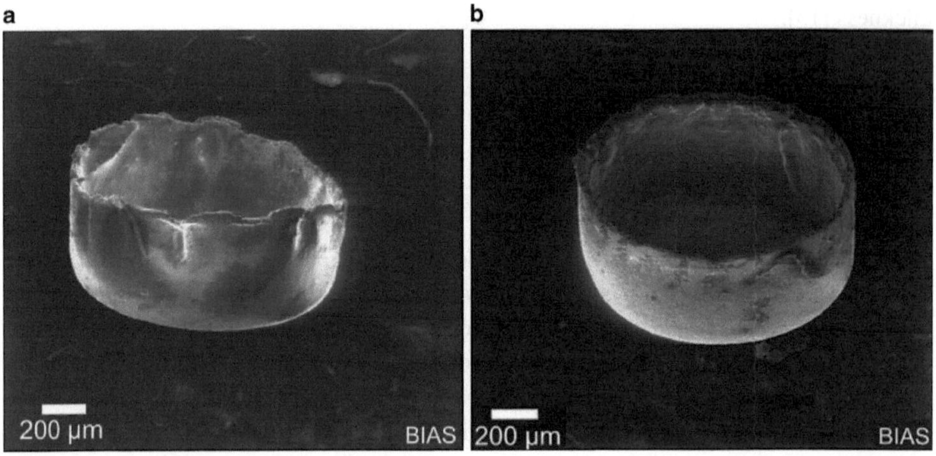

Fig. 2 Deep drawn micro cups with lubricant (**a**) and without (**b**) using a diamond deep drawing die (tool material (blank holder and drawing die): X155CrVMO121 (62HRC), punch diameter: 1 mm, punch radius: 0.1 mm, sheet material: 1.4301 (thickness: 25 μm), drawing ratio: 1.8, punch velocity: 1 mm/s, blank holder pressure: 2 N/mm^2, lubricant: HBO 947/11)

conventional micro deep drawing process. A reason for wrinkles using a diamond tool and lubricant is that the friction between diamond tool and work piece differs and another lubricant should be used. But in this case a lubricant is not necessary because the micro cup looks as a micro cup deep drawn in a conventional process.

5 Conclusions

In the future dry metal forming could change the forming industry, because without lubricant less resources as petroleum and cleaning chemicals are needed and less process steps are needed. For saving resources and energy dry metal forming exhibits a high potential as a green technology. However, dry metal forming requires an appropriate choice of tool materials to match the requirements of low friction and high wear resistance. Diamond tools have the capacity to overcome the limits of non-lubricated micro deep drawing of stainless steel micro cups. As shown in this work it is even possible to avoid the appearance of wrinkles and thus achieve a higher quality output in dry metal forming as compared to lubricated deep drawing.

6 Acknowledgement

The Priority Programme 1676 "Dry Metal Forming" was established recently by the Deutsche Forschungsgemeinschaft (DFG), and its coordination is supported under Vo530/72-1. Part of the work reported in this paper was funded within the DFG Collaborative Research Centre 747 "Micro Cold Forming" (subproject B3 and A6). The authors would like to thank the DFG for the financial support.

References

1. Flosky, F., Vollertsen F., 2014, Wear behaviour in a combined micro blanking and deep drawing process, CIRP Annals-Manufacturing Technology, 63 (1), 281–284.
2. Vollertsen F, Schmidt F, 2014, Dry Metal Forming: Definition, Chances and Challenges, Int. J. precision engineering and manufacturing – Green technology, 1 (1), 59–62.
3. Weihnacht V., Brückner A., Bräunling S., 2008, ta-C Beschichtete Werkzeuge für die Trockenumformung von Aluminiumblechen, Vakuum in Forschung und Praxis, 20 (3), 6–10.
4. Abele E., Feickert S.,Weigold M., 2004, Symposium zum Thema Innovative Zerspanung – Effizient und Umweltgerecht, TU Darmstadt, Inst. f. Technologie und Werkzeugmaschinen, 4–6.
5. Kataoka S., Murakawa M., Aizawa T., Ike H., 2004, Tribology of Dry Deep-Drawing of Various Metal Sheets with Use of Ceramics Tools, Surface and Coatings Technology 177–178, 582–590.
6. Tamaoki K., Manabe K. I., Kataoka S., Aizawa T., 2010, Electroconductive Ceramic Tooling for Dry Deep Drawing, Journal of Materials Processing Technology, 210 (1), 48–53.

7. Alsmann M., Behrens B. A., 1998, Gebrauchs- und Verarbeitungseigenschaften von Organisch Vorbeschichteten Feinblechen fur die Automobilindustrie, Tagungsband zum Industriekolloquium Fertigen in Feinblech, Clausthal-Zellerfeld, 66–70.
8. Carlsson P., Olsson M., 2006, PVD Coatings for Sheet Metal Forming Processes – a Tribological Evaluation, Surface and Coatings Technology, 200 (14), 4654–4663.
9. Osakada K., Matsumoto R., 2000, Fundamental Study of Dry Metal Forming with Coated Tools, CIRP Annals-Manufacturing Technology, 49 (1), 161–164.
10. Taube K., 1998, Carbon-based Coatings for Dry Sheet-Metal Working, Surface and Coatings Technology, 98 (1–3), 976–984.
11. Hirvonen J. P., Koskinen J., Jervis J. R., Nastasi M.,1996, Present Progress in the Development of Low Friction Coatings, Surface and Coatings Technology, 80 (1–2), 139–150.
12. Reisel G., Steinhäuser S., Wielage B., 2004, The Behaviour of DLC Under High Mechanical and Thermal Load, Diamond and Related Materials, 13 (4–8), 1516–1520.
13. Löhr M., Spaltmann D., Binkowski S., Santner E., Woydt M., 2006, In situ Acoustic Emission for Wear Life Detection of DLC Coatings During Slip-Rolling Friction, Wear, 260 (4–5), 469–478..
14. Engel U., Geiger M., Eckstein R., Tiesler N., Engel U., 2001, Microforming, CIRP Annals-Manufacturing Technology 50 (1), 445–462.
15. Schluze Niehoff, Hendrik, 2008, Entwicklung einer hochdynamischen, zweifachwirkenden Mikroumformpresse, BIAS Verlag, Volume 33.

Forming of Tailored Blank

Toshiyuki Takasago and Takao Iwai

1 Background

An automotive body is made up from more than 300 panels, and these panels are mainly assembled by spot welding. It is desirable to integrate several panels by using a large single blank. However, when using a large single blank, it cannot satisfy different requirements such as strength, stiffness and corrosion resistance. As a result, the best material cannot be selected, and material yield is significantly decreased. Therefore, Toyota started to develop tailored blank in the early 1980's. As seen in Fig. 1, several blanks that have different material and thickness are jointed into single blank by butt welding, and make an integrated body panel by pressing. A high-power CO_2 laser was selected as this welding process. Development items are laser butt welding technology and press forming technology [1, 2, 4, 6].

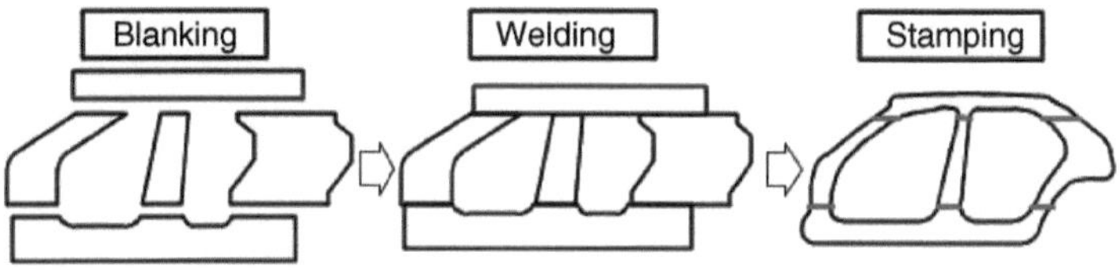

Fig. 1 Concept of the integration procedure using jointed sheets

Toshiyuki Takasago ✉ · Takao Iwai
Toyota Motor Corporation, Aichi, Japan
e-mail: takasago@mail.toyota.co.jp

A. E. Tekkaya et al. (eds.), *60 Excellent Inventions in Metal Forming*,
DOI 10.1007/978-3-662-46312-3_19

2 Laser Butt Welding Technology

CO2 laser has a high energy density. In butt welding, it can make narrow weld bead and reduce heat input to material. Therefore, a welded blank gets small deformation. It can reduce the effect on formability. In addition, it is more competitive than other welding processes in corrosion resistance. On the other hand, gap tolerance of laser butt welding in thin steel sheet is small. In addition, tolerance of misalignment between laser beam and butt joint line is also small. Thus, a high-precision cutting technology for material edge was developed for reducing the gap [6]. Laser welding speed is faster than other butt welding methods. Therefore, higher productivity is secured. Typical sectional shape of weld bead and distribution of hardness are shown in Fig. 2. Width of weld bead is narrow, and hardness is increased by 2 1/2 times. However, there is no soft formation at the boundary of welding area. It was confirmed that there is no functional problem when welding area is hard and thickness secures 80 % or more against matrix thickness (t/t_0 shown in Fig. 2 is 0.8 or more).

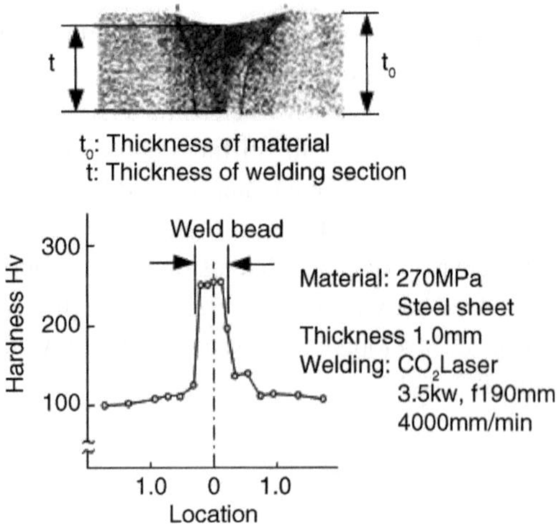

Fig. 2 Sectional shape of weld bead and hardness

3 Study of Formability for Tailored Blank

For applying to panels, press formability of tailored blanks that have different strength and thickness should be figured out, and incorporated into panel design. Forming of pressed parts for automotive body is composed of three typical forming patterns which are stretch forming, stretch flanging and deep drawing. These three basic forming patterns were evaluated by forming test [1, 2, 4].

Stretch Forming

Regarding punch stretchability, materials that have several different strengths were welded by laser beam, and stretch forming was implemented by ball head punch shown in Fig. 3 for evaluating forming height *h* and strain in the top of the test piece before crack. Major strain direction was set by making an arc cut in a round test piece, and strain ratio was changed by W that is the center width of the test piece. Figure 4 shows an example of the

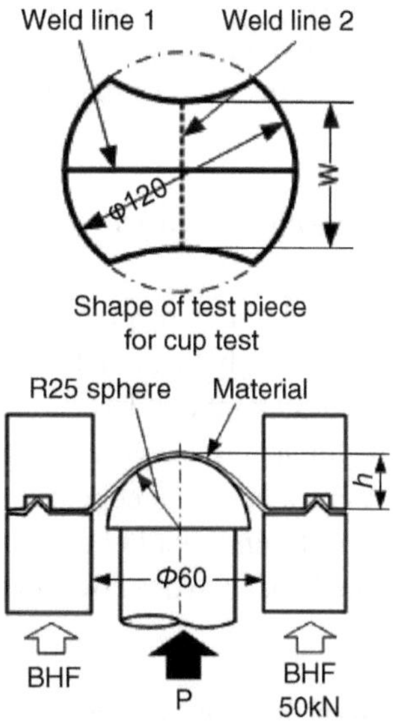

Fig. 3 Ball head expansion test

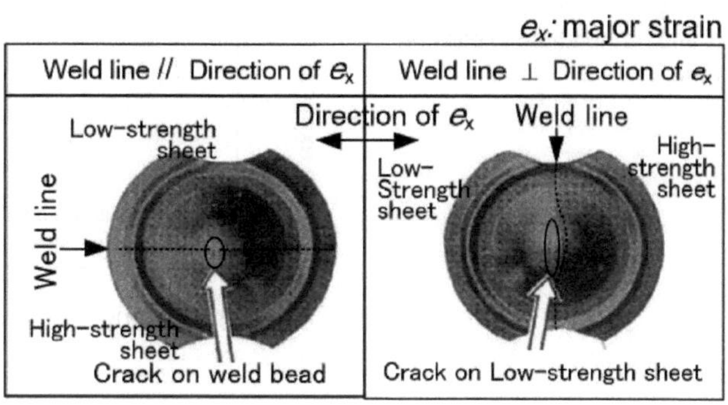

Fig. 4 Crack in ball head expansion test

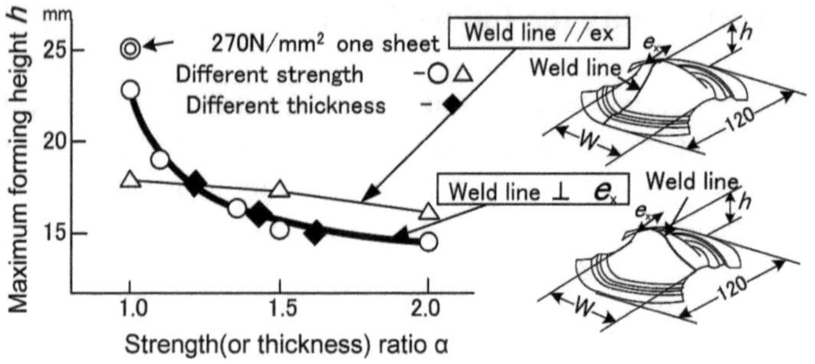

Fig. 5 Relation between Maximum forming height h and strength (or sheet thickness) ratio α

test pieces. As seen in Fig. 5, the relationship between maximum stretch forming height h and strength ratio differs depending on the direction that is perpendicular to weld line or parallel to weld line. In case major strain direction is perpendicular to weld line, the larger strength ratio, the lower maximum stretch forming height. It is because strain at the higher strength side gets smaller when strength ratio becomes larger. In case major strain direction is the same direction to weld line, it is determined based on elongation of weld line. Even though maximum stretch forming height gets significantly lower, it received a small effect from strength ratio.

Stretch Flanging

As seen in Fig. 6, a piercing hole (φ14 mm) was formed by stretch flanging, and diameter of the hole before crack was measured. By the effect of welding area of edge, limited hole expansion ratio of laser welded blank is decreased by 20 % compared to a single blank. In addition, it is decreased in proportion to strength ratio and thickness ratio of welded blank. A crack begins at edge next to weld line in the lower strength side. Therefore, placement of weld line should be set to area that is lower stretch flanging. Additionally, allowance should be increased by lengthening edge of stretch flange in blank shape.

Fig. 6 Relation between strength (or thickness) ratio α and limited hole expansion ratio λ

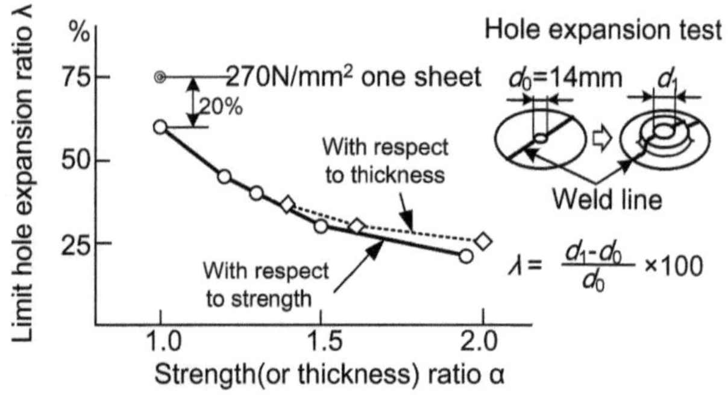

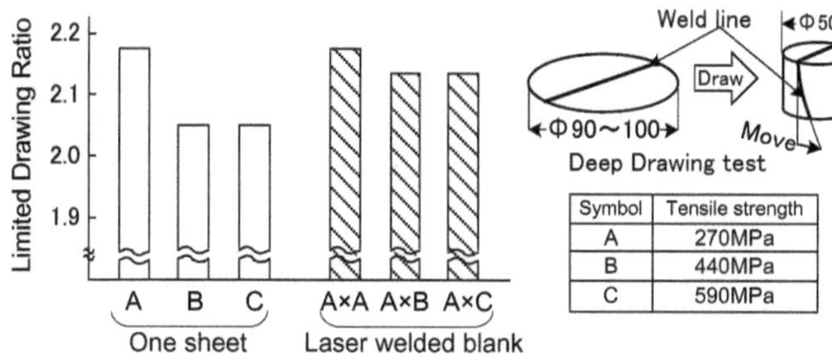

Fig. 7 Limiting Drawing ratio of welded material

Deep Drawing

By using a φ50 mm cylindrical punch type deep drawing testing machine, limiting drawing ratio L.D.R. was found by changing blank hold pressure within drawing ratio from 1.8 to 2.2. As seen in Fig. 7, it maintained the same level with a single blank. Moreover, deep drawability was excellent result. However; weld line will be moved to the lower strength side, the distance is proportional to strength ratio. When different thickness tailored blank applies to actual parts, clearance between upper die and lower die should be set in consideration of this distance.

4 Application to Actual Part Forming

Side Panel of Automotive Body

For integrating several parts, a different-material different-thickness welded blank was examined for application to side panel of automotive body. The panel is shown in Fig. 8.

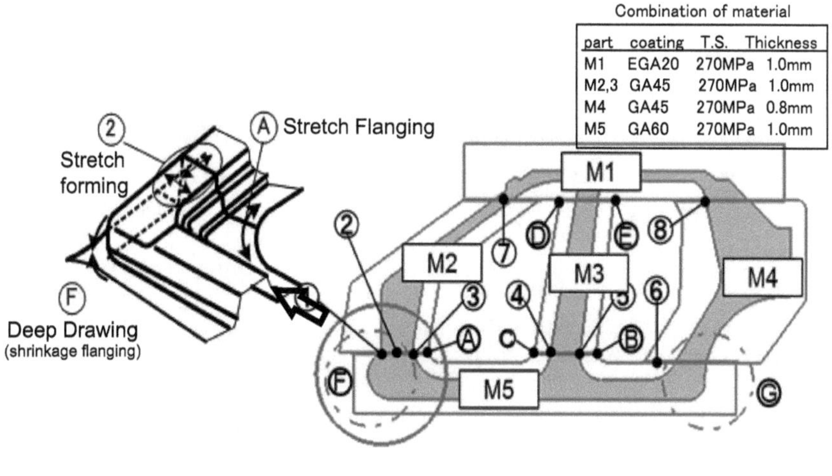

Fig. 8 Example of side panel outer

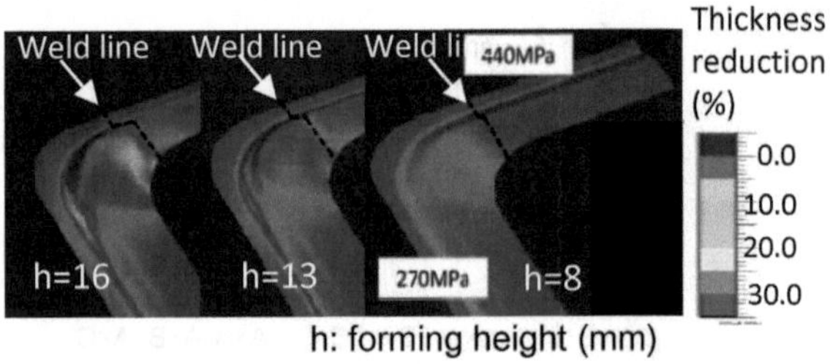

Fig. 9 Example of simulation at Stretch flange

As seen in the figure, No. 1 to 8 are stretch forming. A to E are stretch flanging. F and G are deep drawing. Weld line was set in a horizontal direction because it is easy to process blank and weld by laser beam. Based on the basic evaluation, stretch forming area is changed to reduce strain by considering function of a part from the limit of stretch of weld bead. In deep drawing area, clearance between upper die and lower die was adjusted to respond movement of weld line in different-thickness area. In stretch flanging area, weld line was set not to be maximum stretch flanging, and also allowance was tested by simulation as shown in Fig. 9. Based on the above facts, each laser welded area can be designed within forming limit.

5 Achievement of Development

Examples of parts that are applied by tailored blank are shown in Fig. 10. We have proceeded integration of parts that have a big difference in thickness. Side panel outer was in 1989, and side panel inner was in 1991 and door inner panel was in 1994. At the

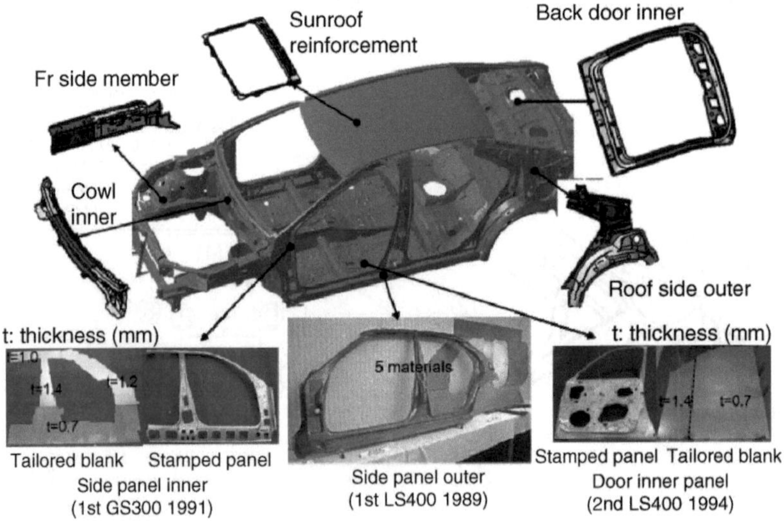

Fig. 10 Integrated body panels by tailored blank

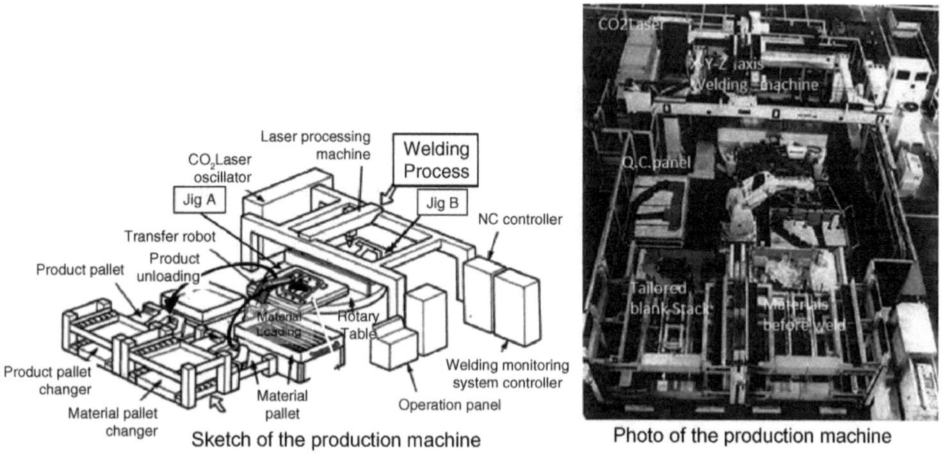

Fig. 11 CO_2 laser welding system

same time, we have developed and launched production equipment shown in Fig. 11. As the result of this innovative development, we could improve body accuracy and reduce production cost [5, 6].

6 Conclusion

1. Regarding different-strength different thickness tailored blanks by laser butt welding, basic forming patterns that are stretch forming, stretch flanging and deep drawing were tested and studied for clarifying each forming limit.
2. Based on the result of this study, tailored blanks by laser butt welding were applied to parts such as side panel, and we achieved press integrated forming to practical use. As the result of this achievement, material yield was developed by 15 % compared to conventional split type in side panel. Variation of body accuracy was improved by half in door opening area.
3. This technology was applied to other parts in automotive body. As the result of this application, press die cost, press working cost and body assembling cost have been cut by reducing number of parts. With this method, an improvement could be achieved in body accuracy, high rigidity and weight saving.

References

1. K. Ikemoto, H. Sugiura, K. Azuma, K. Arima, T. Takasago, Press Forming of Laser Welded Blank, JSTP vol. 132. 370 (1991-11) 1383–1390
2. K. Ikemoto, H. Sugiura, K. Azuma, K. Arima, T. Takasago: Press Formability of Laser Welded Blank: IDDRG (1990-6)

3. K. Yasuda, T.Jinma, T.Onzawa, Formability of Butt welded Stainless Steel Thin sheet, JWS 2–3 (1984) 161–166
4. T. Kobayashi, K. Murata, T. Morimoto, H. Ishigaki, T. Abe, R. Mitsuo, Research about the Press Forming Limit of Steel sheet, JSTP The 19th Conference (1968), 289
5. Y. Sadamura, Press Forming of Laser Welded Blank JSTP, vol. 34, 391, (1993-8) 917–924
6. T. Iwai, K. Sadamura, F. Natsumi, Integrated Automotive Panel Production System using Laser Welding, LANE '94, vol. 1, 135–144

New Forming Technologies Using Screw Type Servo Press

Junichi Endou and Chikara Murata

1 Introduction

There are two types of mechanical servo presses as introduced by K. Osakada et al. [1]: (i) crank type and (ii) screw type. Crank-type servo presses are widely available on the market, especially in Japan, but screw-type servo presses have some inherent advantages over crank-type servo presses including the following;

1. The maximum load and speed are available at any slide position. On the other hands, for a crank type press, the maximum load is possible near the bottom dead center where the ram speed becomes slow.
2. Bottom dead center can be placed at any position by numerical controller, and the positional accuracy of the slide is the same along its entire length.
3. The parallelism of the slide during the entire stroke, especially through forming, can be controlled by numerical controller.

In this article, we introduce new stamping technologies utilizing screw-type servo presses.

Junichi Endou ✉
Kanagawa Institute of Technology, Kanagawa, Japan
e-mail: endo@sd.kanagawa-it.ac.jp

Chikara Murata
Hoden Seimitu Kakou Kenkyusho Co., Ltd., Ueno-cho Kasugai-shi, Japan

A. E. Tekkaya et al. (eds.), *60 Excellent Inventions in Metal Forming*,
DOI 10.1007/978-3-662-46312-3_20

2 One Shot Forming

In sheet metal stamping, several press processes are generally needed to manufacture a finished product. A so-called transfer die and press line are generally used for this purpose, but this requires several presses working in sequence.

In one shot forming [1, 2], however, several dies are assembled lengthwise according to the processing sequence. When using a screw-type servo press with a double action mechanism and hydraulic die cushion, the products can be processed in one shot of the press. Figure 1 shows a screw-type servo press with a double action mechanism and hydraulic die cushion and Fig. 2 shows an example of the process sequence. If we tried to

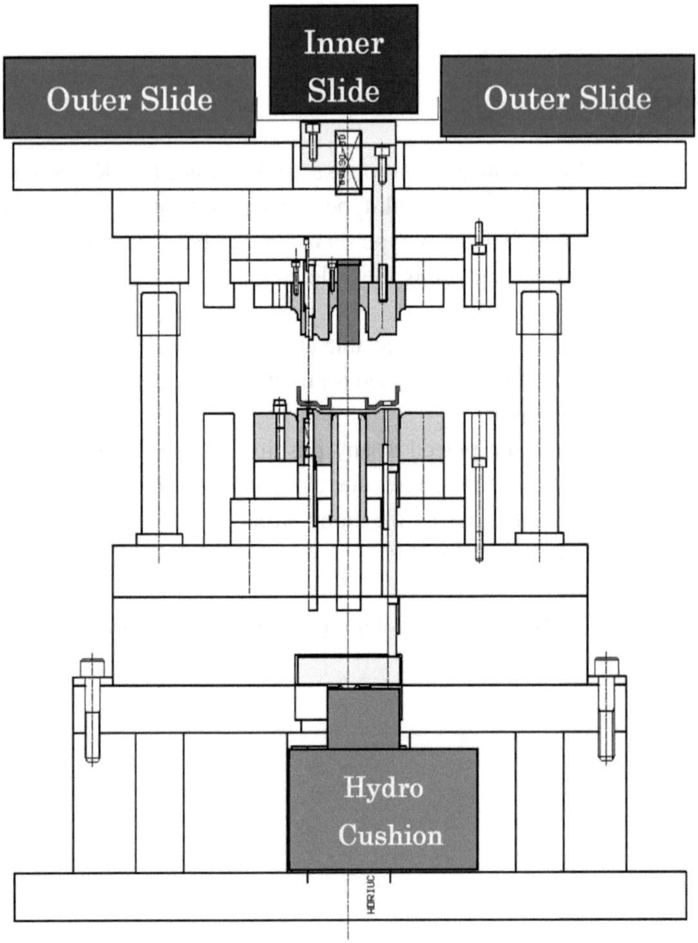

Fig. 1 Double action screw type press

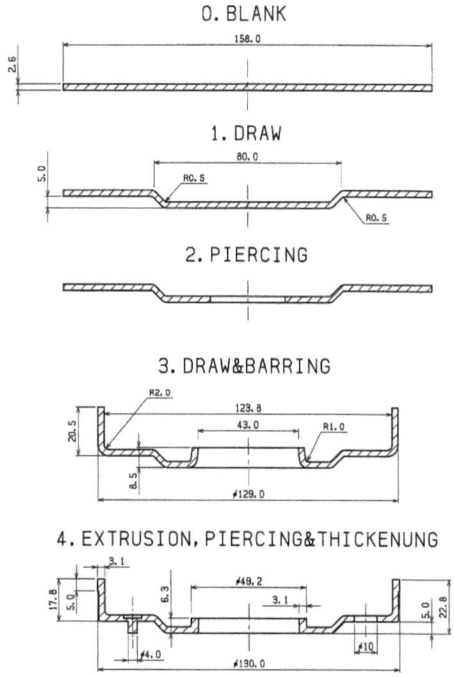

Fig. 2 One shot forming

manufacture the same product using a transfer press line, four sets of dies and presses are needed, but with one shot forming technology, only one multiple motion tooling (dies) and one screw-type press are enough to manufacture the product. The disadvantage of this one shot forming technology is generally the complicated design of multiple motion tooling. However, we can use modern CAD with numerical analysis to alleviate this disadvantage.

3 Slide Parallelism

In order to investigate slide parallelism on a conventional press, we applied eccentric loads to a press having a normal capacity of 300 KN [3]. The loading condition of this test is shown in Fig. 3, and the test results are shown in Fig. 4 [3]. A load difference of 200 KN over a span of 1 m caused an inclination of 0.00035 rad. and a total clearance of 0.3 mm. This total clearance is caused by the so-called lost motion of the press. The reason that the displacement of the slide is not in direct proportion to the load is because of the elastic deflection caused by applied load.

The same loading condition was applied to a screw type servo press of the same capacity (300 KN) [3]. The results are shown in Fig. 5. Total clearance cannot be measured, because the absolute positions of the slides are monitored with linear scales mounted on the bed of the press. The slide did not tilt with the eccentric load, because the servo motors which drive the screws can compensate by measuring the absolute position and using

Fig. 3 Eccentric loading

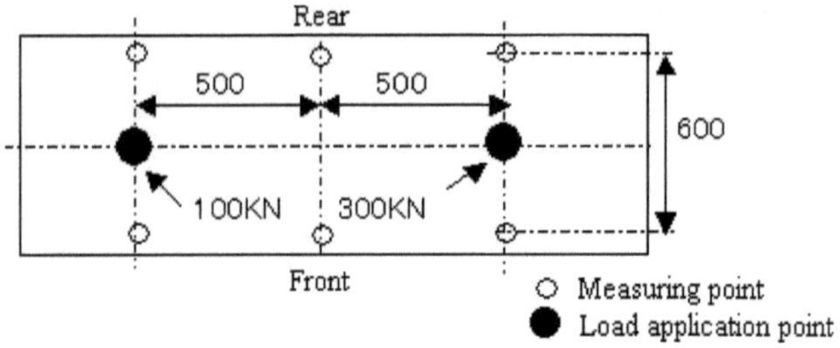

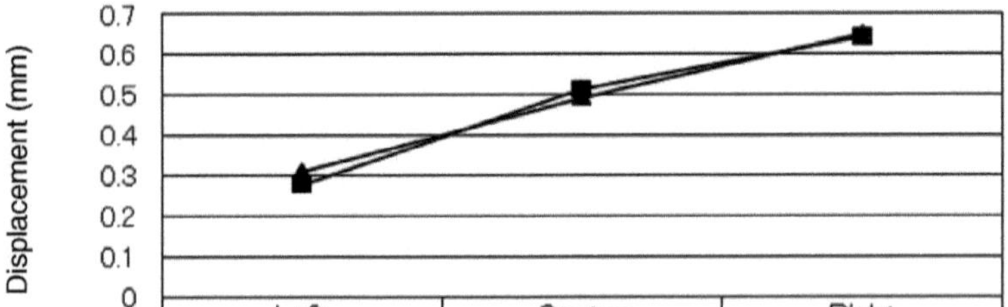

	Left	Center	Right
front	0.309	0.493	0.651
rear	0.276	0.512	0.642

Fig. 4 Slide displacements (conventional press)

feed-back control system. A slight displacement can be seen at the center of the slide, but this is caused by the elastic deflection of the slide itself and can be compensated for by inner slide shown in Fig. 1.

The processing time for stamping sheet metal is usually short and the applied load sometimes changes abruptly, often resulting in blanking or punching due to a so-called break-through. This is caused by micro cracks in the sheet metal that arise from shear deformation and penetrate through to the opposite side of the sheet. The stored elastic energy in the press body is abruptly released resulting in negative loading.

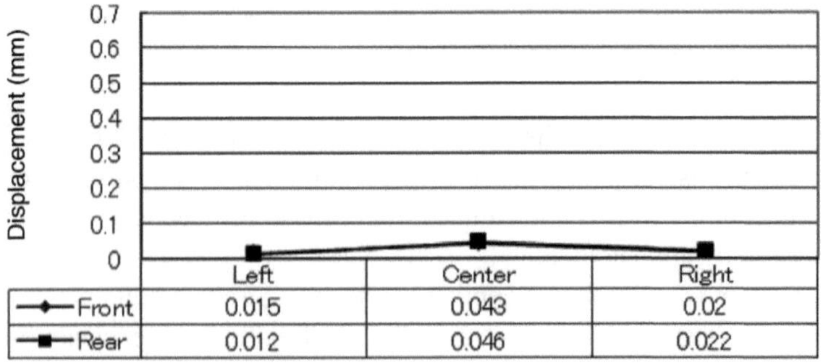

	Left	Center	Right
Front	0.015	0.043	0.02
Rear	0.012	0.046	0.022

Fig. 5 Slide displacements (screw type servo press)

Fig. 6 Displacements of 4 axes without parallel control

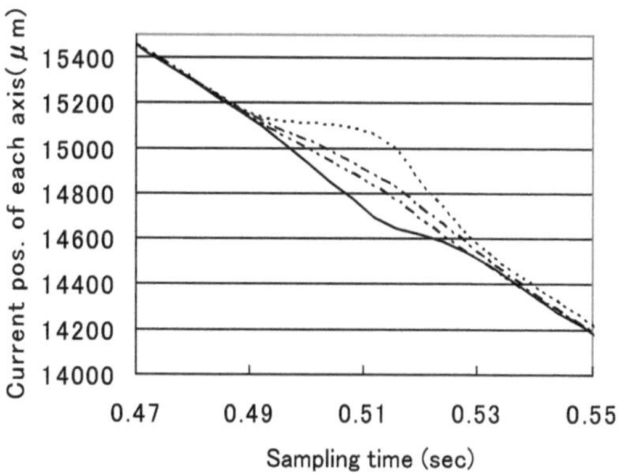

In order to realize the same eccentric load condition, we set an actual die in the same position shown in Fig. 3 and blanking a 4.5 mm thick steel plate were done [4], and the resulting enlarged slide displacement curves are shown in Fig. 6. At the start of the process, the four slide displacement curves come out of alignment, but are then corrected, resulting in the parallel movement of the slide. This correction is caused by the feed-back control system. However, the temporary misalignment did result in an inclination of the slide with a maximum tilt angle of 0.0002 rad. This temporary tilt of the slide is resulted in undesirable effects to the mold.

In order to avoid this temporary tilt of the slide, a new feed-forward control method was developed. The test results of this new control method under the same eccentric load condition are shown in Fig. 7. Again, the slide displacement curves do not perfectly coincide

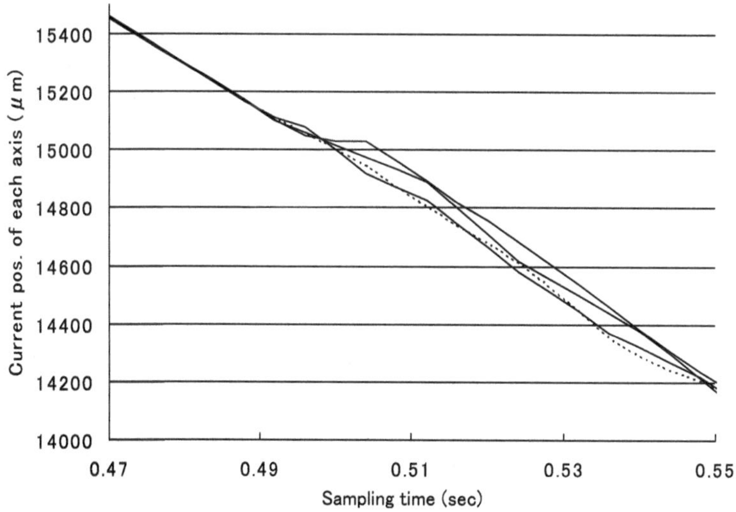

Fig. 7 Displacements of 4 axes with parallel control

with each other during blanking, but this time the differential has been greatly reduced. The inclination of the slide shown in Fig. 7 is 0.00005 rad proving the effectiveness of the newly developed feed-forward control method.

4 Progressive Stair Die Stamping (Dansa Kako)

A new technology called Progressive Stair Die Stamping (*dansa kako* in Japanese) [5, 6], utilizes before mentioned slide parallelism technique in static and dynamic states.

In sheet metal stamping, progressive dies, which combine several press processes into one shot, are often used for mass production. However, the disadvantages of progressive dies include the following;

1. Increased press load. Each process requires its own process load although the processes are conducted simultaneously. This means that the sum amount of each process load is required, often necessitating a large load capacity press.
2. Eccentric load is inevitable. As each individual process load is different, an eccentric load mustnecessarily occur, which means that a high rigidity press must be used.

However, the progressive stair die technique uses a parallel control system as shown in Fig. 8.

The punch length of each process varies and each process is completed sequentially not simultaneously. Therefore, the total stamping load is greatly decreased and a smaller press (which has a load capacity in excess of the load required the most load-intensive process) can be used.

For example, please see Fig. 8, where we have six individual processes. The 1st process is blanking of pilot pin holes which require 910 KN, the 2nd is deep drawing of 200 KN, the 3rd is blanking of 1100 KN, the 4th is coining of 600 KN, the 5th is shaving of 1760 KN and the 6th is blanking of 30 KN. In typical progressive die stamping, all six processes are done simultaneously, requiring 4600 KN. On the other hand, in progressive stair die stamping, the 1st and the 5th processes are paired as Step 1. The same goes for the 3rd and the 6th processes, which become Step 2 and the 2nd and the 4th processes, which become Step 3. The punch lengths for the Step 1 are a little bit longer than Step 2 and the punch lengths for Step 3 are a little bit shorter than Step 2 as shown in Fig. 8. (This

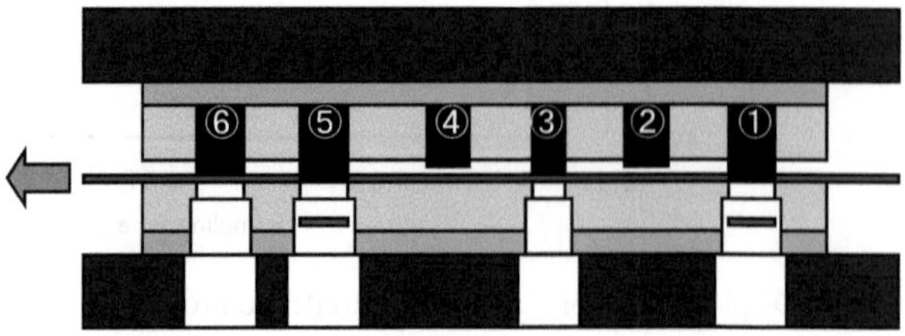

Fig. 8 Progressive stair die stamping (dansa kako)

is why we called it a "stair die"). As a result, the maximum load needed for stamping decreases to 2670 KN from 4600 KN. Using the parallel control system, the press slide is kept parallel during stamping. However, impact unloading due to break-through in the blanking processes results in undesirable effects on the press body. To limit this impact unloading, we modified the punch length of some of the processes. For example, we decreased the punch length of the 5th process in Step 1. As a result, when the 1st process (blanking) achieves break-through, the 5th process is still loading, thereby reducing the undesirable effects on the press body.

5 Conclusion

The screw type servo press and its characteristic features together with its applications are introduced. We believe that screw-type servo presses and crank-type presses still have the potential for significant improvement in sheet metal stamping and forging. One such avenue for improvement is enhanced utilization of press motion, which we have not yet harnessed fully. The second is the development of multiple motion tooling for servo presses as discussed in this article. This development may be enhanced through the use of CAD to better analyze deformation. The third is the potential of unmanned systems. The servo press has numerical control equipment and can be connected to various networks, including LAN, WAN and the internet, which form the basis for unmanned systems.

References

1. Osakada, K., Mori, K., Altan, T., Groche, P. 2011, Mechanical servo press technology for metal forming. CIRP Annals – Manufacturing Technology 60 , 651–672.
2. Murata,C 2011, Possibility of Forming Technology by Multi-slide Servo Press. Press Working, 48(11), 47–49 [in Japanese]
3. Murata, C., Machida, T., Futamura, S., Endou, J., 2004, Development of direct drive digital servo press. Steel Grips 2 (2004) Suppl. (Proc. Metal Forming 2004), 371–374 Nakamura, M., Maki, S., Nakajima, M., Hayashi, K., 1996, Bending of Circular Pipe Using a Floating Spherical Expanding Plug, Advanced Technology of Plasticity 1996: Proceedings of the 5th International Conference on Technology of Plasticity, Columbus, Ohio, 501–504.
4. Murata,C., Machida, T., Futamura, S., Endou, J., 2005, Intelligent Control System for Direct Drive Digital Servo Press. Proc. 5th International Conference on Intelligent Processing of Materials. Monterey, CD-Rom.
5. Murata, C., Machida, T., Futamura, S., Endou, J., 2005, A Proposal of "Progressive Stair Die Working" (Development of Direct Drive Digital Servo Press) Advanced Technology of Plasticity 2005: Proceedings of the 11th International Conference on Technology of Plasticity.
6. Ichikawa, K., Murata, C., Takahashi, T., 2014, Nonsimultaneouse press forming using 4-axes direct drive digital servo press. Procedia Engineering. (Proceedings of the 11th International Conference on Technology of Plasticity), 2014, 1669–1674, CD-Rom

Part IV
Incremental Forming

Non-circular Spinning

Sebastian Härtel and Birgit Awiszus

1 Introduction

Metal spinning is defined by DIN 8584 as a process of sheet metal forming with no intention to reduce the sheet's thickness. By applying forming techniques with a combination of tensile and compression conditions, a flat blank can be reshaped into almost any kind of rotationally symmetric, hollow component, or alternatively, the form of such components can be changed. Sheet metal spinning is an efficient and economical manufacturing process for the production of components in small- to medium-sized lots. Due to low tool costs (only one tool with the shape of the part is required), short set-up times and adaptable machine construction, spinning offers several advantages over conventional forming processes for small production lots, such as deep-drawing (see [1]). Moreover, it is possible to spin components with tight geometrical tolerances and to achieve a high-quality surface finish (see [2]). An opportunity to increase the economical relevance and flexibility of conventional metal spinning is the development of the process of "non-circular spinning" for producing rotationally non-symmetric components (see Fig. 1).

Within the scope of the DFG (German Research Foundation) research project, "Metal Spinning of Rotationally Non-Symmetric Hollow Components", a machine tool and a control system have been developed and presented in [3] that can produce components using this new "non-circular spinning" process. However, it is necessary to optimize the non-circular spinning process in order to produce components with a constant wall thickness, without wrinkles and cracks.

Sebastian Härtel · Birgit Awiszus ✉
TU Chemnitz, Chemnitz, Germany
e-mail: birgit.awiszus@mb.tu-chemnitz.de

A. E. Tekkaya et al. (eds.), *60 Excellent Inventions in Metal Forming*,
DOI 10.1007/978-3-662-46312-3_21

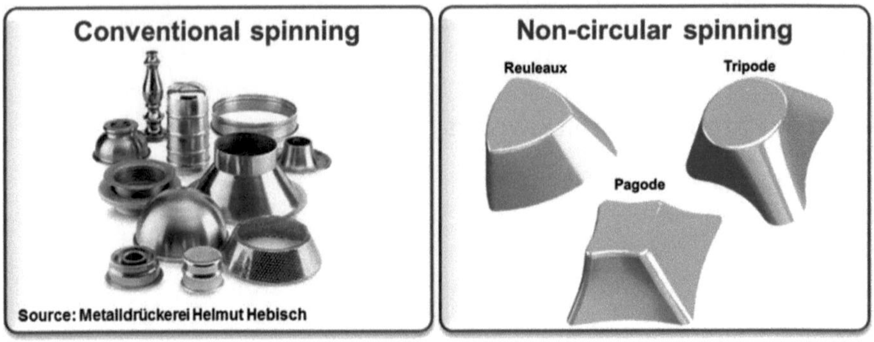

Fig. 1 Components produced by conventional and non-circular spinning

2 Process Idea

In principle, the production of rotationally non-symmetric, hollow components through spinning is possible with the application of force-controlled or motion-controlled roller tools (see Fig. 2).

Initial investigations by Awiszus [4] and Arai [5] showed that a force-controlled roller tool allows the production of rotationally non-symmetric hollow parts. A disadvantage of force-controlled roller tools are the different forming conditions in the concave and convex sections of the component. It is the constant forming force that leads to different forming pressures in the concave and convex areas.

Another possibility for the manufacturing of rotationally non-symmetric hollow parts through spinning is the exact feeding of the roller tool in the axial and radial directions. This ensures a constant gap between the spinning mandrel and the roller at all times during the spinning process. The advantage of motion-controlled rollers in contrast to force-controlled rollers is that the former provide the possibility of actively influencing

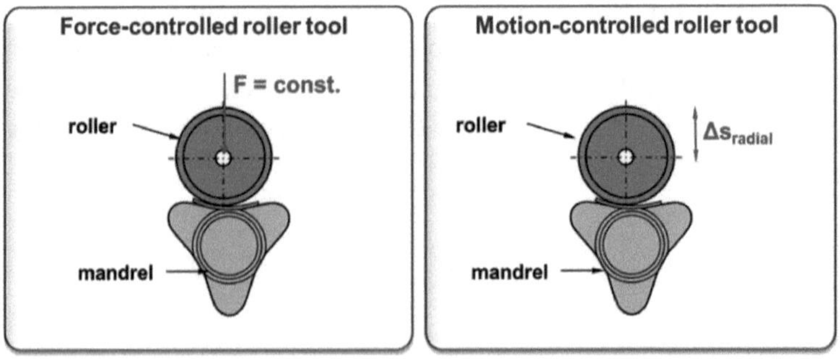

Fig. 2 Principle of force-controlled and motion-controlled roller tool

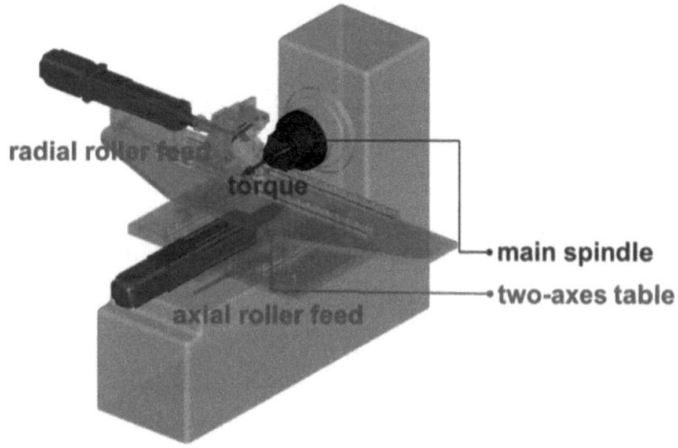

Fig. 3 Main components of the non-circular spinning machine

the resulting blank thickness, due to the constant gap between the mandrel and the roller. The motion-controlled roller method thus creates constant forming conditions in the concave and convex areas.

For the production of rotationally non-symmetric hollow parts, a spinning machine was developed according to the kinematical requirements for non-circular spinning with motion-controlled roller tools. The requirements are the supply of torque at the main spindle and the independent translational movement of the roller in the axial and radial directions. The non-circular spinning machine consists of two main parts, which can be seen in Fig. 3:

- A main spindle as the drive component for the mandrel
- A two-axes table for radial and axial feed of the roller tool

The torque generation at the main spindle, which sets the mandrel and the sheet in rotation, is provided by a direct current motor. The translational movement in the axial and radial directions is carried out by electro-mechanical spindle drives, each of which consists of one electric cylinder (including one screw spindle) and the motor. Roller circulating guidance units were chosen for the axial and radial movements of the roller tool.

The roller tool is controlled by using a master/slave control and the rotating angle of the main spindle is used as the master-axis. Therefore, the axial and radial movement of the roller tool (slave-axis) is dependent on the actual rotating angle of the main spindle. The advantage of this configuration is that rotation speed fluctuations do not influence the slave-positions. An encoder measures the rotating angle of the main spindle incrementally and the roller tool is moved to the predefined slave-positions.

3 Results and Applications

The experimental investigations of the non-circular spinning process are done with blanks (initial sheet thickness 1.5 mm) of pure aluminum Al99.5 H24 and the "Tripode" mandrel type shown in Fig. 4.

To determine the influence of the diameter of the blank, the roller feed rate and the operating angle of the roller on wrinkling and thinning, experiments were done by Härtel and Awiszus [6] with parameters varying according to Fig. 5.

The analysis of the experiments showed an increasing trend of wrinkling by using higher initial blank diameters and higher roller feed rates. Moreover, for each part geometry there is an optimum of the operating angle which leads to minimal wrinkling. Furthermore, the analysis of the sheet thickness showed that lower initial blank diameters and higher roller feed rates lead to a reduction of sheet thinning – independent of the part geometry. There is also a significant influence of the operating angle on sheet thinning for spinning the component "Tripode". For this, a higher operating angle leads to reduced sheet thinning.

Nevertheless, it is possible to produce a wide range of geometries using the newly developed non-circular spinning process as it is shown in Fig. 6. The component "Tripode" is characterized by a variation of concave and convex regions whereas the side walls of the component "Reuleaux" are convex curved and side walls of the component "Pagode" are concave curved.

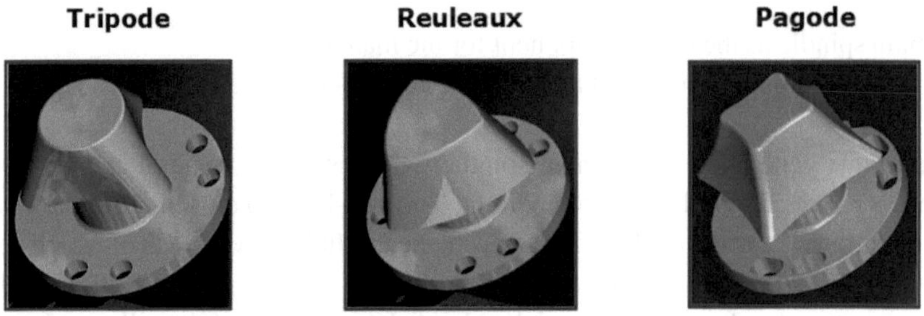

Fig. 4 Shape of the mandrels for non-circular spinning

technological parameters and variation:		
roller feed rate f:	0.5 – 2.0 mm/R	Δ 0.5 mm/R
operating angle α:	50° - 90°	Δ 10°
rotation velocity n:	50 rpm	
blank:		
material:	Al99.5 H24	
diameter d_{blank}:	120 - 140 mm	Δ 10 mm
thickness s:	1.5 mm	

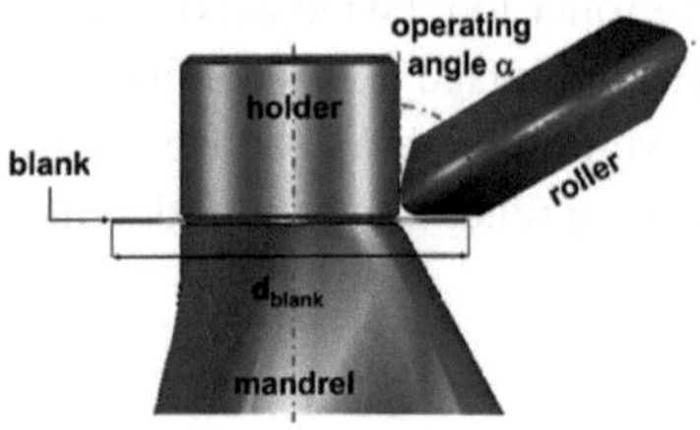

Fig. 5 Technological parameters and variation for the experiments

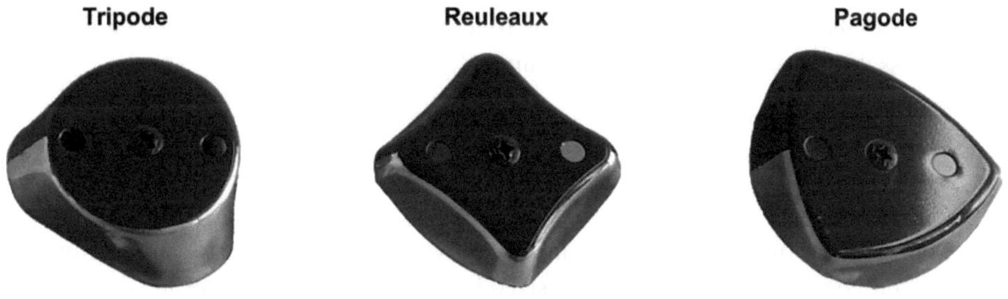

Fig. 6 Produced components by non-circular spinning

4 Optimization

For reducing wrinkling and thinning, a process optimization was done by using the Finite Element Method (FEM). The numerical process analysis shows that a targeted varying of the roller feed rate and the operating angle during the process leads to minimized thinning and also wrinkling. Varying the roller feed rate during the non-circular spinning process can be realized by the existing machine and control conception. But for realizing the swiveling of the roller tool during the process, the existing machine conception has to be extended with a synchronous servo drive and a planetary gearing. The integration of the synchronous servo drive and planetary gearing box within the existing machine configuration is shown in Fig. 7.

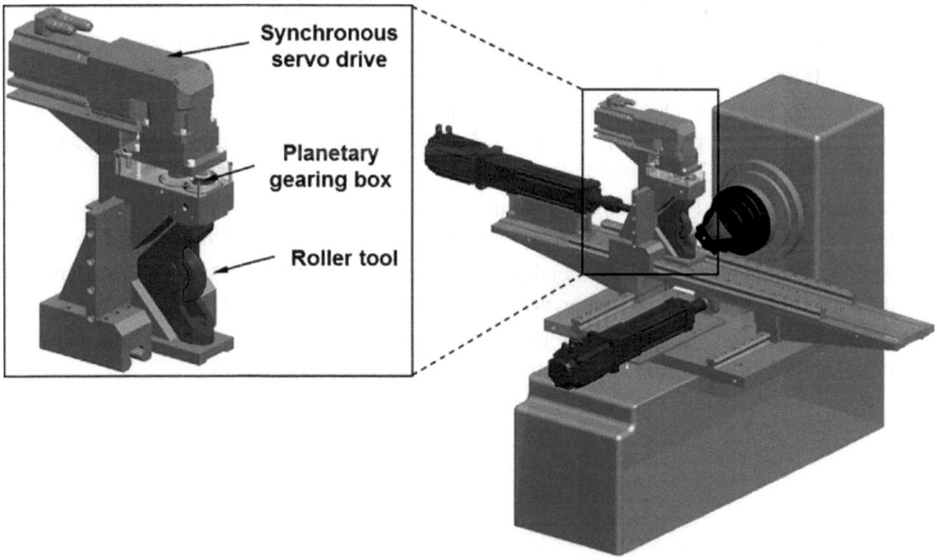

Fig. 7 Integration of the swivel-drive in the existing machine configuration

Thanks to the controlled variation of the roller feed rate and the operating angle during the process thinning could be reduced by up to 25 %.

5 Conclusion

The non-circular spinning process is an innovative forming process for producing non-circular components in small to medium sized lots. Therefore, a machine and control conception was developed to produce the components "Tripode", "Reuleaux" and "Pagode". Furthermore, the newly developed process was optimized by means of the FEM. The optimization approach is to adapt the roller feed rate and the operating angle to reduce wrinkling and also thinning. For varying the operating angle during the process a swivel-drive was integrated in the existing machine configuration.

6 Acknowledgement

This work was kindly supported by the German Research Foundation (DFG). The authors thank the German Research Foundation (DFG) for its financial support of this work.

References

1. Lange, K., 1990, Umformtechnik – Handbuch für Industrie und Wissenschaft, Bd. 3: Blechbearbeitung. Berlin, Heidelberg, New York: Springer-Verlag – ISBN 3-54050-039-1
2. Tschätsch, H., 2005, Metal Forming Practise. Berlin, Heidelberg, New York: Springer-Verlag, 2005. – ISBN-10 3-540-33216
3. Awiszus, B.; Härtel, S., 2011, Numerical simulation of non-circular spinning: a rotationally non-symmetric spinning process. In: Production Engineering. Springer Berlin/Heidelberg, 5 (2011) 6. Doi: 10.1007/s11740-011-0335-9.
4. Awiszus, B.; Meyer, F., 2005, Metal Spinning of Non-Circular Hollow Parts. Proceedings of 8th International Conference on Technology of Plasticity (Verona 2005). – ISBN 88-87331-74-X.
5. Arai, H., 2006, Force-controlled Metal Spinning Machine Using Linear Motors. Proceedings of the 2006 IEEE International Conference on Robotics and Automation (Orlando 2006).
6. Härtel, S.; Awiszus, B., 2012, How to avoid wrinkling and cracking in non-circular spinning. Proceedings of MAPT 2012, 28th–29th of June 2012, Hawaii, pp. 48–53.

Hybrid Sheet Metal Processing Center

David Bailly, Laura Conrads, and Gerhard Hirt

1 Motivation

Due to high tool and facility costs, conventional sheet forming processes such as deep drawing are only profitable in mass production. Incremental Sheet Forming (ISF) is a process with a low demand for geometry specific tooling and thus enables the economic production of small batch sizes. For more than 15 years, ISF and its various process variants are in the focus of research. However, all known process variants show the following drawbacks: long process times depending on the length of the tool path and hence the component size [1]; increasing wall angles lead to increasing sheet thinning. Wall angles greater than 60–70° can only be produced by complex forming strategies [2]; the geometric accuracy is influenced by residual stresses and springback [3].

Accordingly, when aiming for a wide range of applications, all three main process limits need to be expanded. All concepts which are presented here, focus on the application of the ISF process with a partial or full die. This enables an increased geometric accuracy as well as the production of more complex geometries.

2 Setup of the Hybrid Sheet Metal Processing Center

To face the drawbacks of the conventional ISF process, a Hybrid Sheet Metal Processing Center has been developed and installed at the Institute of Metal Forming (IBF) of the RWTH Aachen University. The machine features eight axes for tangential stretch forming with two-sided or four-sided stretching, five axes for milling and ISF and one axis for guiding a laser beam. This way, it allows for milling of geometry specific tooling, stretch

David Bailly · Laura Conrads · Gerhard Hirt ✉
RWTH Aachen University, Aachen, Germany
e-mail: hirt@ibf.rwth-aachen.de

A. E. Tekkaya et al. (eds.), *60 Excellent Inventions in Metal Forming*,
DOI 10.1007/978-3-662-46312-3_22

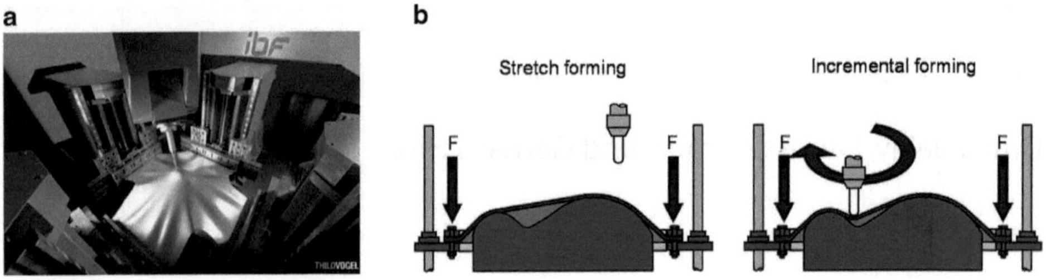

Fig. 1 **a** Impression of Hybrid Sheet Metal Processing Center, **b** Procedure of the process combination of stretch forming and ISF [2, 4]

forming, ISF, laser heat treatment and trimming in one concerted machine setup. The integration of pre-, main- and post-processing in one machine reduces the overall process time and improves the geometric accuracy.

The process combination of stretch forming and ISF is beneficial in terms of process time, sheet thickness distribution and geometric accuracy compared to pure ISF. The process combination starts with the generation of a preform using stretch forming. Afterwards, cavities and remaining features are formed by ISF (see Fig. 1) [2, 4, 5]. Since the process of stretch forming is significantly faster than ISF and only smaller areas have to be formed by ISF, a drastic shortening of the process time can be achieved. Furthermore, stretch forming causes less sheet thinning in steep wall angles than ISF which can lead in total to a more homogeneous sheet thickness distribution. Last but not least, the induced tensile stresses can superpose the residual stresses resulting from the ISF and hence reduce springback.

In the field of small series production, titanium and magnesium components for aerospace applications are a potential market for ISF. Due to the limited formability of these materials at room temperature, components have to be produced by warm forming processes. The two general approaches of warm ISF are the combination with either a global or a local heating. The main expected advantage of local heating is the further localization of the forming zone. While the formability within the forming zone is enhanced, the surrounding area which is not heated can provide the required stiffness to stabilize the formed component [2, 6, 7].

At the Hybrid Sheet Metal Processing Center, laser beam and forming tool operate on the same side of the sheet. This permits the application of laser heating even with additional full or partial dies (see Fig. 2b). The application of dies enables higher accuracy and complexity of the formed geometries. With the developed laser optics the laser beam can be guided in front of the forming tool according to the tool path [2, 7]. Besides the local heating during forming processes, the setup also allows for a local heat treatment prior to forming processes.

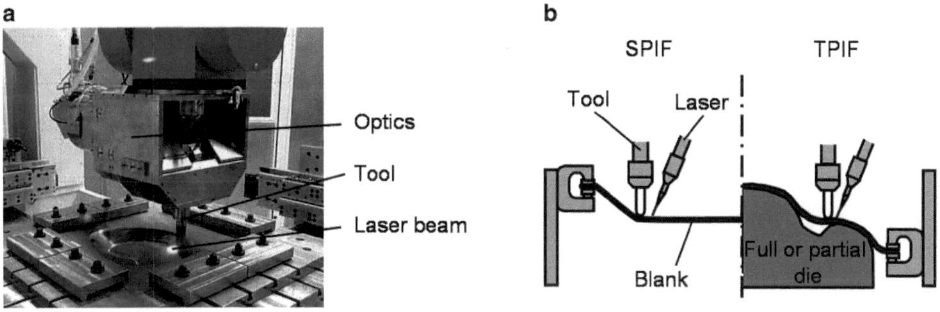

Fig. 2 **a** Laser assisted ISF process, **b** Schematic depiction of laser assisted ISF in case of single point incremental forming (SPIF) and two point incremental forming (TPIF)

3 Applications and Results

At the Hybrid Sheet Metal Processing Center, benchmark parts and industry relevant components have been produced using pure ISF, the process combination of stretch forming and ISF, laser assisted ISF as well as additional pre- and post-processing. Examples are illustrated in Fig. 3.

In order to test the capabilities of the hybrid machine setup, a benchmark component (see Fig. 3a) made from 1.0 mm mild steel (1.0338) has been produced with the process combination of stretch forming and ISF. The evaluation shows a more homogeneous sheet thickness distribution as well as reduced springback compared to the component produced by pure ISF (see Fig. 4).

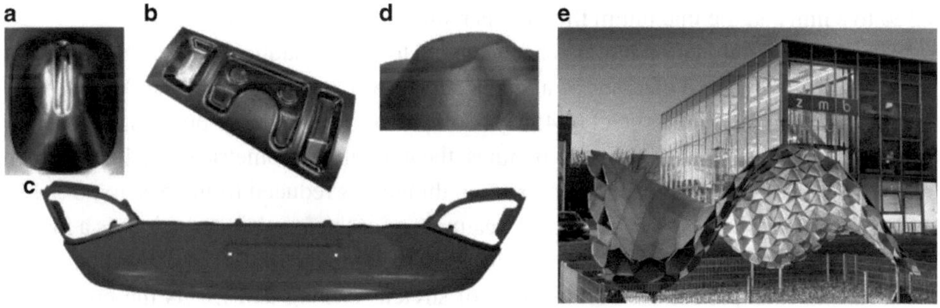

Fig. 3 **a** Benchmark component (four-sided stretch forming + ISF), **b** Airbus A320 inspection cover (two-sided stretch forming + ISF), **c** Audi A4 Avant back door (four-sided stretch forming + ISF), **d** Example component made from TiAl6V4 (laser assisted ISF), **e** self-supporting freeform structure consisting of 140 individual pyramids (pure ISF)

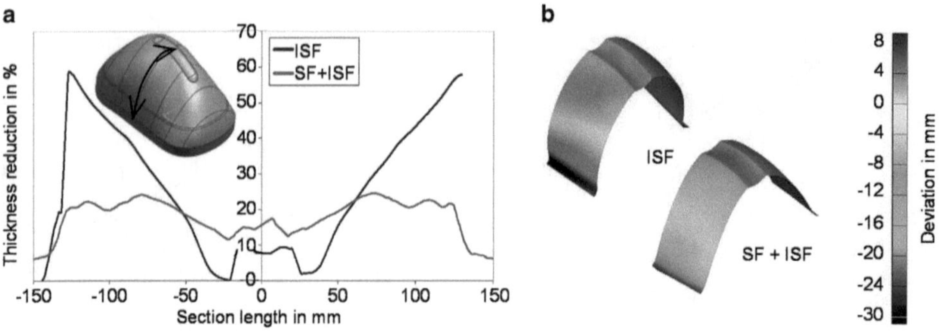

Fig. 4 Comparison of pure ISF to the process combination stretch forming and ISF. **a** Sheet thickness distribution, **b** Deviation of a section of the component from target geometry

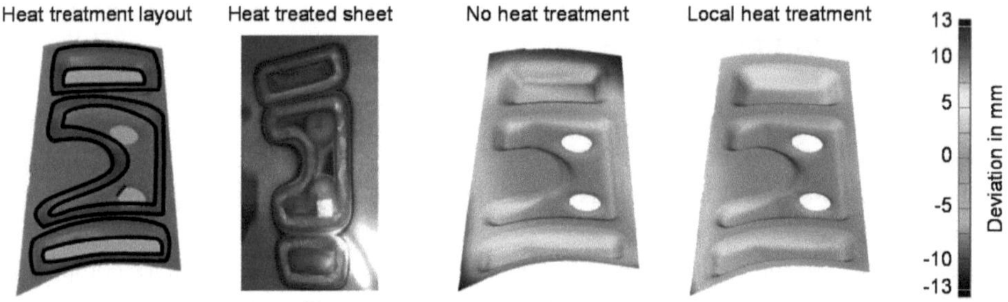

Fig. 5 Local heat treatment prior to forming operation to increase geometrical accuracy

In detail, compared to the pure ISF process, the process time has been reduced by 33 % to 40 minutes, the maximum geometric deviation of the component has been reduced by 63 % to 3 mm and the maximum thinning is reduced from 60 % to 25 % [8].

The process combination has also been applied to an aeronautical part. The part shown in Fig. 3b is a structural element of an inspection cover at the pylon of the Airbus A320 made from 1.0 mm stainless steel (1.4541). Compared to the pure ISF process, the process time has been reduced by 39 % to 65 minutes, the maximum geometric deviation has been reduced by 61 % to 1.3 mm and the maximum thinning is reduced from 35 % to 30 % [4].

Expanding the process by a local heat treatment prior the forming operations is a further option to improve the geometrical deviation from the target geometry (see Fig. 5).

A further example for the combination of stretch forming and ISF is the production of an Audi A4 Avant back door prototype (see Fig. 3c) made from 0.75 mm mild steel (1.0338). Here, the forming and the post-processing trimming and edge flanging have been performed successively in one machine setting. Compared to the pure ISF process, the process time has been reduced by 60 % to 90 minutes [8].

Applying laser assisted ISF, the formability of TiAl6V4 in ISF has been significantly improved (see Fig. 3d). However, using a constant laser output, overheating can be observed in the corners and especially at the end of the process. Due to the narrowing geometry, more heat is stored during the process and less energy would be required to provide the respective temperature. To solve this problem, a closed loop temperature controller has been implemented. The temperature is measured dynamically in the tool tip [9]. As a result, the laser output control leads to a more constant temperature distribution within the sheet.

The self-supporting freeform structure shown in Fig. 3e consists of 140 individual pyramids and 234 individual triangles made from 0.8 mm stainless steel (1.4404) sheets. A software tool has been developed that automatically generates the geometries of all components. Based on these CAD files the pyramids are formed by pure ISF and their edges are subsequently laser trimmed and bent. This example demonstrates the high potential of a production oriented construction concept combined with an automated planning procedure and an optimized manufacturing chain [10].

4 Conclusion

In general, Incremental Sheet Forming is a suitable process for single part and small series production. The main drawbacks of the process are the long process time, the material thinning depending on the wall angle and the limited geometric accuracy. The combination of stretch forming and ISF has the potential to reduce the process time as well as to homogenize the sheet thickness distribution and to improve the geometrical accuracy. Laser assisted ISF allows the application of ISF to materials that need to be formed at elevated temperatures. Furthermore, it is expected that the geometric accuracy can be enhanced since residual stresses and springback might be reduced during warm forming. Both process combinations can be conducted at the described Hybrid Sheet Metal Processing Center in one concerted machine setup. Additionally, preparation processes such as milling of geometry specific tooling and post-processing such as trimming can be carried out integrated in one machine setting.

5 Acknowledgement

The authors would like to thank the German Research Foundation (DFG) for funding received through the Cluster of Excellence EXC128 "Integrative Production Technology for High Wage Countries" and the German Federal Ministry of Education and Research (BMBF) for funding received within the Framework Concept "Research for Tomorrow's Production" (funding nr. 02PU2104).

References

1. Jeswiet J., Micari F., Hirt G., Bramley A., Duflou J., Allwood J., 2005, Asymmetric single point incremental forming of sheet metal, CIRP Annals, 54(2), 623–649.
2. Taleb-Araghi B., Göttmann A., Bambach M., Hirt G., Bergweiler G., Diettrich J., Steiners M., Saeed-Akbari A, 2011, Review on the development of a hybrid incremental sheet forming system for small sizes and individualized production, Production Engineering – Research and Development, 5, 393–404.
3. Bambach M., Taleb-Araghi B., Hirt G., 2009, Strategies to Improve the Geometric Accuracy in Asymmetric Single Point Incremental Forming, Production Engineering – Research and Development, 3, 145–156.
4. Taleb-Araghi B., Manco G.L., Bambach M., Hirt G., 2009, Investigation into a new hybrid forming process: Incremental sheet forming combined with stretch forming, CIRP Annals – Manufacturing Technology, 58, 225–228.
5. Amino H., Matsubara S., Lu Y., 2004, Shaping method and apparatus of thin metal sheet, US Patent 2004/0148997 A1.
6. Duflou J.R., Callebaut B., Verbert J., De Baerdemaeker H., 2007, Laser Assisted Incremental Forming: Formability and Accuracy Improvement, CIRP Annals – Manufacturing Technology, 56, 273–276.
7. Göttmann A., Diettrich J., Bergweiler G., Bambach M., Hirt G., Loosen P., Poprawe R.,2011, Laser-assisted asymmetric incremental sheet forming of titanium sheet metal parts, Production Engineering – Research and Development, 5(3), 263–271.
8. Hirt G., Bambach M., Taleb-Araghi B., Malik A., Vitr M., Bender D., Hieber S. J., Huthmacher V., Zettler J., Nagel G., Krumbacher P., Hofbauer Th., Elvenkemper A., Graef S., Heise F.-J., 2011, SIBUform-Steigerung der Produktivität und des Anwendungsspektrums bei inkrementeller Blechumformung durch Kombination mit Streckziehen, Forschungsvereinigung Stahlanwendung e.V, Düsseldorf, ISBN 978-3-942541-05-3.
9. Göttmann A., Bailly D., Bergweiler G., Bambach M., Stollenwerk J., Hirt G., Loosen P., 2013, A novel approach for temperature control in ISF supported by laser and resistance heating, International journal of advanced manufacturing technology, 67, 2195–2205.
10. Bailly D., Bambach M., Hirt G., Pofahl T., Herkrath R., Heyden. H., Trautz M., 2014, Manufacturing of Innovative Self-Supporting Sheet-Metal Structures Representing Freeform Surfaces, Procedia CIRP 18, 51–56.

Friction-Spinning – Innovative Opportunity for Overcoming Process Limits in Spinning Processes

Werner Homberg and Benjamin Lossen

1 Motivation and the Idea of Friction Spinning

The requirement for a high performance, coupled with ecological concerns, such as for reduced material and energy consumption, calls for innovative lightweight products and hence for efficient manufacturing technologies [1, 2]. In the conventional spinning process, work-hardening is inevitable and significantly restricts the forming limits. This limitation is normally counteracted by burner systems, intermediate heating, or the use of lasers for localized heating [3–6]. All these approaches increase the costs of energy, investment and maintenance. An interesting approach to solving this problem is self-induced heat generation through friction processes, comparable to the friction subprocesses that prevail during the early phases of friction welding. Combining elements from the metal spinning and friction welding processes gives rise to a new process called friction spinning. Through the friction processes, it is possible to achieve defined, local warm-forming. Use can thus be made of semi-finished parts like tubes, sheets and rods in a wide variety of metallic materials. With the appropriate process and tool design, local heat generation and high deformation become possible, which could be used for the production of highly complex part geometries or favorable properties that cannot be produced by conventional forming processes. The defined and locally restricted heat transfer then means that the production of functionally graded properties (e. g. a defined strength or hardness distribution inside the part) becomes possible. Together with the wider range of geometries that can be produced, it becomes feasible to manufacture multifunctional components like complex hollow parts made of tubes, profiles or sheet metals, which can then be employed for lightweight constructions. Apart from this, there is a high poten-

Werner Homberg ✉ · Benjamin Lossen
University of Paderborn, Paderborn, Germany
e-mail: wh@luf.uni-paderborn.de

A. E. Tekkaya et al. (eds.), *60 Excellent Inventions in Metal Forming*,
DOI 10.1007/978-3-662-46312-3_23

tial for savings on manufacturing steps and time, since additional processes like joining operations are no longer necessary.

2 Process Principle

As mentioned above, friction spinning can be used to process a range of semi-finished products. Figure 1 thus shows the working principle of friction spinning for tubular (Fig. 1a–c) and sheet metal components (Fig. 1d–f). To apply the process, it is necessary to achieve a relative movement (rotation) between the (friction) tool and the workpiece (sheet or blank). After applying an axial movement, the rotating workpiece (Fig. 1b) comes into contact with the (stationary) friction tool. This induces friction and hence local heat generation. Consequently, the ductility of the material is increased, causing a flow of material from the contact zone into the outside regions. By additionally using tools, it becomes possible to influence the material flow in a defined way as precondition for the production of defined geometries (Fig. 1c) [7].

Hence, apart from the tool geometry, the process strategy and, in this case, the tool path geometry, in particular, is of importance. That is why the LUF has developed a special machine system (Fig. 2) with the principal components of

- Main drive (application of the rotational movement)
- Two X-Y support units ((one not illustrated) necessary for the tool and workpiece movement) and

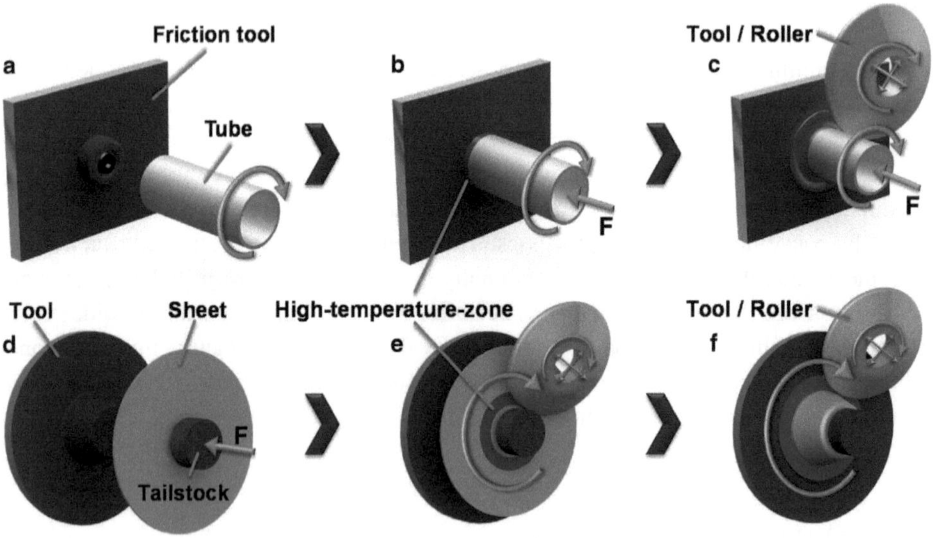

Fig. 1 Process principle of incremental tube and sheet forming by friction spinning [7]

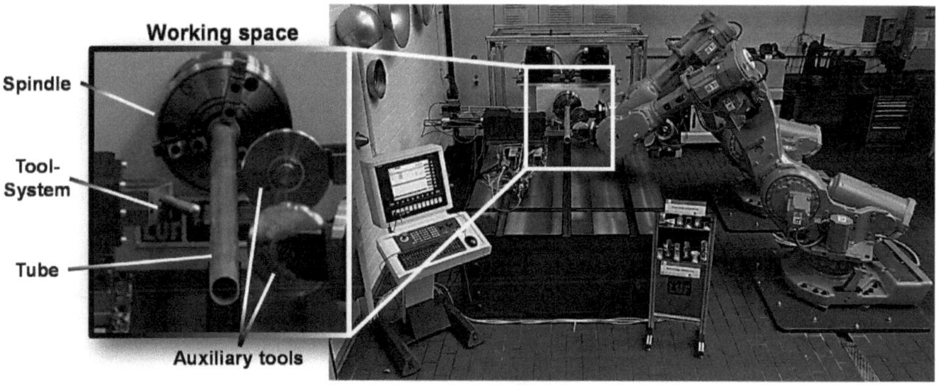

Fig. 2 Prototyping test machine for the manufacture of tubular and sheet metal components by friction spinning

- Two industrial heavy robots (necessary for the movement or positioning of auxiliary tools)

3 Results and Applications

Extensive technically-oriented research work was performed at the LUF for purposes of conducting a detailed analysis and, on the basis of this, of developing an adapted process and tool design for friction spinning. In addition to demonstrating the feasibility of the process, the research work was able to show that the process limits could be extended, for example. Process and tool developments now allow greater formability – it is possible, for example, to produce flanges with a flange diameter ratio of 500 %, expressed in terms of the initial tube diameter, in a single forming operation, compared to a ratio of less than 200 % for conventional spinning. Further promising results were found in respect of a defined influence on the grain structure (see Fig. 3), the hardness distribution [8, 9], the

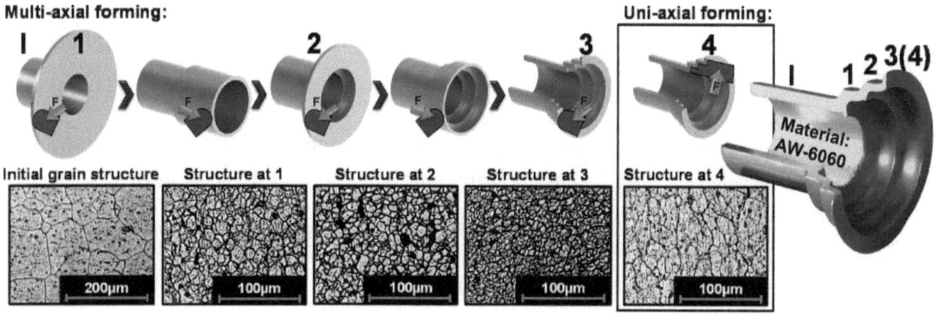

Fig. 3 Influence on grain structure during the friction spinning process with two different forming strategies

Fig. 4 A selection of geometries that can be produced by friction spinning [10]

residual stress situation [10], the attainable surface quality [11] and the high degree of deformation. This makes it possible to produce geometries that cannot be achieved with conventional processes, like thickenings, flanges in the middle regions of parts, hollow tubular structures, and so on [10].

In-depth knowledge of the process was achieved through the fundamental technological investigations related to friction spinning. On the basis of the findings obtained, it proved possible, during the next stage, to mark out the application range by producing more complex profile and sheet parts, as is illustrated by Fig. 4, which shows a number of parts from the geometry range. A large number of materials can be formed by means of friction spinning. The most common materials are non-ferrous metals, aluminum alloys and low-alloy steels, extending right through to ultra-high-strength steels. Even titanium and magnesium alloys have been successfully formed in initial tests.

4 Potential

The innovative incremental forming process of friction spinning has significant process characteristics which make this process suitable for a large number of forming applications and materials which are scarcely formable:

1. Production geometries that cannot be produced with conventional forming operations, e. g. tubular workpieces with flanges, thickening in the middle section or hollow and asymmetrical structures.
2. Manufacture of load-adapted tubes and sheets with functionally graded properties (hardness, strength, elongation, and so on).

3. A high degree of deformation without use of heating systems: The reduction of grain size and hardness avoids the need for external heat strategies to remove work hardening. These characteristics make the process eminently suitable for further processing steps.
4. Defined grain refinement: It is possible to adjust the grain size in the deformed material through the use of the process strategy and an according tool design, for instance.
5. A big reduction in forming forces: The reduction result from the heat treatment due to self-heating during the process.
6. Forming of a wide variety of alloys (aluminum, steel, stainless steel, magnesium, ...).
7. A direct influence on the surface quality during the process: By setting defined temperature profiles through suitably adapted process parameter and tools.

5 Acknowledgement

The authors from the Subproject A7 Centre of the Transregio 30 collaborative research would like to thank the German Research Foundation (DFG) for supporting this project.

References

1. Kleiner, M., Geiger, M., Klaus, A., 2004, Manufacturing of Lightweight Components by Metal Forming, Annals of the CIRP, Vol. 52/2/2003, 521–542
2. Finckenstein, E. von, Dierig, H., 1990, CNC-Drücken, Annals of the CIRP, Vol. 39/1: 267–270
3. Neugebauer, R., Altan, T., Geiger, M., Kleiner, M., Sterzing, A., 2006, Sheet Metal Forming at Elevated Temperatures, Annals of the CIRP 55/2, 793–816
4. Awiszus, B., Meyer, F., Meyer, L.W., Hahn, F., 2005, Erweiterung der Formgebungsgrenzen durch inkrementelle zyklische Umformung am Beispiel des Drückwalzens (Abstreckdrücken). Tagungsband zum Abschlusskolloquium des DFG Schwerpunktprogramms 1074 „Erweiterung der Formgebungsgrenzen bei Umformprozessen“, 3. Mai 2005, Aachen
5. Klocke, F., Brummer, C. M., 2014, Laser-assisted metal spinning of challenging materials. Procedia Engineering of the 11th ICTP Vol. 81, Elsevier, Nagoya, Japan, 2385–2390
6. Music, O., Allwood, J.M., Kawai, K., 2010. A review of the mechanics of metal spinning. Journal of Materials Processing Technology, Vol. 210, Issue 1, pp. 3–23
7. Homberg, W., Lossen, B., 2013, Forming Technology. In: Homberg, W.; Biermann, D; Heim, H.-P. (Hrsg.), Functionally Graded Materials in Industrial Mass Production | Fundamentals. Verlag Wissenschaftliche Skripten, Auerbach, ISBN 978-3-942267-92-2, pp. 43–76
8. Hornjak, D., 2013, Grundlegende Untersuchung der Prozess- und Werkzeugparameter und ihrer Wechselwirkungen für das thermo-mechanisch unterstützte inkrementelle Umformverfahren des Reib-Drückens; Paderborn Univ. Diss., Aachen: Shaker
9. Lossen, B. & Homberg, W., 2014, Friction-Spinning – Interesting approach to the manufacture of complex sheet metal parts and tubes. Procedia Engineering of the 11th ICTP Vol. 81, Elsevier, Nagoya, Japan, 2379–2384

10. Homberg, W., Lossen, B., 2013, Thermal assisted incremental forming of tubes and sheets with process-integrated heat generation. In: Heim, H.-P.; Biermann, D.; Homberg, W. (Hrsg.), Functionally Graded Materials in Industrial Mass Production | Volume 2. Verlag Wissenschaftliche Skripten, Auerbach, ISBN 978-3-942267-91-5
11. Hess, S., Lossen, B., Biermann, D., Homberg, W. & Wagner, T., 2014, Analysis of the surface roughness obtained in a friction spinning process based on empirical models. The International Journal of Advanced Manufacturing Technology Vol. 74 (Issue 9–12), pp. 1655–1665

Single Point "Dieless" Incremental Forming

Masaaki Amino, Masashi Mizoguchi, Yuji Terauchi, and Trent Maki

1 Introduction

Early metalforming required hammering and heat treatment cycles to incrementally shape the material, and metals are still commonly hand crafted to produce prototype parts. Hand crafted parts can be created with relatively little equipment or tooling investment, but it is very time consuming and the consistency of the parts can vary even from skilled workers. Professor Matsubara of Polytechnic University initiated the development of modern incremental forming technology beginning in 1988, to develop a quicker, more consistent method of producing prototype parts. Professor Matsubara applied for a Japanese patent for a single point incremental forming method in 1993. In 1996, a partnership was created between Professor Matsubara and Amino Corporation to accelerate the development of incremental forming technology and to build an incremental forming machine. As well, a research project was sponsored by the Japan Science and Technology Agency (JST) to foster new and original ideas. Amino Corporation developed the first prototype machine in 1996. Amino also developed the CAM (Computer-Aided Manufacturing) software for dieless NC forming in 1998. In 2002, Amino Corporation began providing specialized incremental forming equipment to industries around the world. The total number of machines in operation is over 30 units, and Amino Corporation alone has made over 3000 unique parts.

Masaaki Amino ✉ · Masashi Mizoguchi · Yuji Terauchi · Trent Maki
AMINO Corporation, Misonodaira, Japan
e-mail: mamino@amino.co.jp

A. E. Tekkaya et al. (eds.), *60 Excellent Inventions in Metal Forming*,
DOI 10.1007/978-3-662-46312-3_24

2 Process Idea

The Matsubara method (1) of dieless incremental forming is shown in Fig. 1a. In this method, CAD data for the part to be formed is converted into a series of two dimensional contours that represent tool paths. The panel is formed incrementally from top to bottom as the spherically tipped single point forming tool completes a two dimensional contour, descends a Z pitch, follows the next contour, and so on. The fundamental principle of incremental forming is that material of distance "A" is formed at angle "θ" relative to the vertical axis, and this material is stretched to distance "B", as shown in Fig. 1b. The volume "A1" (before forming) is the same as volume "B1" (after forming). The formed length B is longer than the initial length A, thus the material thickness decreases as the length increases. The relation of forming angle and material thickness is governed by the formula:

$$\text{Formed material thickness } \boldsymbol{t1} = \text{Original material thickness } \boldsymbol{t0} \times \mathbf{sin\theta}$$

The forming process is termed "dieless" since traditional cast dies are not required because the forming force is significantly reduced compared to stamping. Softer, more readily available and lower-cost materials such as plastics and wood can be used, which greatly reduces part cost and lead time. Further, if the part profile has a constant angle (θ) relative to the vertical axis (Z direction) from top to bottom, the tooling costs are virtually eliminated because a master model is not required (Fig. 2).

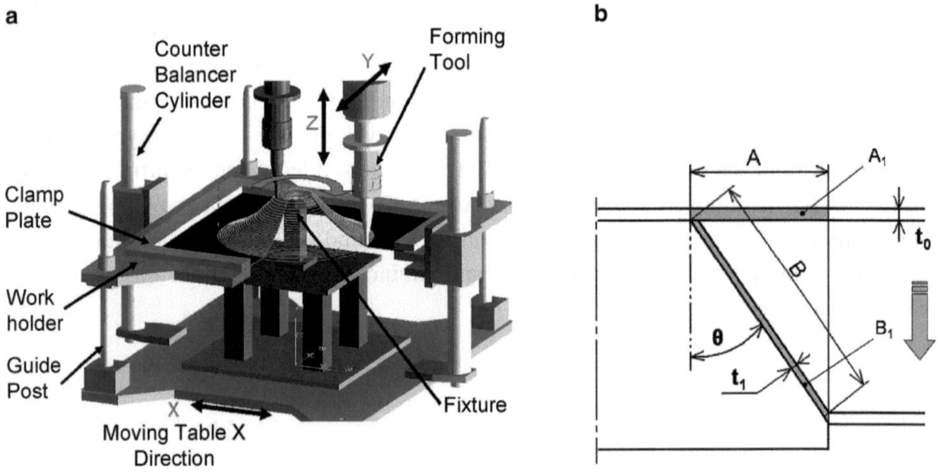

Fig. 1 **a** The incremental forming equipment, **b** Cross-sectional view of the process principles

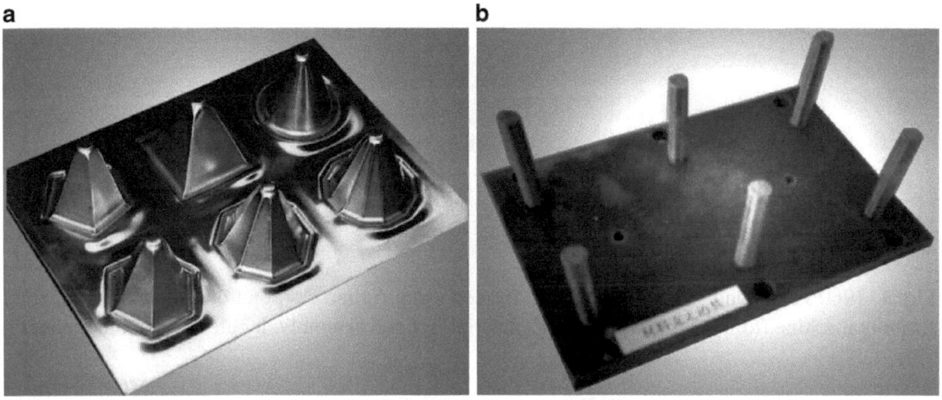

Fig. 2 Six different shapes were formed in a single 1 mm thick A1050 blank using a very simple die made from a base plate and six cylindrical posts

3 Results and Applications

Incremental forming can achieve very complex geometries that have large material elongation, in some cases up to 300 %, because the material is stretched in small, repeating deformation increments. Dieless incremental forming is best suited for low part quantities because the forming time is much longer compared to stamping, and the forming tool creates tool path marks on the panel surface that must be metal finished on exposed surfaces.

Automotive Service Parts

The tool storage and maintenance required to produce small quantities of automotive service parts is very costly and time consuming. Dieless incremental forming can be used to reduce the challenge and cost of storing dies for producing low volumes of service parts because the bulky production tools are not required. In 2000, Amino Corporation partnered with Honda to create hood assembly service parts consisting of an outer skin and inner panel for the 1960's era Honda S800 (Fig. 3a–c).

Additional Styling Details for Existing Panels

Incremental forming can also be used to add additional styling details to existing production panels, which could otherwise not be formed by conventional tools or would be cost prohibitive (4). Amino Corporation recently partnered with Toyota Motor to use incremental forming to add features and styling details on existing stamped panels for special edition Toyota IQ-GRMN vehicles (Fig. 4a–c).

a b c

Fig. 3 1960's era Honda S800 service parts formed using Amino Corporation's dieless incremental forming machine from 0.7 mm thick SPCE material. **a** hood assembly, **b** outer skin, **c** inner panel

a b c

Fig. 4 Dieless incremental forming applications on the Toyota IQ-GRMN. **a** logos on the door, **b** reduced corner radii and enhanced feature lines on the door frame, **c** boss pockets for attaching a fender over the existing part

Thick Material and Tube Forming

Dieless incremental forming is not limited to thin sheet materials. The dieless incremental forming machine has the strength to create and add features to automotive frame parts that are thicker than traditional sheet materials or are made of high strength materials. The dieless incremental forming machine can also be used for tube applications (Fig. 5a–c).

a b c

Fig. 5 Tube applications of dieless incremental forming include. **a** changing the cross-section, **b** adding an embossment, and **c** flaring the end of the tube

4 Conclusion

Dieless incremental NC forming has been advancing since the early 1990's due to the partnership of Professor Matsubara and Amino Corporation. Initially developed for prototype parts and small quantities of production parts, the practical use of dieless incremental forming technology has gradually advanced to include applications such as producing automotive service parts, adding styling to existing panels, forming thick materials, and forming tubes. The low initial investment for equipment and tooling makes dieless incremental forming a cost effective method for producing parts.

References

1. M. Amino, M. Mizoguchi: Practical applications of the dieless NC forming for sheet metal, SOKEIZAI, Vol. 52 (2011), No11, 2–7.
2. H. Amino: Development of the dieless NC forming machine for a sheet metal, J.JSTP, Vol. 51, No599, (2010-12), 1119–1123
3. S. Matsubara: Incremental backward bulge forming of a sheet metal with a hemispherical head tool – A study of a numerical control forming system II, J.JSTP, Vol. 35, No406, (1994-11), 1311–1316
4. M. Amino, M. Mizoguchi, Y. Terauchi, T. Maki: Current status of "Dieless" Amino's incremental forming, 2014 ICTP Conference, Nagoya, Japan

TwinTool

Lukas Kwiatkowski and A. Erman Tekkaya

1 Introduction

Since several years the incremental sheet metal forming (ISF) process is fascinating users and developers of metal forming processes [1]. This is based on two factors: 1) high tooling flexibility and 2) increased forming limits [2]. Nearly arbitrary contoured parts can be manufactured from metals or even polymers [3]. But up to now, the ISF process is only suitable in sheet metal prototyping. The reason for this is inherent in the process set-up. The ISF process set-up consists of a finger like tool which proceeds continuously on a predefined path. ISF is often applied on conventional 3-axis milling machines, featured with a blank holder and optionally with a part geometry specific die. Based on the small forming zone the part is formed slowly step by step. As a concept study the authors summarized several ideas in [4] in order to decrease the process time showing different process set-ups with multiple tools being in service on one workpiece in parallel. In this paper these concepts are assessed. After that one tool setup is realized and a first preliminary part is generated.

2 Tool Design

Using a systematic development approach presented in [5] the invention process exhibits three tool modules, which can be implemented in conventional forming machines used for ISF. The first solution supplies an individual kinematics for each tool by introducing two additional linear axes. It is called "TwinTool" (Fig. 1a). Each forming tool is attached to a slide, which is supported by two guides and actuated by a spindle. Using an inde-

Lukas Kwiatkowski ✉ · A. Erman Tekkaya
TU Dortmund, Dortmund, Germany
e-mail: Lukas.Kwiatkowski@icloud.com

A. E. Tekkaya et al. (eds.), *60 Excellent Inventions in Metal Forming*,
DOI 10.1007/978-3-662-46312-3_25

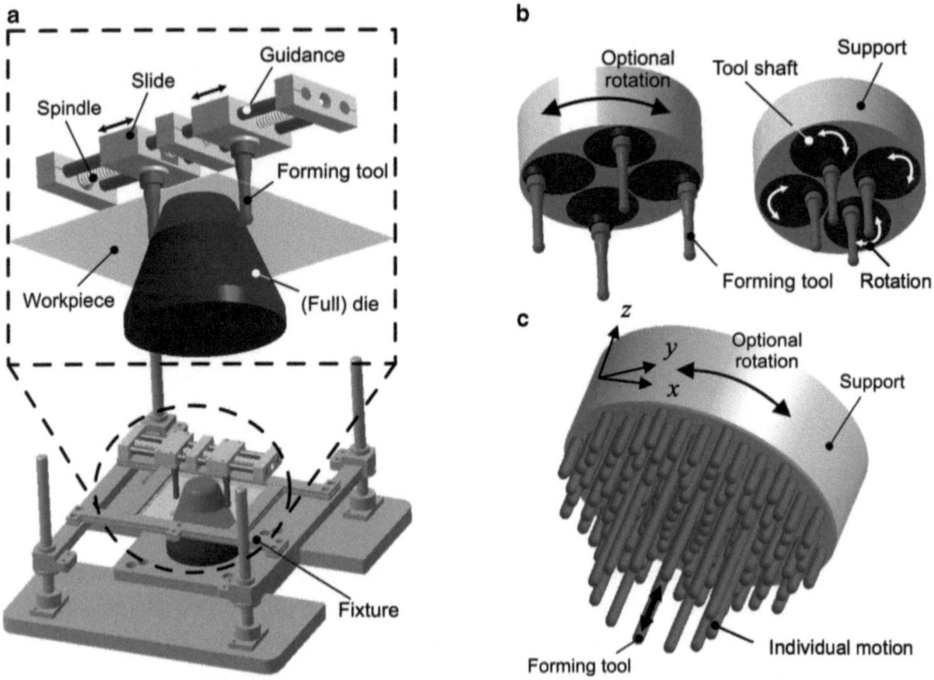

Fig. 1 Invented tool set-ups. **a** TwinTool, **b** RotaryTool, **c** HedgehookTool [5] (Copyright of Photograph: Springer Verlag 2010)

pendent spindle for each slide, asymmetric parts can be formed. Linking both spindles by a coupling allows the manufacture of symmetric parts. While this variant provides one additional degree of freedom for each tool it has to be considered for the tool path programming that all movements, except that of the additional linear axes, are directed in the same direction. Figure 1b shows the established design of the so called 'RotaryTool'. The forming tools are attached to a revolving actuator, which is intended to position each tool. This concept can be realized as a planetary gear set, where the rotational motion of every tool is coupled to each other. An alternative is an individual control of each rotational motion by separate electrical engines. An optional rotational motion of the whole setup can be integrated to support tool positioning. The global tool positioning is carried out by the CNC machine, where this module is attached to. The third solution, shown in Fig. 1c, consists of a tool matrix being vertically adjusted to each incremental pitch as presented in [6] as a setup for rapid prototyping tools used for deep drawing. This concept was called by the authors 'HedgehogTool'. Doing so, each individual tool in the matrix requires an individual actuator. Suitable drives for such a concept are hydraulic or electric motors. Furthermore, a motion of the tools in xy-plane is necessary. This can be done by the CNC machine for the complete assembly or individually for each tool. An optional feature of this concept is the rotation of the whole support.

3 Tool Assessment

The achieved solutions presented in the previous section were compared to a stand-alone alternative using robots as given in [7]. In order to assess all the tool concepts a cost-utility analysis (CUA), as explained in detail in [5], was performed. The CUA classifies the concepts according to three major categories, namely function, potential, and cost. Each category is subdivided into several criteria, as shown in Table 1. Additionally, a positive and negative result for each criterion is listed. To consider the different impacts of each criterion, a weight factor has been introduced following the rules in [6]. Next, each variant is assessed individually for every single criterion to exhibit its individual degree of performance. This is done on a scale ranging from 1 (representing a failure to this criterion) to 10. The results are listed in Table 1 as well. Finally, the degree of

Table 1 Criteria and result of assessment of tool design concepts

Category	Criterion	Positive	Negative	Fac. (%)	Robot	Twin Tool	Rotary Tool	Hedge hook
Function	Forming head fixture	Fixed	Free rotation	2	7	4	7	3
	Control	Simple	Complex	16	3	9	8	1
	Robustness	High	Little	21	7	9	6	3
	Accuracy	Error < 1,5 mm	Error > 1,5 mm	25	5	7	7	6
	Friction, wear	Small	High	14	9	8	6	6
	Weight	Low	High	9	9	9	9	2
	Lubrication	Not required	Recommended	2	8	6	5	3
	Required space	Small	Large	11	1	10	10	7
Potential	Geom. Flexibility	High	Low	19	10	10	3	6
	Speed-up	> 50 %	< 50 %	44	6	5	7	10
	Surface quality	High	Low	6	5	7	5	3
	Complexity	High	Low	31	9	7	1	9
Cost	Production costs	Low	High	24	1	8	7	2
	Assembly	Simple	Time consuming	4	6	9	8	2
	Technical feasibility	Simple	Complex	32	10	10	9	2
	Die	Not required	Required	28	8	6	7	5
	In-house production	Few parts	Most parts	12	10	8	8	4

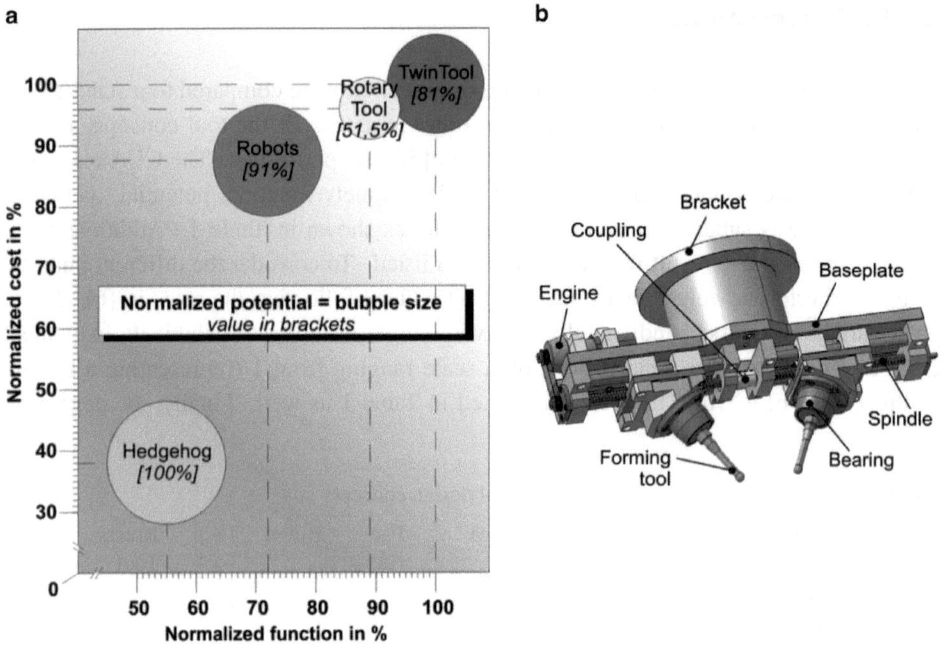

Fig. 2 **a** Tool design assessment by cost-utility analysis, **b** CAD layout of the TwinTool (Copyright of photograph: Springer Verlag 2010)

performance is linked with the predefined weight factor. This methodology establishes a ranking of the invented conceptual designs. Considering the functional aspects first in Table 1, the precision in positioning the tools under load and a robust (fault-free) operation have the greatest impact. Assessing the potential of each design, main emphasis is on minimizing the overall process time. Additionally the feasible geometrical complexity is the most vital aspect. Finally, the cost of each design is evaluated. Each design is intended to be a prototype rather than a massproduced tool. The ranking is visualized in a diagram, as shown in Fig. 2a. The graph shows the normalized function (abscissa) over the normalized cost-efficiency (ordinate). Furthermore, the potential is indicated by the size of each bubble.

The TwinTool has proven to be the simplest and cheapest concept for a speed-up of AISF by the use of more than one tool. Hence, this design is finalized by the authors for further investigations (Fig. 2b). To allow a forming operation of the two tools close to each other, an inclination of the two pins is considered. An additional bearing in each tool clamping will lead to a rolling motion of the tool on the parts surface and therefore reduce wear and friction. The TwinTool system can be attached to the spindle of a conventional milling machine using a bracket. In that way path programming of the tool motion can easily be achieved by integrating the additional linear axis into a given machine control.

4 Conclusion and Outlook

Due to the result in Fig. 2a a prototype TwinTool has been build by the authors for testing which is finally illustrated in Fig. 3d. The produced prototype TwinTool as well as the developed modular concepts have been filed for a patent by the IUL [8]. First preliminary trials with the TwinTool were focussed on forming a symmetric pyramidal shape. Hence, the two additional linear axis were fixed together by a coupling to work in parallel. While this tool setup exhibits two separated forming zones the part was separated either at its center line. Intermediate steps of the CNC cycle are presented in Fig. 3a–c. The tool motion is generated in y- and z- direction by the milling machine while the motion of both tools in x-direction is provided by the additional linear axis. Consequently an area is needed where both halves of the part intersect each other. This overlap area is marked blue in Fig. 3b. One corresponding demonstrator part made of Aluminium EN AW 1050 with a thickness of $t = 2$ mm is shown in Fig. 3e. Here the two overlap areas, one at the front an one at the back of the part, are clearly visible by eye. As a consequence future work will be focussed on the materials behaviour in that overlap area by experimental analysis and FE simulations. Additionally, some effort will be spent on improved CNC programming to manage two or more tools with one forming machine.

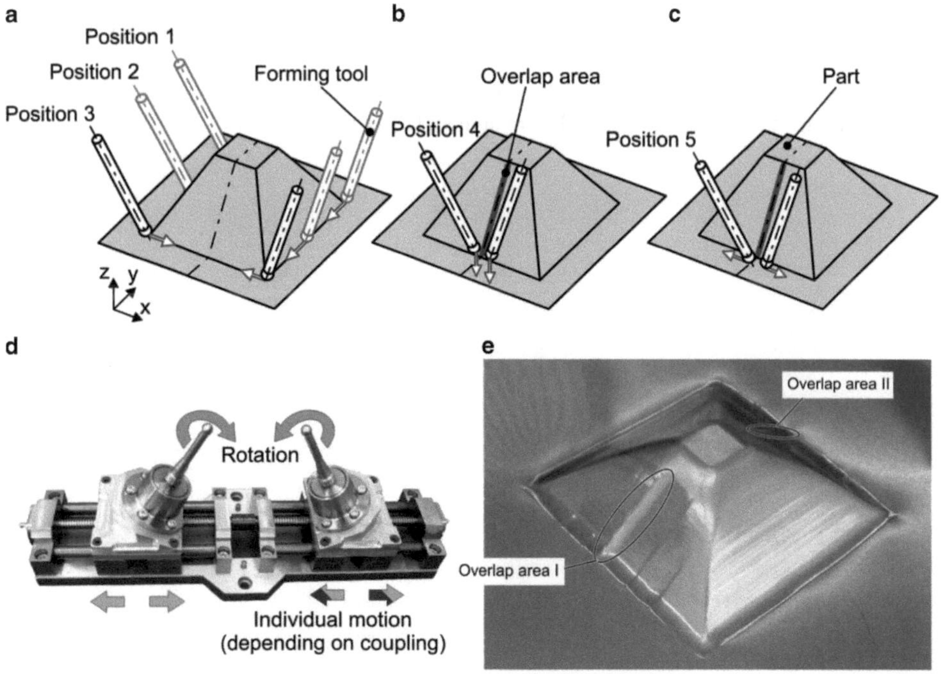

Fig. 3 **a-c** Tool cycle of TwinTool applied to form a symmetric pyramid, **d** TwinTool prototype, **e** achieved part manufactured by TwinTool

5 Acknowledgement

The authors thank the German Research Foundation (DFG) for their kindly support of the work under the grant number TE 508/26–1.

References

1. Amino M., 2014, Current status of "Dieless" Amino's incremental forming, 11th international Conference on Techonolgy of Plasticity (ICTP).
2. Jeswiet J., Micari F., Hirt G., Bramley A., Duflou J., Allwood J., 2005, Asymmetric single point incremental forming of sheet metal. CIRP Ann Manuf Technol. 2, 623–649.
3. Martins P., Kwiatkowski L., Franzen V., Tekkaya A., Kleiner M., 2009, Single point incremental forming of polymers, CIRP Ann Manuf Technol 58 1, 229–232.
4. Kwiatkowski, L., Urban, M., Sebastiani, G.,Tekkaya, A. E., Tooling Concepts to Speed up Incremental Sheet Forming, German Academic Society for Production Engineering (WGP), 2009.
5. Pahl G., Beitz W., 1993, Konstruktionslehre: Methoden und Anwendung, 3rd edn. Springer, New York.
6. Smatloch C., 1996, Entwicklung eines numerisch einstellbaren Werkzeugsystems für das Tief- und Streckziehen. Ph.D. thesis, Technische Universität Dortmund.
7. Tuominen T., 2004, Method and apparatus for forming three-dimensional shapes in a sheet metal, Patent number WO2004030843A1.
8. Tekkaya, A. E., Kwiatkowski, L., Urban, M., Abel, H.-J., 2009, Verfahren und Vorrichtung zur inkrementellen Blechumformung", Patent number DE102009025726.8.

Laser Adjustment Using Actuators

Hinnerk Hagenah and Manfred Geiger

1 Motivation

Experts from both science and industry agree that micro systems technology is one of the key technologies. The aim of micro systems technology is the integration of electronic, optical and mechanical functionality in micro components and workpieces. Besides the ever increasing pressure on prices the reason for the importance of this technology are the raising demands of the customers. Customers demand small products with high functionality and low prices. Therefor a necessary condition to exploit the potentials of micro systems technology is the development of comparably inexpensive but reliable concepts and technologies for the manufacture and assembly of hybrid micro systems. Special attention has to be paid to the adjustment processes. Due to the miniaturisation it is becoming very difficult and expensive to assemble workpieces within tolerances. Often an adjustment step is needed to ensure product quality. The adjustment process should be flexible to be applicable to a large variety of products. In addition it is desirable to use a process that has the potential to be automated. Laser beam forming of metallic material is a process that satisfies these demands. The integration of laser beam forming into the process chain of adjusting preassembled systems leads to a new technology called laser adjustment that allows the highly accurate and inexpensive production of high precision parts [1].

Hinnerk Hagenah ✉ · Manfred Geiger
Friedrich-Alexander-Universität Erlangen-Nürnberg, Erlangen, Germany
e-mail: Hinnerk.Hagenah@fau.de

A. E. Tekkaya et al. (eds.), *60 Excellent Inventions in Metal Forming*,
DOI 10.1007/978-3-662-46312-3_26

2 Process Idea

The fundamental idea was to use the potentials of laser forming as a contactless, repeatable, reversible forming process in order to modify the process chain for high precision assemblies in micro system technology as shown in Fig. 1.

While the conventional process chain has lowest accuracy in the final step being tool removal, the chain incorporating laser adjustment has this in the very end. Though there are more process steps the process is faster and more accurate. The basic mechanisms of laser forming were known and well researched. Applications in forming of prototypes and geometric correction were being investigated. Micro adjustment was a promising field of application for this new forming technology. However, a closer look at the applied mechanism revealed challenges that required a fundamental approach to the idea. The required geometries to be formed in order to carry out the requested adjustment needed to be designed. After the applicability had been shown by a sample application this was the remaining central challenge.

3 Mechanisms of Laser Forming

A comprehensive and structured review on laser forming mechanisms is given by Vollertsen in [3]. Fundamental works incorporated into that book were among others carried out by Holzer [4] focussing on the contactless forming itself and Kraus [5] applying the mechanisms to profiles. The ongoing investigations on laser forming of metal by thermally introduced stresses were documented in the books written by Hennige [6] emphasizing the flexibility and Merklein [7] focussing on the effects of laser forming on aluminium material properties. The two most important mechanisms applied in both micro and macro

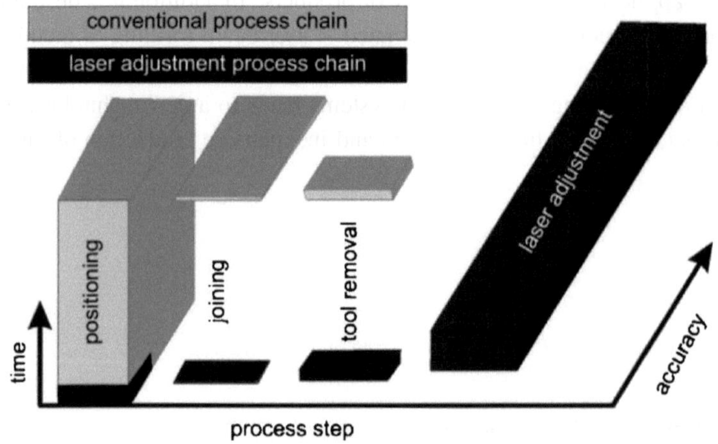

Fig. 1 Effect of laser adjustment on the process chain [2]

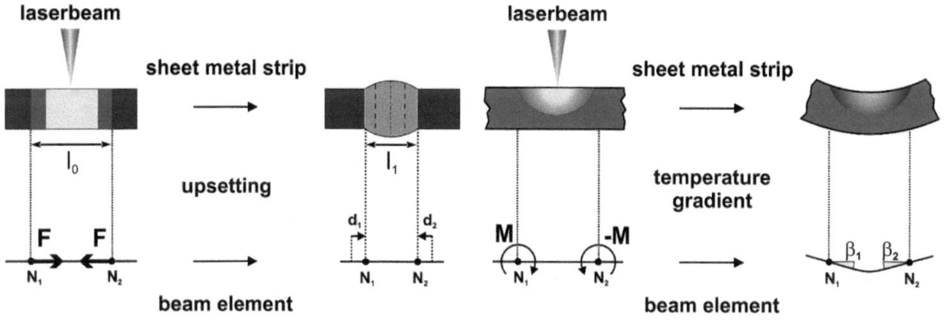

Fig. 2 Modelling the upsetting mechanism (**a**) and the temperature gradient mechanism **b** with beam elements [8]

laser adjustment are the temperature gradient and the upsetting mechanism depicted in Fig. 2. In more recent work these mechanisms have been modelled using beam elements to allow for a faster computation of the resulting forming effects [8].

The forces F induced by laser forming by means of a homogeneous heating of the specimen are causing a displacement d_1 and d_2 in the nodes N_1 and N_2 of the beam element (Fig. 2a). This results in the upsetting mechanism shortening the heated part of the specimen. If a temperature gradient is applied the resulting moments of force M applied at the nodes N_1 and N_2 cause bending of the beam element by an angle of $\beta_1 = -\beta_2$ (Fig. 2b).

4 Basic Actuator Geometries and Fundamentals of Laser Micro Adjustment

The fundamental idea of laser micro adjustment was the fixture of the element to be positioned with high accuracy to specially formed sheet metal parts called actuators. These were then formed by means of laser forming in order to move the effector into the desired position. This process might well be applicable for the adjustment of the lens position e. g. in cd-players. The major challenge in this approach is the design of actuators that do provide the mandatory stiffness to hold the effector in the desired position while at the same time assuring all needed adjustment movements can be carried out. Initial research was carried out by Müller [9] and Huber [10] who identified basic actuator elements able of realising specific movements. The basic actuators were distinguished by their ability to move in plane or out of plane. Furthermore some of them were designed to be able to move in two directions while others could only move in one direction. The basic actuator types used for all following more complex designs are shown in Fig. 3 [11]. First steps towards a computer assisted system for the challenge of actuator design were taken by Wurm [12]. Using Denavit-Hartenberg matrices he verified the ability of combinations of basic actuator elements to achieve the required adjustment movement from a given initial positioning of the effector element to the desired target position.

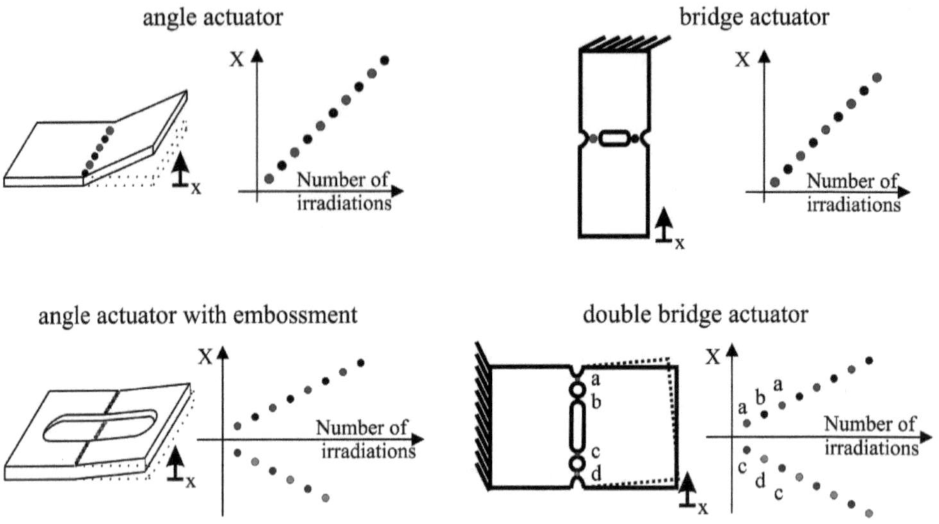

Fig. 3 Basic actuator types and possible movement [11]

5 Actuator Design as Multi-Objective Optimisation Problem

Plettke understood the design of actuators as a multi-objective problem [13] and identified 5 goals partially contradicting each other: simplicity of design, fitting in the given space, reliability of adjustment, accessibility for the laser and suitability for the task. Using a given initial actuator design and a mathematical description of the adjustment task, a non-dominated sorting genetic algorithm was used to compute the best compromises for the 5 objectives using Pareto dominance.

The optimisation process involves (Fig. 4) the evaluation of the beam model of laser forming (Fig. 2) for a fast computation of the adjustment taking place in closed kinematic chains. In a sample application part of the sigma actuator was modelled (Fig. 5) with distinguished degrees of freedom and the target to make an adjustment move in z-direction.

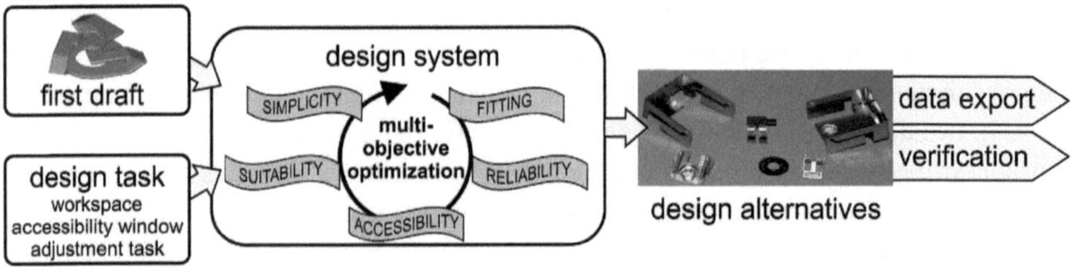

Fig. 4 System structure of the computer aided design system for actuators [14]

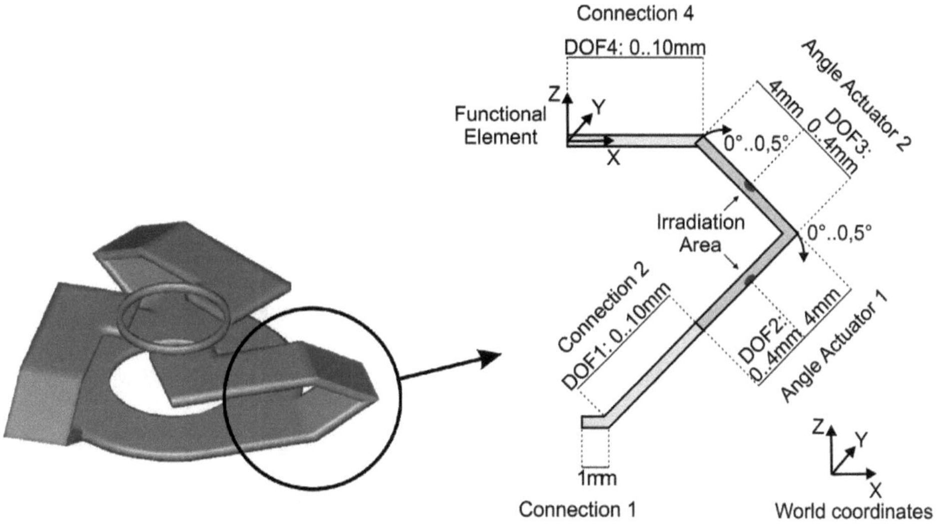

Fig. 5 Sigma actuator and test actuator system [15]

The system quickly reduced the degrees of freedom 1 and 2 (Fig. 5: DOF1, DOF2) to a length of 0 as they do not affect the adjustment and their removal increased the simplicity of the design. Furthermore some actuator elements were reduced as the effect on the possible adjustment movement was minimal. The algorithm was able to cover the entire field of possible solutions also focussing only on single criteria neglecting the others. Compromises reaching acceptable adjustment capabilities while being very simple were also developed by the optimisation system.

At this stage the required understanding of the laser forming mechanisms applied for adjustment tasks, the mandatory fast computable models for the forming taking place using a certain irradiation to a given sheet metal geometry and the needed optimisation procedures searching for compromises in the multi-objective search space describing the world of actuators for micro adjustment are available to design actuators for given adjustment tasks to successfully apply laser forming in micro systems technology.

6 Acknowledgement

The presented results are a summary of the research work of numerous colleagues at the LFT. The authors wish to thank these researches as well as the involved funding institutions in Germany and the EU and the partners from industry for their support and trust.

References

1. W. Hoving. "Laser applications in micro technology.", Laser Assisted Net Shape Engineering 2, Proceedings of the LANE '97, M. Geiger, F. Vollertsen, Bamberg: Meisenbach, 1997, pp. 69–80.
2. M. Schneider: Laseranwendungen in der Automobilelektronik. Aachener Kolloquium für Lasertechnik'98. Aachen: Fraunhofer ILT, 1998
3. F. Vollertsen. "Laserstrahlumformen – Lasergestützte Formgebung: Verfahren, Mechanismen, Modellierung", Bamberg: Meisenbach, ISBN 3-87525-071-0, 1996
4. S. Holzer. "Berührungslose Formgebung mit Laserstrahlung", Bamberg: Meisenbach, ISBN 3-87525-079-6, 1996
5. J. Kraus. "Laserstrahlumformen von Profilen", Bamberg: Meisenbach, ISBN 3-87525-094-X, 1997
6. T. Hennige, "Flexible Formgebung von Blechen durch Laserstrahlumformen", Bamberg: Meisenbach, ISBN 3-87525-140-7, 2001
7. M. Merklein, "Laserstrahlumformen von Aluminiumwerkstoffen – Beeinflussung der Mikrostruktur und der mechanischen Eigenschaften", Bamberg: Meisenbach, ISBN 3-87525-156-3, 2001
8. R. Plettke, H. Hagenah, M. Geiger. "Application of genetic algorithms for the optimisation of actuator systems for laser micro adjustment", International Journal of Microstructure and Materials Properties (IJMMP) 4(2009)6, pp. 615–625
9. B. Müller. "Laserstrahljustieren mit Excimerlasern – Prozeßparameter und Modelle zur Aktorkonstruktion", Bamberg: Meisenbach, ISBN 3-87525-159-8, 2001
10. A. Huber. "Justieren vormontierter Systeme mit dem Nd:YAG-laser unter Einsatz von Aktoren", Bamberg: Meisenbach, ISBN 3-87525-153-9, 2001
11. H. Hagenah, T. Wurm, M. Geiger. "Computer assisted design of actuators for high precision adjustment in micro technology" In: Bramley, A.; Brissaud, D.; Coutellier, D.; Mcmahon, C (Hrsg.): Advances in Integrated Design and Manufacturing in Mechanical Engineering, Springer, Dordrecht, The Netherlands, 2005, pp. 403–416
12. T. Wurm. "Lasertrahjustieren mittel Aktoren – Entwicklung von Konzepten und Methoden für die rechnerunterstützte Moellierung und Optimierung von komplexen Aktorsystemen in der Mikrotechnik", Bamberg: Meisenbach, ISBN 3-87525-206-3, 2004
13. R. Plettke. "Mehrkriterielle Optimierung komplexer Aktorsysteme für das Laserstrahljustieren", Bamberg: Meisenbach, ISBN 978-87525-315-3
14. M. Geiger, R. Plettke, H. Hagenah. "Multi-Objective Optimization of Actuator System Design for Laser Micro Adjustment", Production Engineering 3(2009)2, pp. 181–188
15. R. Plettke, H. Hagenah, M. Geiger. "Computer assisted design of actuator systems for laser micro adjustment", Key Eng. Mater. 344(2007), pp. 807–814

Flexible Asymmetric Spinning

Omer Music and Julian M. Allwood

1 Introduction

Over the past sixty years, there has been a large volume of research on metal spinning. Metal spinning refers to a group of forming processes that allow production of hollow, axially symmetric sheet metal components. The basic spinning technique involves clamping a sheet metal blank against a mandrel on a spinning lathe and gradually forming the blank onto the mandrel using a roller (Fig. 1). The process has evolved from the art of pottery using a potter's wheel, which dates back to ancient Egypt.

Spinning has developed over time. The demand by the industry for shorter production times and components with tight dimensional tolerances has driven improvements in the

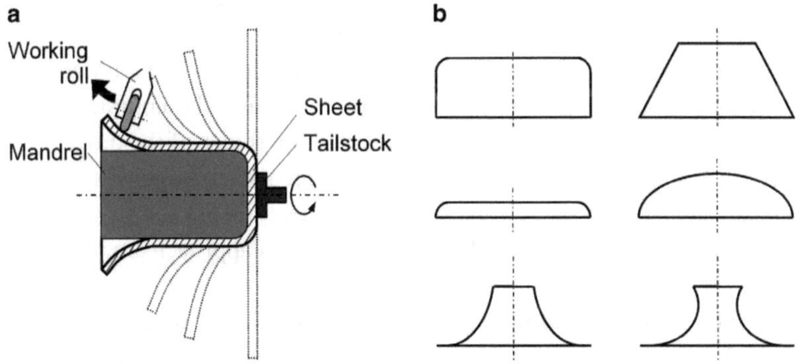

Fig. 1 Conventional metal spinning. The process (**a**) and example of achievable geometries (**b**) (Copyright Elsevier 2010)

Omer Music ✉ · Julian M. Allwood
University of Cambridge, Cambridge, UK
e-mail: omerm@cantab.net

A. E. Tekkaya et al. (eds.), *60 Excellent Inventions in Metal Forming*,
DOI 10.1007/978-3-662-46312-3_27

process, including automation and increased power in spinning machines, resulting in a major improvement in both process automation and the quality of produced parts.

In spite of its age, the process has seen almost no change from the original configuration. In the last 20 years, together with the evolution of numerical models, the understanding of the process and its mechanics has developed significantly. Music et al. [1] report on academic work on the analysis and application of the mechanics of spinning: novel process configurations, attempts to understand the process mechanics and application of the mechanics of spinning. They conclude that this knowledge is still limited, and a relatively simple task of making a given product by spinning is still largely based on the skill of the operator – the process still remains an art. Perhaps more importantly, the process has two main limitations; since a specific mandrel is required for each product, the spinning process is not flexible; in addition to being inflexible, it is limited to production of axially symmetric parts. Both limitations have been addressed separately in academic literature. However, the application has been limited to production of relatively simple shapes, and has not yet been accepted in industry.

This study provides new insight into the role of the mandrel in spinning and presents a process that overcomes limitations of traditional spinning – flexible asymmetric spinning.

2 Role of the Mandrel in Spinning

Spinning is inflexible because of the mandrel that defines the product geometry. Therefore, the key to exploring options for making spinning more flexible is to examine the interaction between the mandrel and the sheet – to examine the role of the mandrel in spinning.

To study the interaction, a numerical model was developed using Abaqus and validated against both physical trials and results published by Wang et al. [2]. A case study was the setup for spinning a simple can from a 1 mm thick aluminium blank (AlMg3 – 500 mm diameter) with a spinning ratio of 2. Instead of simulating the complete process from beginning to the end, which would require a week, three stages of the process were simulated: early, mid- and late stage. The shape of the sheet at each stage was estimated based on the toolpath and taken as the starting point for the analysis of each stage. At each stage, three rotations of the sheet were simulated. Details of this approach are reported in Music and Allwood [3].

Figure 2 shows the key and perhaps most surprising result of the analyses. The red arrows show the contact forces between mandrel and the sheet. The contact is clearly limited to three areas, and no significant pressure occurs on the opposite side of the mandrel; areas B and C are at the limit of where the cup has been formed onto the mandrel, but offset to either side of the working roll; area A is at the corner radius of the can where the base blends into the wall. Intuitively, these areas arise because the working roll force tends to crush the cup, bending it at A, and the contacts B and C oppose the crushing and are sufficient to ensure no other contact around the circumference. It appears that the effect of the mandrel in spinning is limited to these three small areas of contact.

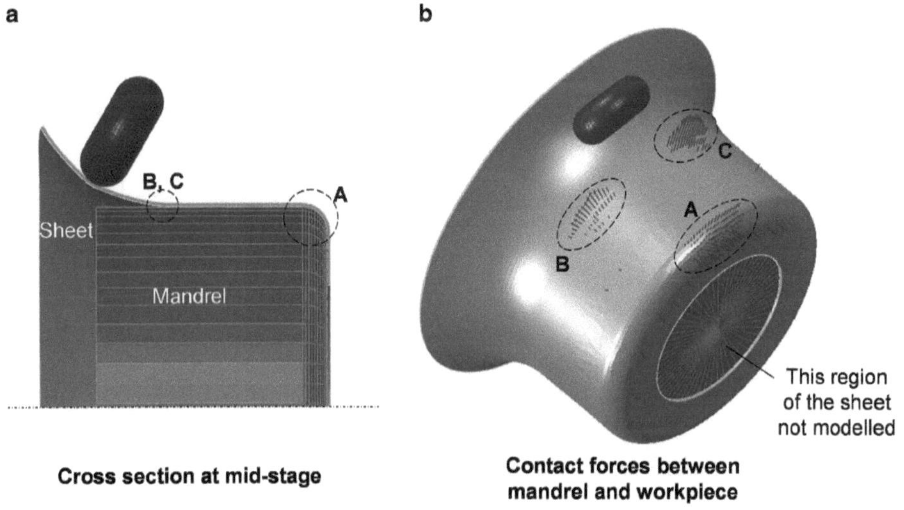

Fig. 2 – Simulated process stage (**a**) and contact forces between mandrel and workpiece (**b**)

The implication of this brief analysis is that the mandrel could be replaced, by three rolls: one at the base of the spun product (the 'blending roll') and two 'support rolls' placed on each side of the working roll. If the three rolls were allowed to move independently, asymmetric spinning would be possible. However, further investigation is necessary and the main question that needs to be answered is: How predictable and consistent is the contact pattern?

3 Design of Flexible Asymmetric Spinning

To investigate the interaction in more detail and explore options for designing the process and a prototype machine, a second, more detailed study was performed using the same set of process parameters. Three main parameters were investigated: tool position, stage in the process and mandrel radius. For each parameter three values were investigated, leading to a total of 9 analyses (Fig. 3a). Figure 3b shows the variation in α, the separation between contact areas A and B. The angle between the two contact areas varies between 10° and 33°, but more importantly, the pattern of contact remains consistent.

Further analysis of spinning over an asymmetric mandrel revealed the same pattern. Figure 4 shows the extreme case of a 'kidney-bean' mandrel. Although slightly different, the same pattern is still clearly visible.

These analyses confirm that the mandrel could be replaced, in both axisymmetric and asymmetric spinning, by three rolls: one at the base of the spun product (the 'blending roll') and two 'support rolls' placed on each side of the working roll, and moving along the axis of rotation as the final diameter is reached.

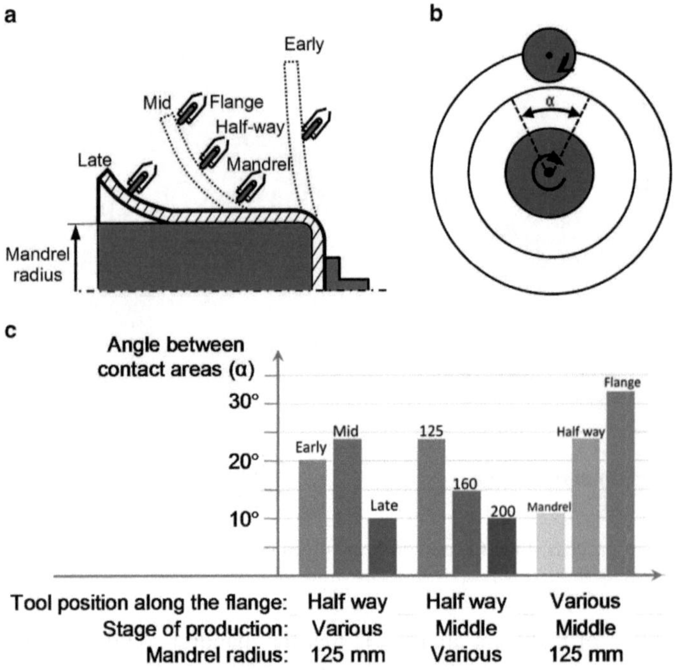

Fig. 3 Case study parameters and results: stage and tool location (**a**), angle between contact areas (**b**) and variation in radial separation of contact areas during spinning (**c**) (Copyright Elsevier 2011)

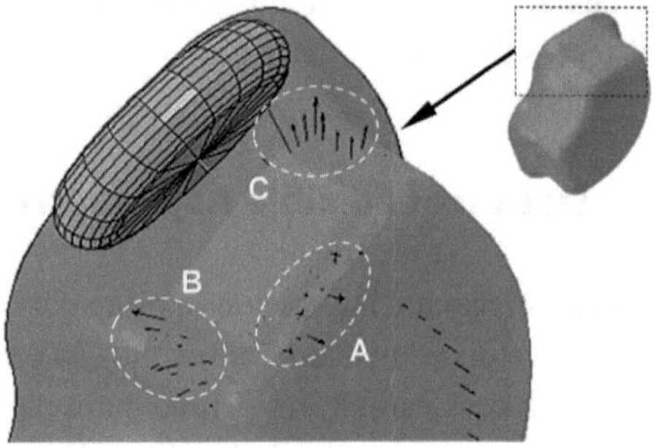

Fig. 4 Contact forces in the extreme asymmetric case of a 'kidney-bean' mandrel (Copyright Elsevier 2011)

4 Prototype Machine and Physical Trials

Investigation presented in the previous section has shown that the mandrel could be replaced by a set of three carefully controlled tools, which make the process mandrel-free and allow for production of asymmetric parts. In order to evaluate machine capability, a set of target parts was specified. Rather than having a set of real parts, a set of target parts and wall profiles demonstrating machine capability have been selected. The parts show increasing complexity; a circular, elliptical and a rectangular cup and as the most

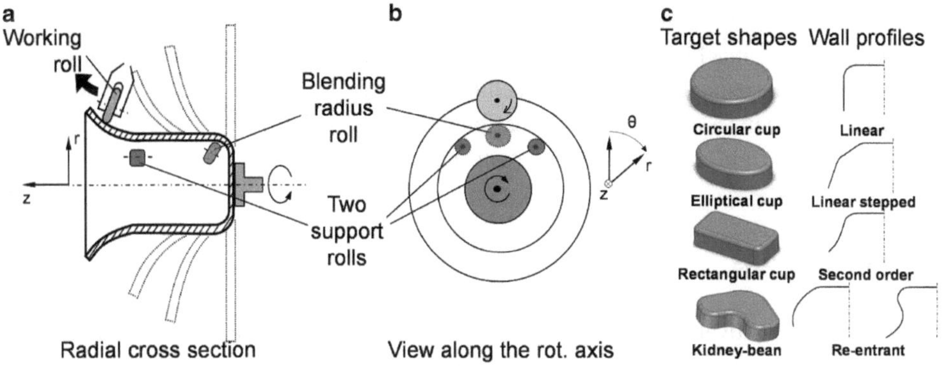

Fig. 5 Flexible asymmetric spinning process design (**a**, **b**) and target geometries (**c**) (Copyright Elsevier 2011)

challenging shape a 'kidney-bean' asymmetric shape having a combination of both stretch and shrink flanges. Similarly, range of wall profiles range from a linear, linear-stepped and a second order profile, with the most challenging shape being a second-order re-entrant profile.

Realising the schematic of Fig. 5 requires several automated degrees of freedom, increasing the cost of the machine. In order to limit cost, support roller motion was limited to planar motion parallel to working roller, leading to a seven axis machine design. To allow production of asymmetric shapes, each roller can move independently in *r*-direction. Details of the machine design are given in [4].

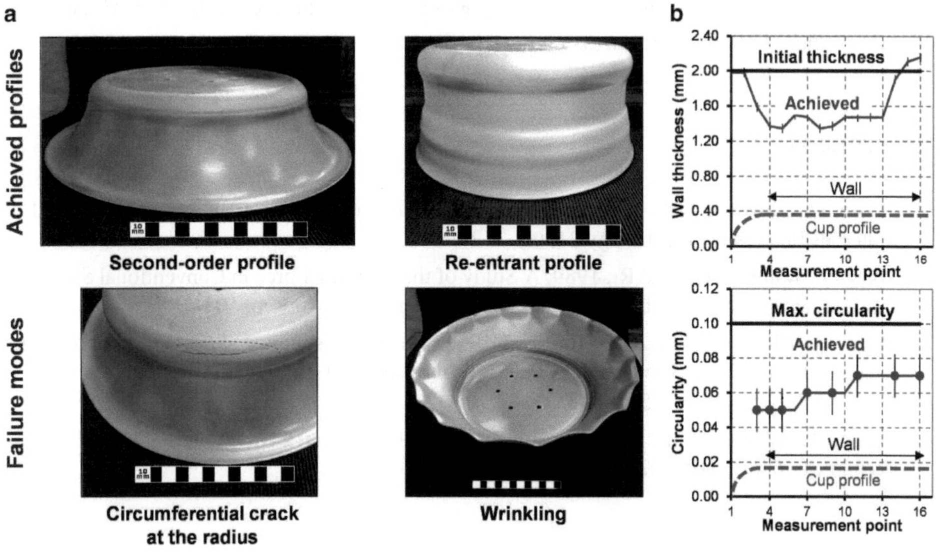

Fig. 6 Physical trials: achieved profiles (**a**) and part measurements (**b**)

Figure 6 shows the results of the physical trials, aimed at evaluating the mandrel-free capability of the machine. Figure shows achieved second-order profile and re-entrant profiles produced using the prototype machine and the failure modes observed. The failure modes (bottom left) are similar to conventional spinning and dimensional accuracy of the produced parts (right) is comparable to conventional spinning.

5 Conclusions

This study was motivated by the need for flexible net-shape sheet forming. Analysis of the sheet-mandrel contact has led to a key discovery; the function of the mandrel in spinning is limited to support in only three regions, suggesting that the mandrel could be replaced by three rolls. A detailed analysis of sheet-mandrel interaction has led to design of a novel process – flexible asymmetric spinning.

A prototype flexible asymmetric spinning machine has been designed and built to test the proposed process through physical trials. Physical trials demonstrated the mandrel-free capability of the machine and showed that the process is capable of producing components with tight geometrical tolerances and high surface quality and confirmed the design of the process and the machine itself.

6 Acknowledgements

The first author, and the costs of the machine, were supported by the Engineering and Physical Sciences Research Council through the Dorothy Hodgkin Postgraduate Award, Ford Motor Company through its University Research Programme, Siemens VAI, Novelis, Metal Spinners Ltd. and Cummins Generator Technologies.

References

1. Music, O. Allwood, J.M., Kawai, K.-I., 2010, Review of mechanics of spinning, Journal of Materials Processing Technology 210(1), 3–23.
2. Wang Q., Wang T., Wang Z.R., 1989, A Study of the Working Force in Conventional Spinning, Proceedings of 4th International Conference on Rotary Metalworking,103–108.
3. Music O., Allwood J.M., 2011, Tool-path design for metal spinning, Special edition of Steel Research International (Proceedings of the 11th International Conference on Technology of Plasticity, Aachen, Germany), 542–547.
4. Music, O., Allwood, J.M., 2011, Flexible asymmetric spinning, CIRP Annals – Manufacturing Technology 60(1), 319–322.

Part V
Shear Cutting

Micro Hole Piercing with a Slant Angle

Tomomi Shiratori and Takafumi Komatsu

1 Introduction

Development of lower fuel consumption vehicles is one of most crucial issues in automobile industry and the competitions have been accelerating year by year. Gasoline fuel injection system was developed by BOSCH in about 40 years ago, and the system is improved with technologies developments. Figure 1[1] shows two fuel injectors. The injector has an orifice plate and there are a lot of types of orifices (Fig. 2). Figure 3 shows a type with layout of 12 holes. As Fig. 4 showing, the hole has a slant angle which is determined by spray specifications. The size of micro holes are from 80 to 300 μm and the material thickness is normally equal to hole diameter. Angle of micro pierced hole is from 0 degree (vertical) to over 45 degree. It is produced by specially fabricated punch and die, and by using vertical press machine.

Fig. 1 Gasoline injector

Tomomi Shiratori ✉ · Takafumi Komatsu
Komatsuseiki kosakusho Co., Suwa-City, Japan
e-mail: shiratori@komatsuseiki.co.jp

A. E. Tekkaya et al. (eds.), *60 Excellent Inventions in Metal Forming*,
DOI 10.1007/978-3-662-46312-3_28

Fig. 2 Several types of orifice plate

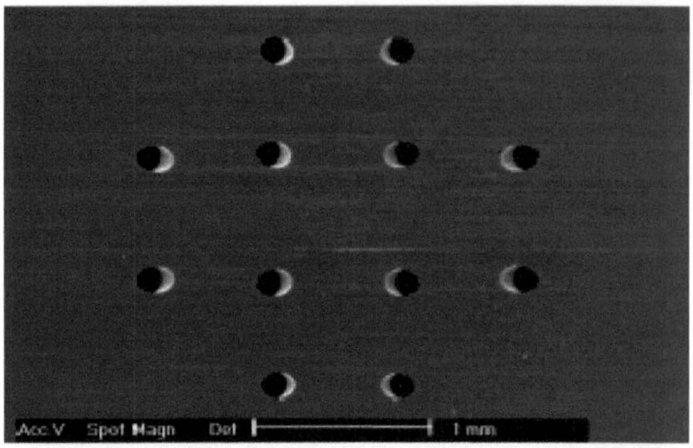

Fig. 3 Multi holes orifice plate

Fig. 4 Cross section of micro piercing with slant angle

The development of micro piercing with a slant angle started in the late 1980's. Since that, The piercing technologies are replaced from electric discharge machine, and the cost of production reduced to 1/100. The developed technologies won awards and patents as shown in Table 1.

Table 1 Main awards and patents on production of orifice plate by angled micro piercing

Year	Topics
2005	Ministry of economy, trade and industry: 1st Monodukuri-Award, Special award.
2006	Small and medium enterprise agency: Best 300 of energetic small and medium-sized enterprises in japan.
2007	Japan patent No. 3934547: Method for manufacturing injection member.
2008	The medium and small business research institute: Good company award, Grand prize.
2012	Japan society for technology of plasticity: SME Technology development award.

2 Piercing Method for Micro Hole Piercing with a Slant Angle

There are 2 types of piercing method when pierced the slant angle hole. One is the piercing by vertical movement of the punch with slanted-cut stripper and die plate. The material settles on slanted plate as shown in Fig. 5. Another is the slanted punch movement and material is settles on horizontally as shown in Fig. 6. In mass production, selection of material handling is important for the productivity. Horizontal material transferring and progressive die system with several small die system is a suitable for high productivity of micro piercing with slant angle (Fig. 7).

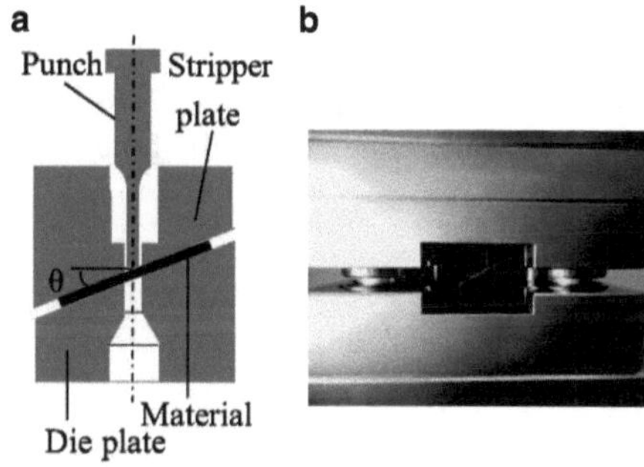

Fig. 5 Slanted material and vertical movement of the punch. **a** Figure of the die, **b** Picture of the die

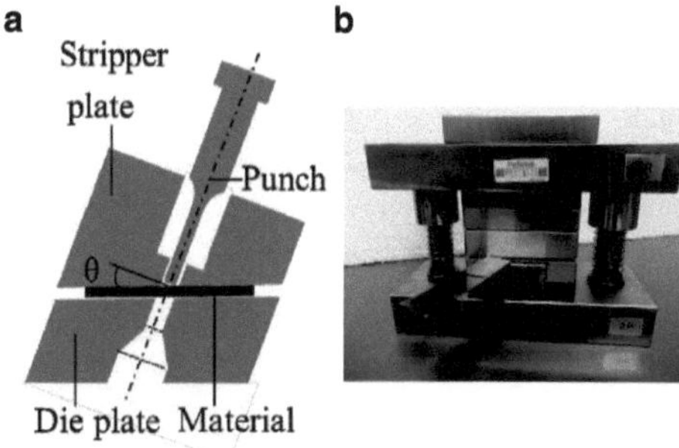

Fig. 6 Slanted movement of the punch. **a** Figure of the die, **b** Picture of the die

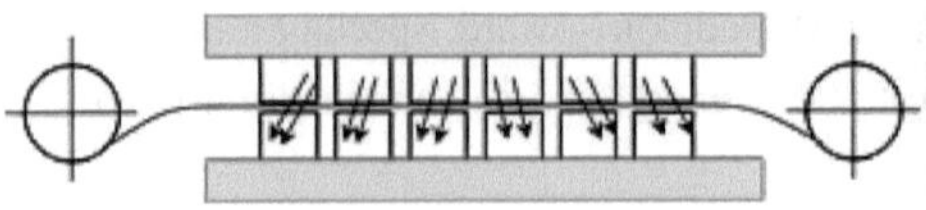
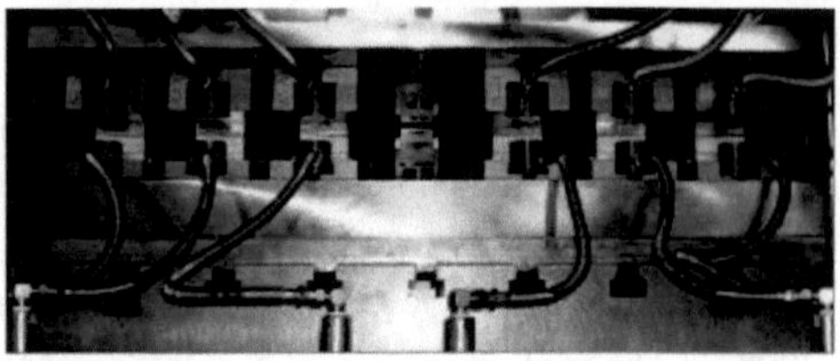

Fig. 7 Progressive die systems for orifice plate. **a** layout of die systems, **b** Picture of die systems

3 Piercing Characteristics of Micro Piercing with a Slant Angle

Micro hole piercing with a slant angle has many characteristics. Figure 8 shows an example of piercing a slant angle of 30 degree for a material with thickness of 0.1 mm by using a punch with diameter of ϕ0.15 mm. When the punch starts the piercing, only left side of punch contacts the material and shears the left side of hole. Then, the shearing spreads to whole areas. On the contrary, while the punch stroke is of 0.115 mm to 0.202 mm, only the right side of hole was sheared with right side of punch.

Figure 9 [2] shows the piercing diagram of piercing load and punch strokes for the austenitic stainless SUS304 that condition is similar to Fig. 8. From this figure, shape of diagram is look like triangle that because of one side shearing within punch stroke of 0.202 mm. Furthermore, maximum piercing force is monitored at point of "a" and it volume is 38.6 N. And then, left side of hole became fractured (see Fig. 8); as a result, piercing force reduced to point "b" at the stroke of 0.111 mm. From point "b" to point "c" right side of the punch shear the hole.

Fig. 8 Coordinate of 30 degree slant angle piercing

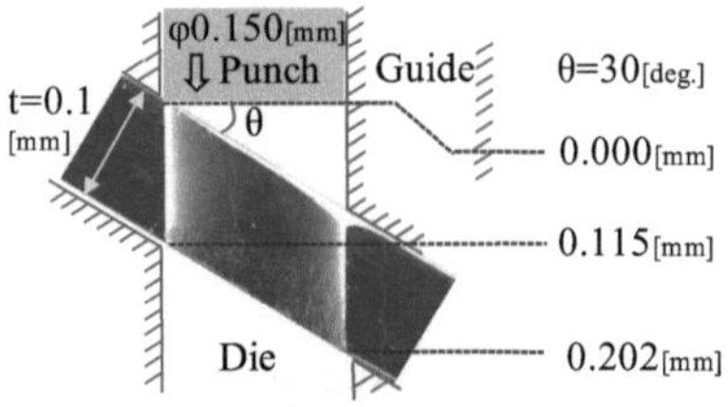

Fig. 9 Piercing force-punch stroke diagram of 30 degree slant angle piercing

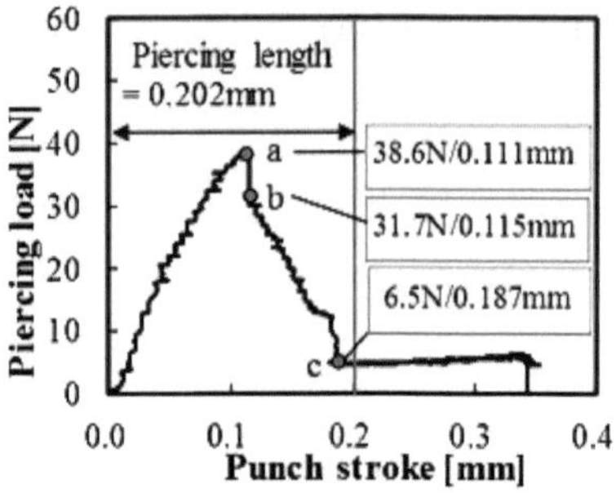

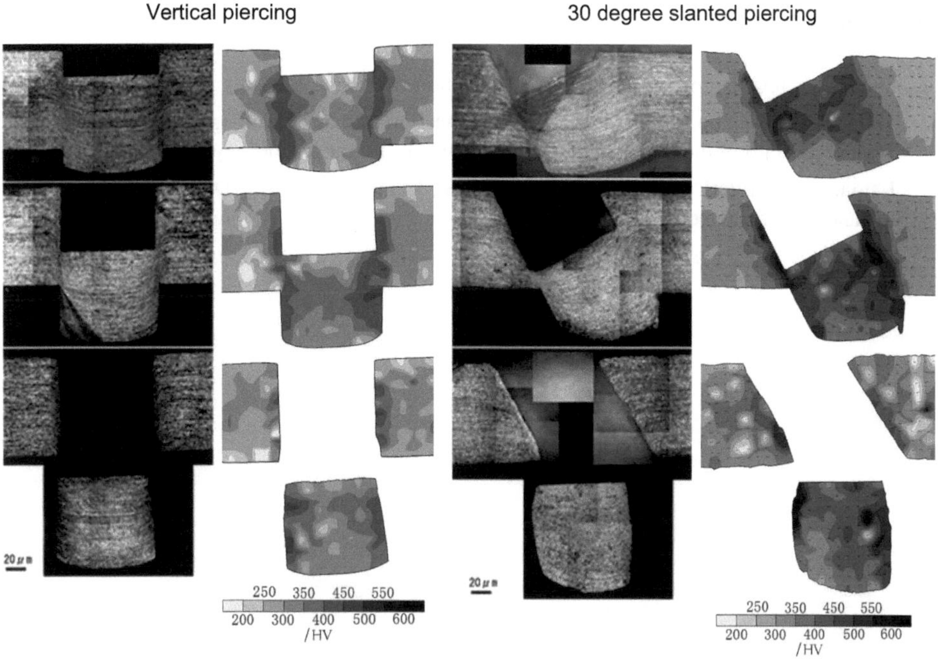

Fig. 10 Optical microstructures and map of Vickers hardness in vertical and 30 degree slant angle piercing

Figure 10 [3] shows pierced hole and pierced scrap by the vertical and slanted piercing processes for comparison at start point of burnished surface, end point of burnished surface, which are investigated by optical microstructures observation and map of Vickers hardness (HV). Experimental conditions are of punch diameter of $\phi 94\,\mu m$, die tool diameter of $\phi 100\,\mu m$, austenitic stainless SUS304 as material with thickness of $100\,\mu m$, and the slant angle of 0 and 30 degree. In vertical piercing, shearing strain becomes larger when the piercing proceeds. Therefore, HV became higher around edge of the punch and die. Inner side of pierced hole work hardened and Vickers hardness become higher. Pierced scrap has higher hardness than pierced hole. Both side of hole and scrap have similarly characteristics.

On the other hand, slantingly piercing process results are clearly difference from vertical piercing. In start stage of piercing, left side of punch is piercing the material, but right side of punch still does not start the shearing. Therefore, HV become higher along the piercing line of left side. In end point of burnished surface of left side, both ends of the punch shear the material. HV became higher than start stage of piercing. Hardness of side wall of pierced hole is over 600 HV partially. Difference of Vickers hardness in left side and right side of pierced hole is one of the features for micro piercing with a slant angle.

References

1. http://www.boschautoparts.com/port-fuel-injection.
2. Shiratori, T., Nakano, S., Suzuki, Y., 2011, Detection of blank jamming in fine slant piercing, The proceedings of the 62nd Japanese joint conference for the technology of plasticity, 443–444.
3. Katoh, M., Shiratori, T., Suzuki, Y., Nakano, S., Komatsu, T., 2014, Deformation of material in punching of slanted fine hole in SUS304 sheets with fine-grained microstructure, J. of JSTP, 55(638), 223–227

Fine Blanking of Helical Gears

Andreas Feuerhack, Daniel Trauth, Patrick Mattfeld, and Fritz Klocke

Fine blanking is a well-established process for the production of near net shape components with high quality. The produced parts are characterized by a smooth sheared edge up to 100 %, excellent surface properties with good flatness and little burr as well as close tolerances for near net shape manufacturing. These process characteristics are suitable for the efficient production of spur gears with large batch size. In this work, the application of fine blanking was extended for the production of helical gears. Therefore, the fine blanking process was modified with an additional rotary movement of the dies to realize the manufacturing of helical gears. In this contribution the process idea, experimental and numerical work as well as the potential of fine blanked helical gears is presented.

1 Motivation

In the automotive industry gears are an essential part for the efficiency of gear boxes. These gears are characterized by high precision with close tolerances, high balance, and application of high strength material. The general demand on thin helical gears in Europe has a volume of about 20 million components, trend rising. These thin helical gears are used in gear boxes for reduction of backlash and to achieve high balance. The conventional process chain for the production of thin helical gears is time-consuming because of many manufacturing steps. The motivation is the reduction of this process chain by application of fine blanking. Fine blanking has been successfully used for the production of spur gears directly from coil. Fine blanked parts in general are characterized by a smooth sheared edge of nearly 100 %, excellent surface properties with good flatness and little burr as well as close tolerances for near net shape manufacturing [1].

Andreas Feuerhack · Daniel Trauth · Patrick Mattfeld · Fritz Klocke ✉
RWTH Aachen University, Aachen, Germany
e-mail: F.Klocke@wzl.rwth-aachen.de

A. E. Tekkaya et al. (eds.), *60 Excellent Inventions in Metal Forming*,
DOI 10.1007/978-3-662-46312-3_29

One approach for the manufacturing of helical gears is the application of shear cutting and the usage of round sheet metal blanks [2]. The use of round sheet metal blanks is uneconomical and time-consuming and shear cutting does not fulfill the needed near net shape properties. Therefore, kinematics of the fine blanking process was adapted by an additional rotary movement to manufacture helical gears directly from coil with enhanced near net shape properties. By performing preliminary tests the rotary movement of the punch was linearized and the sheet metal blank was cut with a defined angle to simulate the real manufacturing process. Thereby, the main influencing parameters were identified and its correlations described [3]. Based on these insights a tool concept for fine blanking of helical gears was developed and verified under industrial conditions. The successful development by WZL and project partners was patented and is applied in the industry [4]. The innovative fine blanking process for the production of helical gears will be presented here.

2 Process Idea

Fine blanking is characterized by a linear vertical movement of the punch. Typically an underneath drive press is used, where the cutting plate realizes the movement. The decisive advantage of fine blanking is the achievable high quality of the blanked part. To reach this high quality in rotary fine blanking it is necessary to ensure a small die clearance. The typical die clearance is approximately 0.5 % of the sheet thickness. The interaction of the die clearance, v-ring and the superposition of compressive stresses are responsible for the achieved part quality [5]. To realize the fine blanking of helical gears tool kinematics had to be redesigned. The linear movement of the active parts gets superimposed with a rotational movement (Fig. 1). Thereby, the punch and counter punch are mounted pivoted and axially fixed. The upward movement of the cutting plate causes the punch to the rotational movement that leads to a helical cut. The PhD thesis of Zimmermann gives a detailed process description for the fine blanking of helical gears made of 16MnCr5 [6].

Technological challenging was the development of the tool redesign. The rotation angle β is only 1.4° and thereby the realization of a pivoted bearing is difficult to implement. The use of rolling bearings was not possible due to process related limitations. According to feasibility study of bearing designs a hydrostatic bearing with an oil chamber was applied. This tool design in combination with the upward movement of the cutting plate initiates the fine blanking process (Fig. 2). Depending on the movement of the cutting plate, the oil chamber pressure build-up and in consequence the punch moves nearly frictionless [7].

Beside the tool development the calculation of the cutting force was adapted. The calculations of the cutting force F_s [N] is based on the sheet thickness s [mm], cutting length l_s [mm], tensile strength R_m [MPa] and shear strength factor c_s [–], see Eq. 1 [7].

$$F_s = s l_s R_m c_s \quad (1)$$

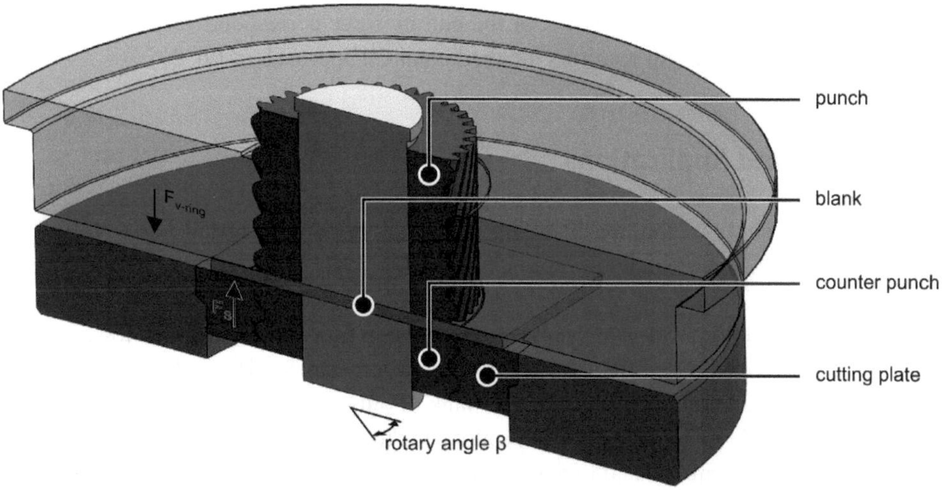

Fig. 1 Principle tool concept for the fine blanking of helical gears

Fig. 2 Tool in industrial application

Taking into account the helix angle β [°] the cutting path increases in comparison to the sheet thickness s. The calculation of the cutting force is modified as shown in Eq. 2:

$$F_s = \frac{s}{\cos\beta} l_s R_m c_s. \tag{2}$$

The determined experimental data of the cutting force correspond very well with the calculated cutting force and confirms the mathematical approach of Eq. 2.

3 Results and Applications

The feasibility study of the fine blanking of helical gears was carried out with a case-hardened steel 16MnCr5 and a sheet thickness of 3.2 mm. In Table 1 the helical gear is specified in detail.

The fine blanked helical gears showed a very good quality, see Fig. 3. However, the quality of the tooth flank differs for the right and left flanks. The right flank shows a smooth sheared edge of 100 % and the left flank of only 81 %. Furthermore, the analysis of the die roll height gives different results along the tooth profile. For the left flank a value of 0.12 mm was determined and in comparison to the right flank (0.65 mm) the difference is nearly by factor 5. These results indicated a change of the die clearance during the fine blanking process [7]. The change of the die clearance is undesired and therefore the experimental work was supported by the application of the finite element method to determine the cause-and-effect relationships between the part quality and fine blanking parameter.

Figure 4 shows a fully elasto-plastic simulation model, tangential expansion of the punch and the contact normal stress distribution at one tooth. The tangential displace-

Table 1 Specification of the fine blanked helical gear

parameter	value
helical angle β	26.74°
normal pressure angle	20°
normal module	3 mm
number of teeth	40
pitch diameter	134.37 mm
addendum modification coefficient	0

Fig. 3 Fine blanked helical gear and detailed view of the 100 % smooth sheared tooth flank

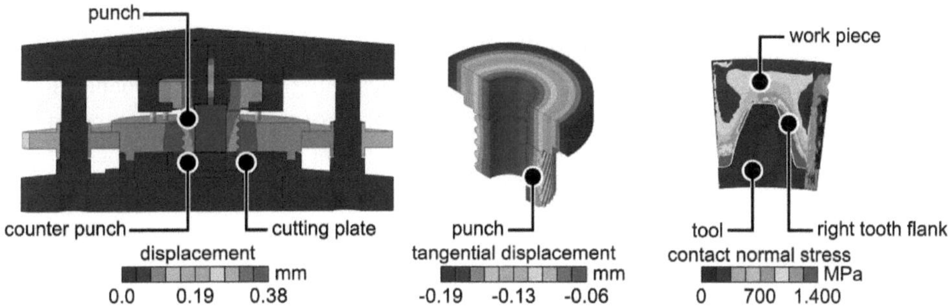

Fig. 4 Fully elasto-plastic FE model with a detailed view of the tangential displacement of the punch as well as the contact normal stress distribution at one tooth

ment of the punch as well as the contact normal stress distribution correlates well with the experimental work. The shifting of the die clearance is related to the tangential displacement and finally leads to a non-uniform stress distribution at the tooth. This leads to a flank depending cutting quality [8]. Current research work is focusing the adaption of the die clearance to achieve a homogenous stress distribution and thereby a uniform 100 % smooth sheared edge for the right as well as for the left tooth flank.

4 Potential

Fine blanking of helical gears is a promising technology and has the following potentials:

1. The helical gears have a good part quality with a smooth sheared edge and requiring less refinishing operations.
2. Manufacturing of helical gears directly from coil compared with conventional machining is very economical.
3. Processing of a wide range of materials and sheet thicknesses is possible.
4. Achieving of a homogenous 100 % smooth sheared edge by an optimization of the die clearance.

5 Conclusion

Fine blanking of helical gears is an innovative process to manufacture helical gears directly from coil. The major challenge in tool design was the bearing of the punch to realize the rotary movement of the punch. The approach with hydrostatic bearing was successfully implemented and ensured a reproducible manufacturing of helical gears. The experiments showed a different smooth sheared edge for the left and right flank of the tooth. Experimental investigations were supported by means of the finite element method.

The tangential displacement of the punch leads to a displacement and thereby to a shifting of the die clearance. The shifting of the die clearance is represented by different contact normal stress distribution at the tooth and correlates with the flank dependent smooth sheared edge. Current research is focusing on the adaption of the die clearance to achieve a homogenous 100 % smooth sheared edge on both the right and left flank of the tooth.

6 Acknowledgement

This process has been investigated in several projects. The work has been kindly supported by the German Research Foundation (DFG) under the grant number KL 500/126-1 and the Industrial Collective Research under the grant agreement number [KF 2120301PK8].

References

1. Klocke, F., König, W., 2006, Fertigungsverfahren 4, Umformen, 5. Ausgabe, Springer Verlag Berlin.
2. Belcredi, L., 1986, Procedimento per la tranciatura di ruote dentate a profile elicoidale e prodotti cosi' ottenuto. Patent IT 1137113.
3. Klocke, F., Zimmermann, M., Bäcker, V., Wegner, H., 2011, Finite Element Simulation of an Analogy Process for the Fine Blanking of Helical Gears, Proceedings of International Symposium on Assembly and Manufacturing ISAM 2011, Tampere, Finland, 1–6
4. Zimmermann, M., Watermann, M., Trump, K., 2011, Verfahren und Vorrichtung zum Feinschneiden von Werkstücken, Patent EP2208552 B1.
5. Hoffmann, H., Neugebauer, Spur, G., 2012, Handbuch Umformen, Hanser Verlag
6. Zimmermann, M., 2015, Feinschneiden von Schrägverzahnungen, Apprimus Verlag
7. Klocke, F., Zimmermann, M., Mattfeld, P., Feldhaus, B., Trump, K., Watermann, M., 2011, Feinscheiden schrägverzahnter Stirnräder. Innovatives Vorverzahnen schmaler Schrägverzahnungen, wt Werkstattstechnik online, Volume 10 (10), 668–672.
8. Zimmermann, M., Klocke, F., Schongen, F., Feldhaus, B., 2011, Fine Blanking of Helical Gears – Finite Element Simulations and First Experimental Results, ICTP 2011, International Conference on Technology of Plasticity, 10, 581–585.

Edge-Fracture-Tensile-Test

Martin Feistle, Michael Krinninger, Isabella Pätzold, and Wolfram Volk

The *edge-fracture-tensile-test* is an innovative examination method, which offers the possibility to characterise high strength steels with respect to edge crack sensitivity. Manufacturing processes, such as expansion and stretch flanging, which generate an uniaxial tensile stress on the components shear cut edge, lead to edge fracture. In particular, deformation and material damage of the component's edge, caused by the shear cutting process, reduce the residual formability and prefer edge fracture initiation.

A modular shear cutting tool generates *edge-fracture-tensile-samples* with different shear cutting process parameters. The *edge-fracture-tensile-test* investigates one-sided shear cut *edge-fracture-tensile-samples* with regard to the residual formability of the shear cut edge.

1 Motivation

In modern car body manufacture, the application of lightweight components is a central challenge. Due to the economical and ecological reasons as well as to improve product properties, a mass reduction is necessary. In addition, an improved crash performance and an enhanced driving comfort can be realised by applying high performance materials, for example high strength steels with yield strengths above 600 MPa and tensile strengths above 800 MPa. For crash-relevant components, higher material-strengths allow to reduce material-thickness and thus the required weight reduction of the body in white [1].

However, the edge fracture sensitivity of many steel sheet materials represents a challenge in the usage of high strength steels. In addition to material selection, the phenomenon of edge fracture is influenced significantly by the cutting process. Due to the

Martin Feistle · Michael Krinninger · Isabella Pätzold · Wolfram Volk ✉
Technische Universität München, Munich, Germany
e-mail: Martin.Feistle@utg.de, Michael.Krinninger@utg.de, Isabella.Paetzold@utg.de, Wolfram.Volk@utg.de

A. E. Tekkaya et al. (eds.), *60 Excellent Inventions in Metal Forming*,
DOI 10.1007/978-3-662-46312-3_30

in series production mostly used shear cutting process, high strains get initiated to the shear affected zone [2, 3]. This deformation at the edge of the component causes work hardening, which comes along with an immense reduction of residual formability. A continuation of stress of shear cut edges due to follow-up operations, which usually generate a nearly uniaxial stress at the component's edge, often lead to edge fracture [4].

The assessment of edge fracture sensitivity can be carried out using a variety of examination methods such as the hole-expansion-test according to ISO 16630 [5], the collar-drawing-test according to VDI 3359 [6], the Diabolo-test or the open-hole-tensile-test. In most of these methods, which detect edge fracture, the investigated edge of the sample is at least at the beginning of testing in contact with the tool, whereby the effects of friction may influence the test results. In addition, the examined component edge is usually not exposed to tensile load, although this is a common edge load in forming processes.

The *edge-fracture-tensile-test*, which was developed at the Institute of Metal Forming and Casting, therefore represents a friction-free process. It allows characterising the edge fracture tendency of a steel sheet material quickly and its determination of forming limit under tensile stress.

2 Process Idea

Several studies determined the main influences on the phenomenon of edge fracture, for example: separation method, cutting surface parameters, shear affected zone, tool wear and structure [4, 7, 8].

To determine these influence factors and develop a new *edge-fracture-tensile-sample*, a modular cutting tool was constructed and manufactured. The cutting tool for the experimental runs is shown in Fig. 1. It allows cutting *edge-fracture-tensile-samples* with an open or a closed cutting line and different cutting parameters. To realise the different cutting line geometries, it is only necessary to adapt the geometry of the initial blank. Different cutting parameters, for example the die clearance, can be adjusted with different cutting active elements. Moreover, it is possible to realise a single and multi-cutting process. The cutting tool, which is used for the experimental investigation, has a very high stiffness. Due to the high tool stiffness, there are no displacements of the cutting elements caused by deflection or shifting during the cutting process. The blank holder plate guides the punch; therefore, a minimal guiding tolerance can be achieved. The initial positioning of the blank holder plate and the die is realised with alignment pins. Due to gas pressure springs, different blank holder forces for the different sheet materials and blank thicknesses can be applied. The sample is positioned by locators. This kind of positioning ensures a high repetitive accuracy in the orientation of the samples. At the beginning of the experimental runs, the edge radii of the cutting elements are sharp edged. This is measured with a profile measurement device manufactured by Mahr GmbH, Germany.

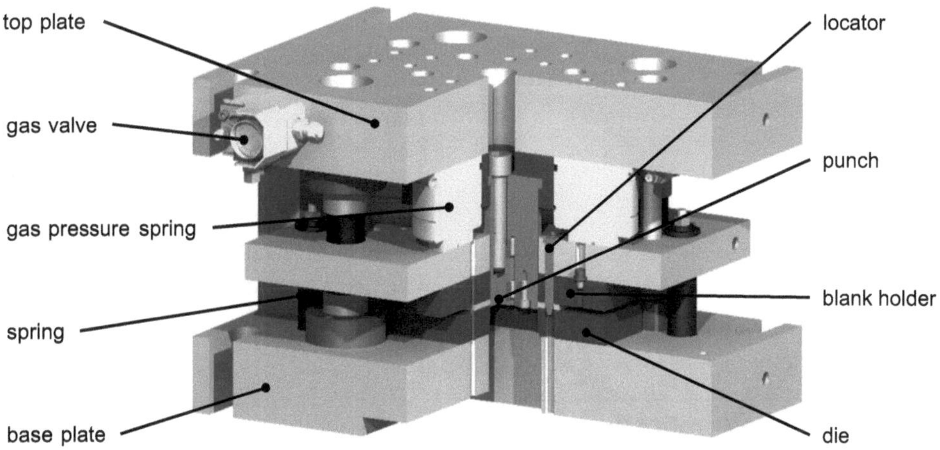

Fig. 1 Used tool for shear cutting of the *edge-fracture-tensile-samples*

Figure 2 shows the manufacturing steps from the initial sample to the *edge-fracture-tensile-sample*. The left image shows the initial sample geometry with the locating holes. The upper side of the *edge-fracture-tensile-sample* is manufactured by a milling process. The sample in the middle shows the geometry of the shear cutting process with an open cutting line, the outline of the punch and the resulting scrap. The right picture shows the final *edge-fracture-tensile-sample* with its dimensions. Due to the shear cutting process and the consequent residual formability, the milled side of the sample must have a higher formability than the shear cut edge. This ensures that the crack initiation starts at the shear cut edge.

For each experimental run, three samples of the material HCT780X (sheet thickness 1.2 mm) are tested. The results are averaged. The created cutting surfaces are also determined with the profile measurement device (see Table 1).

During the sheet forming process, the cutting edges of the components are typically loaded with a plane strain and an uniaxial stress. Therefore, the *edge-fracture-tensile-*

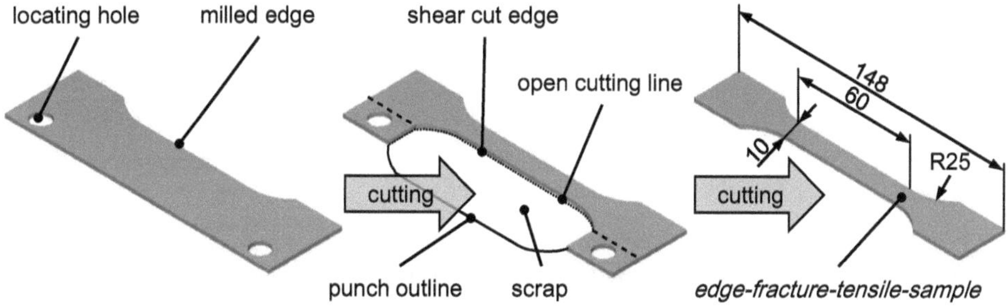

Fig. 2 Schematic draft of sample geometry

Table 1 Cutting surface parameters according to VDI 2906-2 [9]

u [%]	rollover height [mm]	clean-shear height [mm]	fracture height [mm]	burr height [mm]
10	0.10	0.10	1.00	0.02
20	0.16	0.12	0.92	0.04

samples are tested with the tensile test according to DIN EN ISO 6892-1 [10]. The accessibility to the sample in the tensile testing machine allows detecting the elongation and the crack initiation with an optical measuring system.

3 Results and Applications

The key parameter of the *edge-fracture-tensile-tests* for the experimental runs is the residual formability of the shear cut edge until fracture initiation. The beginning of local necking characterises the critical limit of failure. This definition is chosen, because such a local distortion leads to rejected components in the production process. The strains at the beginning of necking are determined by the method according to [11]. The photogrammetric measuring system by GOM – Gesellschaft für Optische Messtechnik mbH, Germany enables the local detection of the degree of deformation caused by tensile stresses. Figure 3 shows the local concentrated deformation of the shear cut edge of the sample at the beginning of necking. The diagram shows the obtainable major principal strains of the both side milled sample and single side shear cut sample at this point.

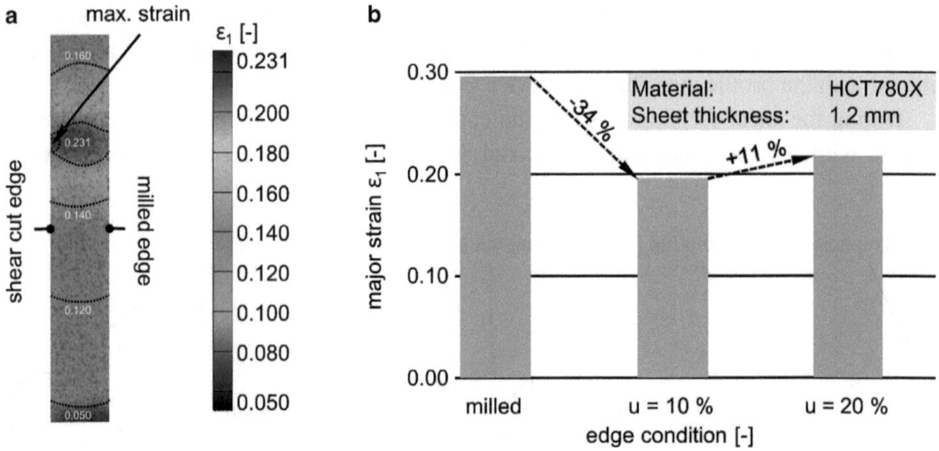

Fig. 3 Visualisation of the strain distribution in the sample (**a**), comparison of the forming capability of the edges of milled and shear cut samples (**b**)

Because of the shear cutting process, the edge of the sample is deformed extremely. This leads to a work hardening of the material in an area along the cutting line. Therefore, a reduced residual formability in this area is the consequence. Compared to the milled sample, the cutting process with a relative die clearance of $u = 10\ \%$ induces a reduction of the major principal strain up to 34 %. An increased die clearance causes a higher residual formability with this steel grade. A die clearance of $u = 20\ \%$ rises the major principal strain about 11 % compared to $u = 10\ \%$. Table 1 shows the characteristic cutting parameters according to VDI 2906-2 [11] for two different die clearances. An increased die clearance causes a higher rollover and clean-shear height. The fracture height decreases at the same time.

4 Potential

The advantages of the *edge-fracture-tensile-test* allow an evaluation of the *edge-fracture-sensitivity* of conventional and high strength steels with less cost of time and material. The cutting parameters can be adjusted in order to utilise the formability of the material in the best way. A new investigation method was developed, which enables the steel and sheet metal processing industry to characterise new steel grades with little effort and time.

5 Conclusion

The realised test shows that the forming potential of high strength steels is limited by edge fractures caused by the shear cutting process. Based on this knowledge, it is necessary to select suitable shear cutting parameters especially for cutting blanks. Therefore, a high residual formability can be maintained. Furthermore, continuing investigations should focus on the influence of die clearance variation, cutting edge geometry and so on.

6 Acknowledgement

This *edge-fracture-test* has been investigated in some projects at the Institute of Metal Forming and Casting. The work has been kindly supported by the Stiftung Stahlanwendungsforschung im Stifterverband für die Deutsche Wissenschaft e. V. under the grant number P 1072/07/2014.

References

1. Kleiner M., Geiger M., Klaus A., 2003, Manufacturing of Lightweight Components by Metal Forming, CIRP Annals – Manufacturing Technology, Volume 52, Issue 2, p. 521–542, Elsevier B.V., Amsterdam, Netherlands
2. Lange K., 1990, Umformtechnik – Handbuch für Industrie und Wissenschaft, Band 3: Blechbearbeitung, 2. Auflage, Springer-Verlag, Berlin, Heidelberg
3. Doege E., Behrens B.-A., 2010, Handbuch Umformtechnik, 2. Auflage, Springer-Verlag, Berlin, Heidelberg
4. Kardes N., Altan T., 2008, Examining edge cracking in hole flanging AHSS, Stamping Journal, FMA Publication, Rockford, IL, USA
5. N. N., 2009, Metallic materials – Sheet and strip – Hole expanding test, ISO 16630, ISO copyright office, Geneva, Switzerland
6. N. N., 2013, Flanged holes – Flange forming, VDI 3359, Association of Engineers, Beuth Verlag GmbH, Düsseldorf
7. Konieczny A., Henderson T., 2007, On Formability Limitations in Stamping Involving Sheared Edge Stretching, SAE Paper No. 2007-01-0340, p. 41–50, SAE International, Warrendale, PA, USA
8. Liewald M., Gall M., 2013, Experimental investigation of the influence of shear cutting parameters on the edge crack sensitivity of dual phase steels, IDDRG 2013 Conference, Zurich, Switzerland
9. N. N., 1994, Quality of cut faces of (sheet) metal parts after cutting, blanking, trimming or piercing; shearing, form of sheared edge und characteristic values, VDI 2906-2, Association of Engineers, Beuth Verlag GmbH, Düsseldorf
10. N. N., 2014, Metallic materials – Tensile testing – Part 1: Method of test at room temperature (ISO/DIS 6892-1:2014), DIN EN ISO 6892-1, DIN German Institute for Standardization, Beuth Verlag GmbH, Düsseldorf
11. Krauer J., 2010, Erweiterte Werkstoffmodelle zur Beschreibung des thermischen Umformverhaltens metastabiler Stähle, Dissertation, Technische Hochschule Zürich, Switzerland

Reduction of Vibrations in Blanking by MR Dampers

Andrea Ghiotti, Paolo Regazzo, Stefania Bruschi, and P. Francesco Bariani

1 Introduction

Blanking and, more generally, shearing operations are the most common operations in the process chains for sheet metal parts. Looking at a typical punch force vs. penetration curve in blanking (Fig. 1), two main phases can be distinguished. In the former (from A to B in Fig. 1a), the punch penetrates the material causing the sheet metal to deform until the shearing starts. In this phase, the press and die set store up elastic energy. In the second phase (from B to C in Fig. 1a), the shear strength of the material is exceeded and a fracture starts causing the sudden release of the energy that dissipates through the high frequency oscillations (the tail end of the curve in Fig. 1a). This break through shock generates un-

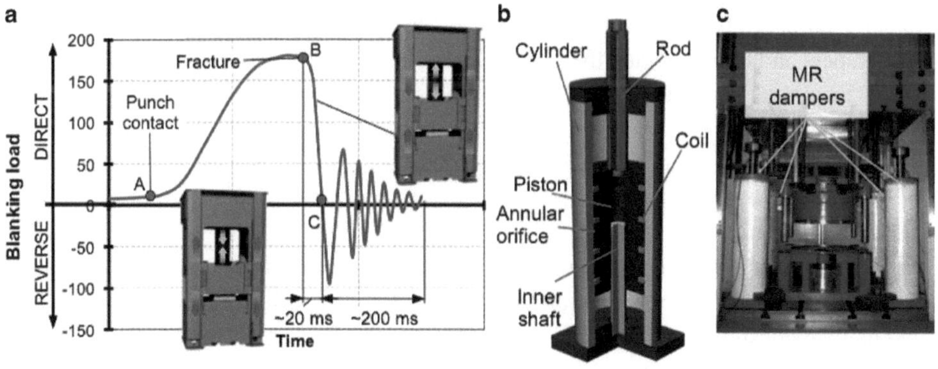

Fig. 1 **a** The reverse load phenomenon in blanking, **b** Sketch of MR damper, **c** MR fluid dampers

Andrea Ghiotti ✉ · Paolo Regazzo · Stefania Bruschi · P. Francesco Bariani
University of Padova, Padova, Italy
e-mail: andrea.ghiotti@unipd.it

A. E. Tekkaya et al. (eds.), *60 Excellent Inventions in Metal Forming*,
DOI 10.1007/978-3-662-46312-3_31

controlled high reverse loads, mechanical vibrations and loud noise that may cause serious problems such as fatigue cracks in the ram, drive linkage, crown, housings and, in the bed of the press, premature wear in punches and dies and great discomfort for press operators. The break through shock in blanking operations has been considered by a number of researchers, most notably for modelling the ductile fracture that causes surface separation and the associated energy release [1–3] or for correlating the stiffness of the system components (press, stripper and die set) and their vibrations during the break through [4–6]. However, few contributions in the scientific literature have dealt with new possible solutions to effectively reduce the reverse load, vibrations and noise. Murakawa [7] developed a "hydraulic inertia damper" that proved to be effective in lowering the punch force reduction rate, thereby causing a reduction in vibration and noise. Osakada [8] demonstrated that the noise level accompanying the break through shock can be significantly reduced if an accurate control of the punch motion – such as that allows by servo controlled presses – is combined with a "continuous two-steps blanking", where the punch is stopped just before the fracture starts. In this research, the application dampers based on magneto-rheological (MR) fluids to reduce the shock response of press systems during blanking operations is considered. Compared to most conventional dampers, which are passive hydraulic devices, MR dampers are semi-active devices where the damping capacity can be modulated either gradually or in real-time based on the system response. MR fluids are suspensions of micron-sized ferromagnetic particles in a liquid carrier that presents controllable and reversible changes in its rheological properties, from free flowing linear viscous liquids to semi-solids, having controllable yield strength when activated by an external magnetic field. Thanks to these features, they offer relevant performances in the typical working conditions of metalworking presses, where long working strokes are required and the vibrations caused by break through shocks present high frequencies and small-amplitude vibrations. The aim of the investigation described in the paper was to evaluate, through full-scale experiments, the feasibility and practicability of implementing MR dampers and to understand the potential benefits when they are used in a semi-active manner in comparison with conventional dampers.

2 MR Dampers

The MR dampers designed and developed by the Authors are presented in Fig. 1b and c (details of the design and manufacture in [9] and [10]). The prototypes consist of single-hand dampers designed to withstand the break through shock generated during blanking in a long-stroke hydraulic press. A 1 mm thick annular orifice between the piston rod and the liner allows the fluid to flow through the entire annular region as the piston moves. The inner shaft prevents possible misalignments from arising due to high-frequency vibration and reverse load and, at the same time, works as a volume compensator to balance the volume change during the rod stroke. The electrical coils embedded into the piston generate a uniform magnetic field across the 240 mm long gap and perpendicularly to the

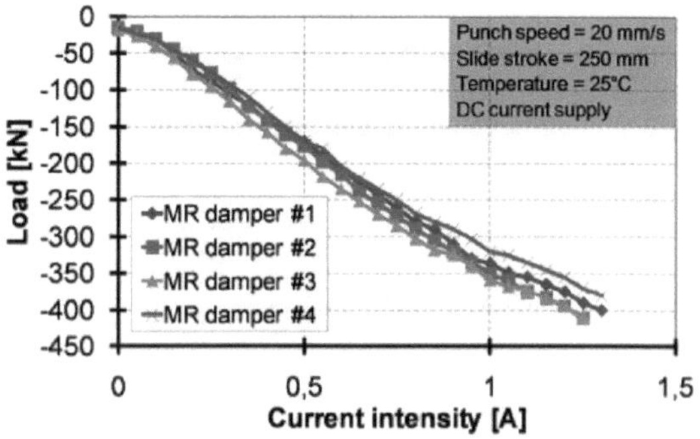

Fig. 2 Force vs. current intensity curves of the MR dampers for a fixed value of the punch speed

direction of the MR fluid flow. Figure 1c shows the four dampers in the working area of the press.

The damping force of the four MR dampers for different values of the current intensity was measured through laboratory experiments. Figure 2 shows the curves of the force vs. current intensity of the MR dampers for a fixed value of the punch speed. For the four dampers, the damping force is constant along the entire testing stroke and presents a good linearity with the current intensity.

3 Experiments

In the experiments three damping systems were compared: (a) the MR dampers above described, (b) conventional hydraulic cushions, and (c) commercially available hydraulic dampers. The damping system based on hydraulic cushions consisted of four hydraulic cylinders whose chambers were connected in parallel to a common manifold. The working stroke and maximum counter force were 40 mm and 80 kN, respectively. The system, based on hydraulic dampers, consisted of four shock dampers providing a counter pressure generated by an internal orifice and a flow control valve. The working stroke and maximum counter force were 20 mm and 200 kN, respectively. The experiments were carried out on a four-column double-effect 2500 kN hydraulic press. Figure 3a–c shows the machine and the details and the sketch of the tooling.

Data regarding the pressure and acceleration were acquired during the experiments: the former was monitored inside the manifold of the upper actuator of the press, the latter measured on four points of the frame, respectively located on the die plate, on the ram and on one of the columns of the press. The blanking experiments were conducted on 2 mm and 3 mm thick C40 steel sheets with a constant punch-die clearance of 0.2 mm and two different speeds of the press ram (20 mm/s and 50 mm/s). The experiments were repeated for four different configurations of the press: with no damping device, with MR dampers, with hydraulic cushions and with hydraulic dampers. The experiments with the MR dampers were carried out using four levels of current intensity: 0.4 A, 0.6 A, 1.1 A and 2 A. All the experiments were replicated to assure the repeatability of results.

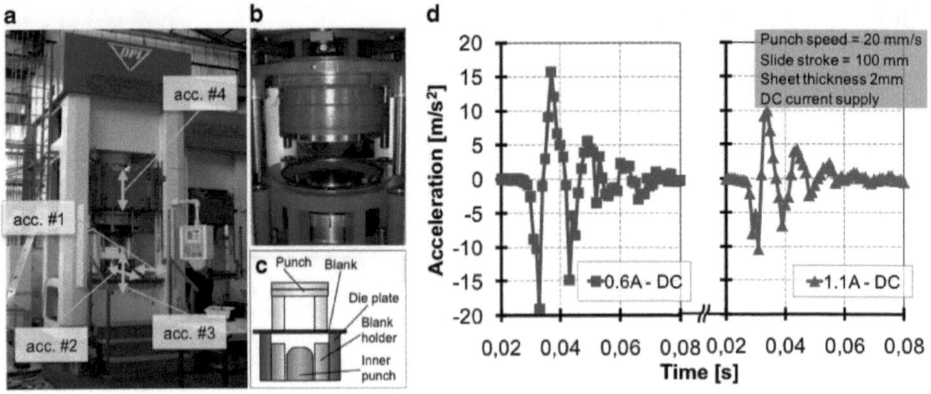

Fig. 3 **a** The 2500 kN hydraulic press with indication of the accelerometers, **b** Detail and **c** sketch of the tooling, **d** Accelerations for different electrical currents with the MR shock dampers

4 Results

Figure 3d shows the effect of the current intensity in the MR damper circuits on the response of the press system to the break through shocks. Damping performance increases with the current intensity until a critical value of 3 A generates the saturation of the magnetic circuit and the permanent magnetization of the particles in the MR fluid. A value of 1.1 A proved to be a good compromise between damping performance and system stability. Figure 4 shows the acceleration-time response over the duration of the break through shock when the press system is equipped with the hydraulic cushions (Fig. 4a), the hydraulic dampers (Fig. 4b) and the MR dampers (Fig. 4c).

Each of the three responses is compared to the one measured on the press without any damping device. Reductions in the peak acceleration are achieved by using both the hydraulic cushions and the MR dampers. Compared to the response of the system equipped

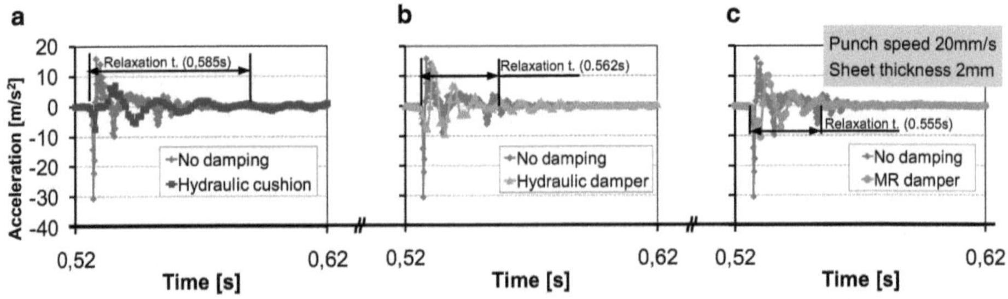

Fig. 4 Accelerations measured at the press ram with **a** the hydraulic cushions, **b** the hydraulic shock dampers and **c** the MR shock dampers

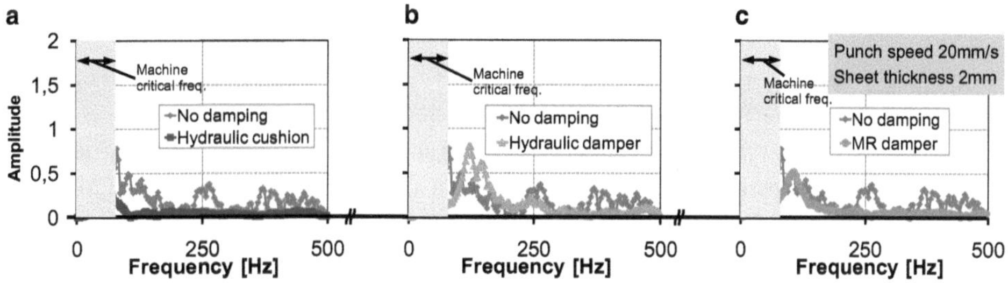

Fig. 5 Amplitude spectra of accelerations at press ram with **a** the hydraulic cushions, **b** the hydraulic shock dampers and **c** the MR shock dampers

with the conventional dampers, the decay of residual vibrations that is observed for the system with the MR dampers is significantly faster (approximately a reduction of 30 % compared with hydraulic cushions and 10 % compared with hydraulic dampers). This is because the damping effect of the highly viscous MR fluid (when the magnetic field is activated) predominates over the unavoidable elasticity of the fluid pressurized in the conventional dampers. Figure 5 shows the amplitude spectrum of the ram acceleration in the frequency domain for the press system equipped with the hydraulic cushions (Fig. 5a), the hydraulic dampers (Fig. 5b) and the MR dampers (Fig. 5c). Each of the three spectra is compared to the one measured on the press without any damping device. As expected, the spectrum for the press without any damper shows a high number of excited frequencies, with the highest values close to 25 Hz. Significant reductions of the number of the excited frequencies and the acceleration amplitudes are observed for the three kinds of dampers. However, the hydraulic cushions behave worse than the other two shock dampers typologies since they are not able to damp the low frequencies vibrations that are typical of the tested hydraulic press. Among the latter, the amplitude of vibrations measured with the MR dampers is approximately 40 % less than the one obtained with the hydraulic dampers.

5 Conclusions

MR dampers were developed for metalworking presses to withstand the break through shock generated during blanking operations. Full-scale experiments were set up to understand the potential benefits of MR dampers compared to conventional dampers. The results show that MR dampers can be effective in damping the vibrations induced by the break through shock and in limiting the number of excited frequencies. Thanks to their short response time they appear suitable to be implemented in processes where multiple blanking operations are required.

References

1. Breitling J., Pfeiffer B., Altan T., Siegert K., 1997, Process control in blanking, Journal of Materials Processing Technology, 71 (1), 187–192.
2. Stegeman Y.W., Goijaerts A.M., Brokken D., Brekelmans W.A.M., Govaert L. E., Baaijens F.P.T., 1997, An experimental and numerical study of a planar blanking process, J. of Materials Processing Technology, 87 (1–3), 266–276.
3. Stegeman Y. W., Goijaerts, D. Brokken, W. A. M. Brekelmans, L. E. Govaert, F. P. T. Baaijens, 1999, An experimental and numerical study of a planar blanking process, Journal of Materials Processing Technology, 87 (1–3), 266–276.
4. Doege E., Seidel H.-J., 1985, Noise Reduction on Mechanical Punch Presses, CIRP Annals, 34 (1), 507–509.
5. Guo B., Chen W. M., Wang Z. R., 1998, Analysis of blanking vibration with consideration of the break-through state, Journal of Materials Processing Technology, Vol. 75 (1–3), 117–121.
6. Bassiuny, A.M., Li, Xiaoli, Du R., 2007, Fault diagnosis of stamping process based on empirical mode decomposition and learning vector quantization, Int. J. Mach. Tools and Manuf., 47 (15), 2298–2306.
7. Murakawa M., Mo J., Wakatsuki Y., Koga N., 2001, Investigation of blanking noise reduction using a hydraulic inertia damper, J. of Materials Processing Technology, 112 (2–3), 205–213.
8. Otsu M., Yamagata C., Osakada K., 2003, Reduction of blanking noise by controlling press motion, CIRP Annals, 52(1), 245–248.
9. Regazzo P., Ghiotti A., 2008, Internal report in Italian, DIMEG 2008/56.
10. Ghiotti A., Regazzo P., Bruschi S., 2014, Shear surface control in blanking by adaptronic systems, Procedia Engineering, 81, 2014, 2512–2517.

Force Reduction During Blanking Operations of AHSS Sheet Materials

Andreas Mackensen, Matthias Golle, Roland Golle, and
Hartmut Hoffmann

Within the manufacturing process of sheet metals, blanking represents an essential process operation. As the industrial application of high-strength multi-phase steels grows, the blanking process must consider high blanking and shear forces, which are characteristic for processing these materials. This paper presents options for reducing these forces. Experiments were performed utilizing a novel tool concept, which can correlate necessary blanking forces to the punch stroke in three dimensions and in the direct force path. Results from three different AHSS materials are presented showing the variation of decisive blanking parameters such as die clearance, shearing angle and sheet positioning angle.

1 Motivation

Challenges in modern sheet metal manufacturing have become manifold as material strength and quality requirements have steadily increased. Some of these challenges arise from the use of advanced high-strength steels (AHSS) in the manufacturing of sheet metal blanks. These AHSS blanks are being adopted by the automotive industry due to the potential for lightweight vehicle design and improved or at least comparable conformance to crash-relevant safety requirements [1].

AHSS materials belong to the family of multi-phase steels. They are characterized by a complex microstructure consisting of microconstituents such as bainite, martensite and retained austenite dispersed in a ductile ferritic matrix. Compared to the typical ferritic phases in the microstructure of conventional micro-alloyed steels, multi-phase steels possess unique mechanical properties. Of particular interest are the high strength combined with comparatively good formability which is the result of the processing and funda-

Andreas Mackensen · Matthias Golle · Roland Golle · Hartmut Hoffmann ✉
Technische Universität München, Munich, Germany
e-mail: hartmut.hoffmann@utg.de

A. E. Tekkaya et al. (eds.), *60 Excellent Inventions in Metal Forming*,
DOI 10.1007/978-3-662-46312-3_32

mental microstructure of AHSS. The typical AHSS materials are Complex-Phase (CP) steels, Dual-Phase (DP, ferritic-martensitic) steels, TRIP (Transformation Induced Plasticity) steels and TWIP (Twinning Induced Plasticity) steels.

Traditionally, when AHSS are used in shearing processes, cutting tools of high quality in terms of stiffness and dimensional accuracy are usually preferred in order to minimize wear and improve tool life [2, 3]. This method is also generally regarded as a decisive strategy for producing consistently high quality manufactured parts. However, reduced tool life is symptomatic in industrial settings when AHSS sheet materials are used due to the high blanking and shear forces. In this context, the immense cutting shocks that typically occur negatively affect not only the tool life but also the press functionality [4].

2 Process Idea

It is desirable to realize a shear cutting operation with reduced process forces that still results in acceptable quality levels. Thus, the objective is to investigate the influence of standard shear cutting parameters, such as die clearance, cutting angle and sheet positioning angle, on the cutting and in particular on the transverse forces in a blanking process.

The experiments were performed utilizing a novel tool concept which provides possibilities for measuring process forces in ways more extensive than known experimental approaches from literature. The developed cutting tool is able to correlate necessary blanking forces to the punch stroke in three dimensions.

The sheet materials used in this study include a DP800, a TRIP700 and TWIP1000 steel, each showing a unique microstructure as well as different formability and strength characteristics. Additionally, a mild carbon steel (DC04) is used for comparison. For the materials tested, the sheet thickness is 1 mm (±30 μm) and the cutting line length is 70 mm.

The effect of different die clearances, shear angles as well as sheet positioning angles on process forces were investigated. The investigated specific die clearances are 5, 10 and 15 %, investigated shear angles are 0°, 2.5° and 5°, while the sheet positioning angles are 5°, 10° and 15° (see Fig. 1).

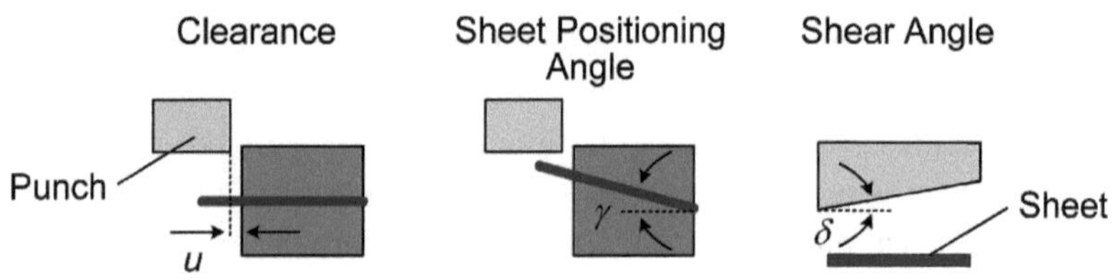

Fig. 1 Schematic illustration of selected blanking process parameters

3 Results and Applications

Figure 2 shows the normalized force-displacement behavior at a specific die clearance of 10 % with respect to the 1 mm sheet thickness and with a punch position parallel-edged to the sheet for the blanking as well as shear force. The force is normalized to a cutting line length of 1 mm. In this context, the displacement is limited to the stroke movement until bottom dead center; therefore, stripper force is neglected in the figure. Due to very small values (≤ 0.4 kN), the process force in the direction parallel to the cutting edge (F_y) is also neglected in the figure. It can be seen, that the material strength has a significant influence on the progression of the curves and their maximum force level, independent of the effective directions F_z and F_x. In addition, the maximum force level for the shear force F_x is significant higher for the AHSS materials compared to standard estimation approaches given in literature. For the TWIP steel the maximum shear force F_x is up to 45 % of the maximum blanking force F_z.

Selected force-displacement-diagrams of the DP steel grade at different clearance ratios for a shear angle of 2.5° are illustrated in Fig. 3. Compared to the results with the punch position parallel-edged to the sheet from Fig. 2, the maximum blanking and transverse forces are significantly reduced when a shear angle is used in the cutting operation. Therefore, this configuration is already widely used in industrial applications. Moreover, the examination shows a considerable dependence of the force progression on specific die clearance for the blanking force F_z. Generally, the process forces present a characteristic plateau-shaped curve progression. However, with increasing clearance ratio, the force level increases significantly in the F_z direction, whereas for the transverse force F_x

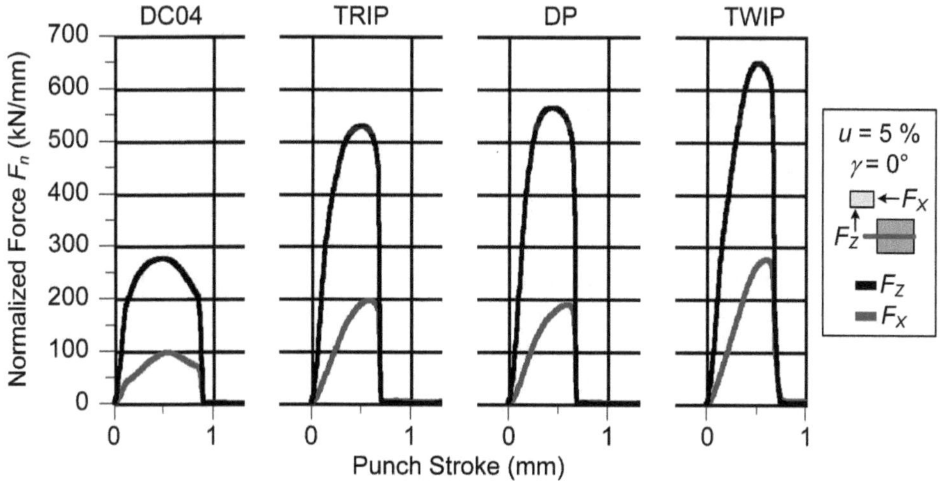

Fig. 2 Selected normalized force-displacement diagrams for the sheet materials investigated with parallel-edged punch position (specific die clearance of 5 %)

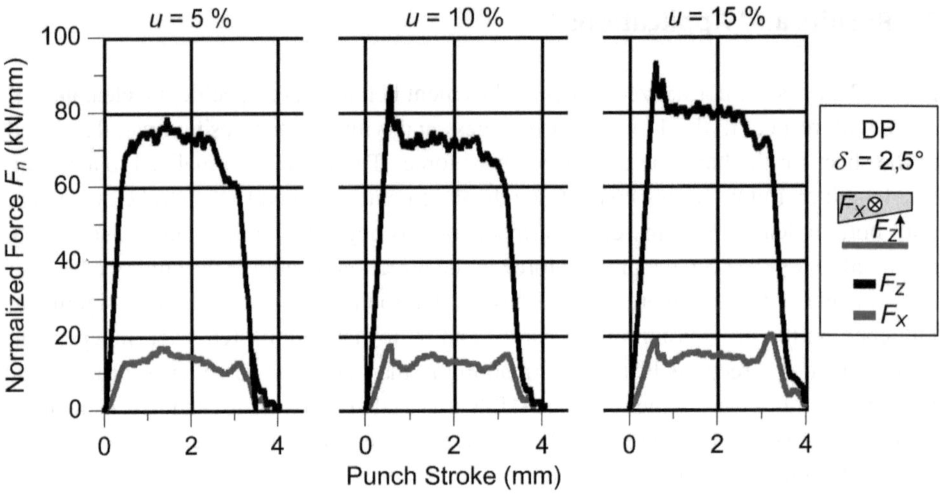

Fig. 3 Selected DP steel grade normalized force-displacement diagrams at different specific die clearances and a constant shear angle of 2.5°

a dominant tendency cannot be identified. An increase of the specific die clearance leads to a characteristic peak formation of the blanking force at the beginning of the cutting operation, which is due to the enhanced blanking force necessary until crack initiation. In contrast, the transverse force also presents a characteristic temporary increase at the end of the cutting process, when the part tears apart. This phenomenon is based on the augmented bending impact along the cutting edge during the cutting operation. This leads to an increased transverse load at the final cutting stage. In this context, the force peaks increase slightly with increasing specific die clearance. The results illustrated in Fig. 3 are representative of the AHSS materials examined, which show that fundamentally different force levels are required depending on the material strength.

As each one of the directional force components, namely F_x, F_y, and F_z, are determined throughout the stroke, it is possible to calculate the resultant force vector f and its magnitude with respect to the specific time step, sheet metal, and specific parameter set. From this, the value of the maximum load impact during the cutting operation can be derived. Figure 4 shows these maximum magnitudes, summarizing the different experimental investigations performed by means of the normalized force. In this context, the values plotted on the vertical axis are normalized using the maximum value of the force magnitude derived from all material specific experiments, which is generally at 5 % specific die clearance, a punch position parallel-edged to the sheet and a sheet position angle of 0°. It can be seen that the application of a shear angle induces a significant reduction of the resultant cutting force, whereas the force reduction is more distinctive for the AHSS materials compared to the DC04 steel grade. In addition, varying the sheet-positioning angle leads to a reduced maximum cutting force, which is due to the superimposed bending impact during the cutting operation.

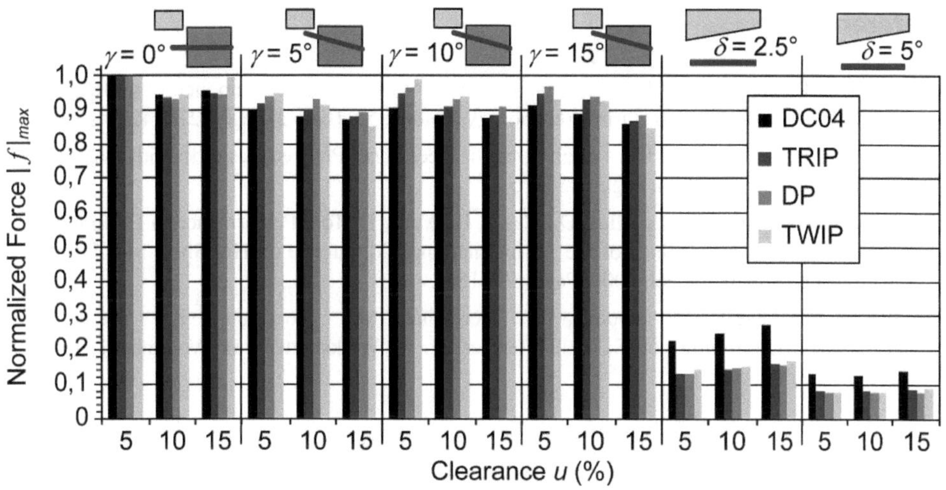

Fig. 4 Normalized force vector f depending on the investigated process parameters

4 Potential

Now it is possible to reduce blanking and shear forces purposively even for AHSS materials to prolong tool life in an industrial setting. In addition, existing presses that were procured for the processing of conventional steel grades may still be used for the processing of AHSS materials.

5 Conclusion

The results of the experimental investigations show that in particular using a shear angle of 2.5 % or 5 % respectively can reduce the blanking and shear forces up to 90 %, especially for AHSS materials. In opposite to that the influence of the specific die clearance on the blanking and shear forces is circumstantial.

6 Acknowledgement

The work has been kindly supported by the Bayerische Forschungsstiftung.

References

1. Kleiner, M.; Geiger, M.; Klaus, A.: Manufacturing of Ligthweigth Components by Metal Forming. CIRP AnnalenManufacturing Technology, Volume 52, Issue 2, Page 521–542
2. Hogman, B.: Blanking of Ultra High Strength Steel Sheet.Proceedings of the 6th International Tooling Conference, 2002, Page 203–216
3. Picas, I.; Hernandez R.; Casellas, D.; Casas, B.; Valls, I.: Tool Performance in Cutting of Hot Stamped Steels. Proceedings of the 1st International Conference on Hot Sheet Metal Forming of High-Performance Steel, 2008, Page 179–189
4. Weck, M.; Krell, M.; Doege, E.; Derenthal, M.; Grossmann, K.; Wiemer, H.: Schnittschlagdämpfung von Schnellläuferpressen. wt Werkstatttechnik, Volume 91, Issue 9, 2001, Page 540–544

Part VI
Rolling

Flexible Rolling

Markus Grüber, Reiner Kopp, and Gerhard Hirt

1 Motivation

In order to implement the idea of light-weight design by using load-aligned structural parts it is obviously necessary to manufacture the required semi-finished products with the desired thickness distribution. The basic idea of adapting a conventional rolling process for this purpose has been thought of a long time ago and has been described in a patent of Mannesmann in 1891 [1]. Nonetheless, an industrial applicable process design for the production of sheets and strips with a variable thickness did not come to reality before the past years. In 1989 the DFG granted a respective research proposal of the Institute of Metal Forming (IBF) [2]. Thereafter, the successful implementation of a continuous rolling process for the manufacturing of semi-finished products with a defined thickness distribution as well as a process chain for further processing of those sheets and strips have been developed at the IBF with kind support of the DFG and the BMBF [3].

This modified rolling process is classified according to Allwood et al. [4] as a flexible rolling process wherein the roll gap may be varied through the process. An alternative production process for strips and sheets with a variable thickness is the welding of so called tailor welded blanks (TWBs). Those TWBs exhibit the drawback of welded seams which define a weak spot during forming operations applied to sheets and strips. Furthermore, the welding of TWBs is a discontinuous process making flexible rolling the more efficient choice by means of mechanical properties and process control [5]. Despite this drawback of TWBs, it is a valuable option for further processing as it allows for a joining of different steel grades and even different TRBs. Within this general survey on the flexible rolling process, the basic process principle, achieved results and potentials are outlined.

Markus Grüber · Reiner Kopp · Gerhard Hirt ✉
RWTH Aachen University, Aachen, Germany
e-mail: hirt@ibf.rwth-aachen.de

A. E. Tekkaya et al. (eds.), *60 Excellent Inventions in Metal Forming*,
DOI 10.1007/978-3-662-46312-3_33

2 Process Idea

The basic idea of the flexible rolling process is a modification of a conventional cold rolling process. During the rolling, the roll gap is adjusted in order to achieve the desired thickness profile in rolling direction (see Fig. 1). Thereby, a material flow in rolling direction is achieved and smooth thickness reductions in longitudinal direction can be produced. These homogeneous thickness reductions outreach comparable tailored blanks produced by joining operations as no welding seams are necessary. Therefore, no material weakening occurs as the flexible rolled sheet still shows a persistent microstructure without a welding seam throughout the whole sheet [6].

Subsequently to the rolling process, the strip is separated into sheets with a respective thickness distribution. Those sheets can be used in further production steps such as deep drawing or stretch forming in order to manufacture load-aligned structural parts.

The adaption of the roll gap during the flexible rolling requires a process control in order to ensure that the desired thickness distribution is achieved. Prior to the process, a first set of control data is generated according to the desired thickness distribution of the final metal strip. As the roll gap and thus the thickness reduction vary throughout the process, the contact length between roll and strip is not constant. In particular, an increase of the roll gap leads to a reduced contact length, while a decrease of the roll gap results in an increasing contact length. Both, an increase as well as a decrease of the roll gap, complicate an analytic description of the process. In consequence, the primarily calculated control data does not necessarily lead to the desired thickness distribution. To optimize the control data, a monitor control is implemented at the outlet of the rolling stand. This monitor control compares the measured thickness profile of the strip to the desired thickness profile and optimizes the control data for the roll gap motion. Therefore, a certain number of profiles are measured at the outlet and the control data are iteratively adapted until the measured thickness distribution agrees with the desired one. Within the

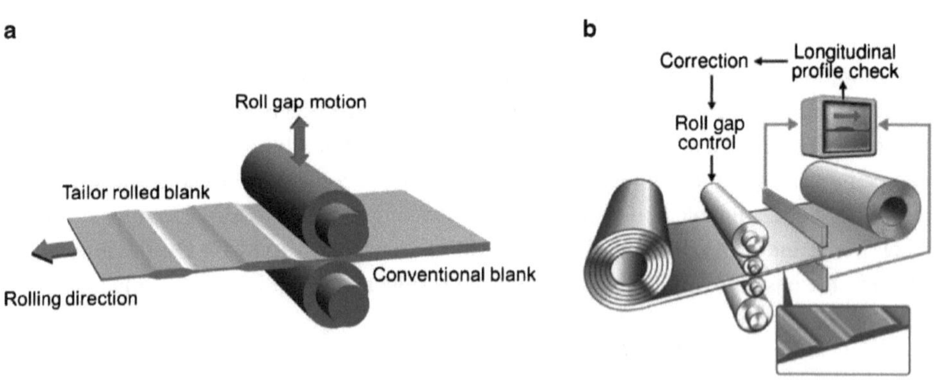

Fig. 1 Schematic view of the process principle for flexible rolling (**a**) [7] and the respective process control (**b**) [8]

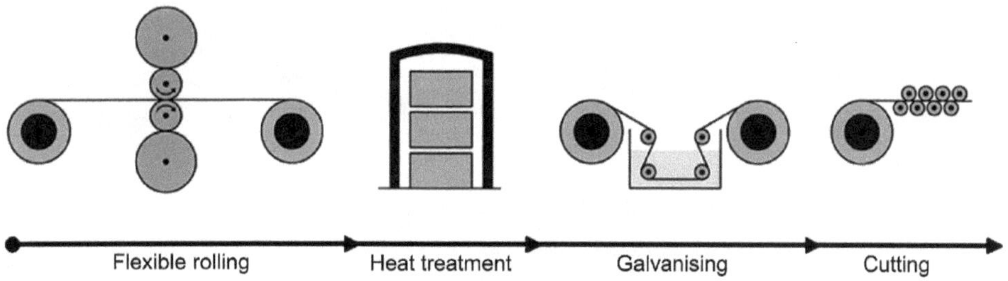

Fig. 2 Process chain for the production of TRBs [8]

optimization-loop the mill stand characteristic is implemented in order to take the machine specific behavior – particularly the elastic deformation of the rolls and the mill stand – into account. A schematic view of the optimization-loop setup is shown in Fig. 1 and illustrates the afore described process control [6].

A general overview of the production chain for TRBs is depicted in Fig. 2. After the flexible rolling process the strips are annealed in a batch-type annealing furnace to homogenize the microstructure of the material after rolling. Subsequently, the surface is galvanized for means of corrosion protection before the final cutting of the strip into individual sheets [9].

3 Results and Applications

The subsequent processes, such as deep drawing and stretch forming, that use TRBs as a starting material, have been investigated and realized successfully. In comparison to conventional blanks, the application of TRBs offers the potential to significantly reduce the weight of the final structural parts in a range of approx. 20 %. As shown in Fig. 3 the

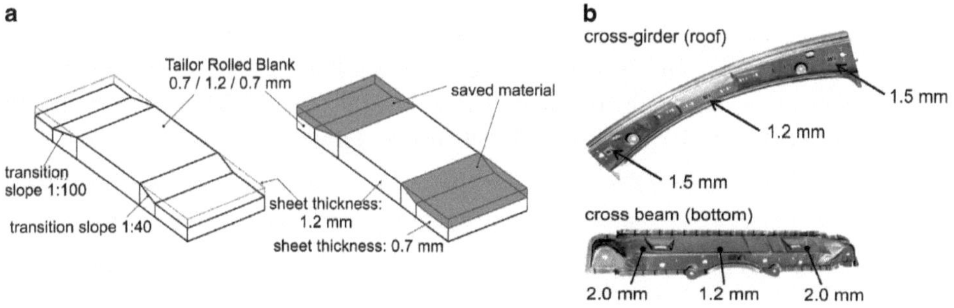

Fig. 3 Schematic view of feasible thickness distributions (**a**) [9] and product examples from automotive applications (**b**) [8, 11]

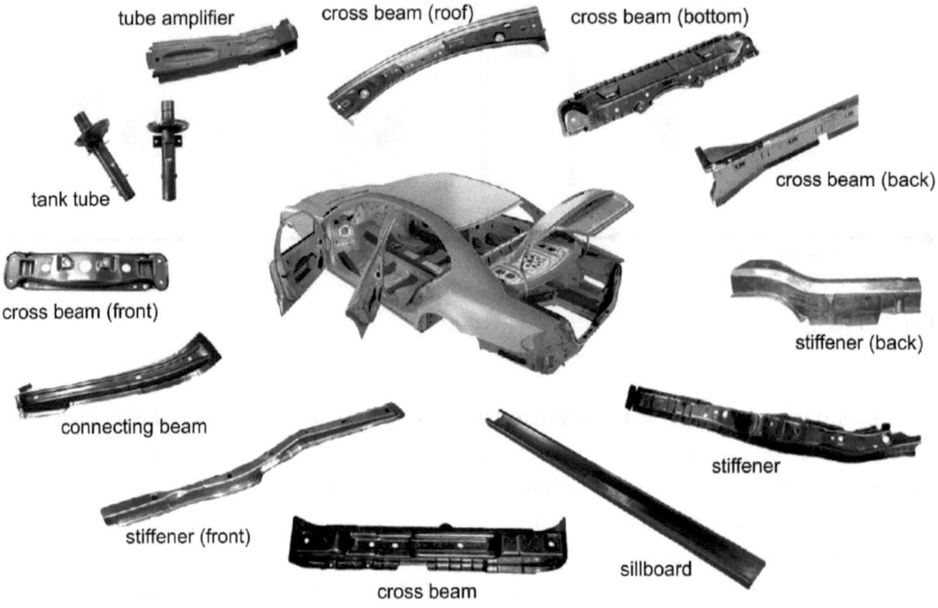

Fig. 4 Car body parts made of TRBs by Mubea [9]

semi-finished sheets can be produced with different transition slopes between the thickness changes allowing for an optimized further processing. To overcome the issue of wrinkling in the deep drawing process of sheets with a non-constant thickness, modified tools with flexible blank holders can be used [9, 10].

Flexible rolling is nowadays a state-of-the-art technology and is used on an industrial scale for the production of load-aligned semi-finished products. The main application of these semi-finished materials is the manufacturing of load-aligned structural parts for the automotive industry. To underline the significance of the flexible rolling, an overview of already produced car parts is given in Fig. 4. The variety of these examples from automotive applications clearly illustrates the great potential of load-aligned profiles

The core technology of this process has been patented in collaboration with Mubea in 2001 [12]. Mubea actually employs the flexible rolling process at the production plants in Attendorn, Weißensee (Germany) as well as Florence (USA) and produces up to 300,000 tons of semi-finished strip and sheet material per year.

4 Potential

Flexible rolling is a promising production process for manufacturing semi-finished products with a variable thickness and thus seizing a suggestion for light-weight design proposed by various authors like Kleiner et al. [13]. The main process potentials and advantages compared to alternate processes can be summarized as follows:

1. Flexible rolling is a continuous process enabling a large scale production of tailor rolled blanks with reduced production time.
2. In contrast to sheets with a thickness distribution manufactured by joining operations no welding seam interrupts the microstructure of the part.
3. The modification of the control data for the rolling process allows for a wide range of slopes in the thickness transitions.
4. Investigations regarding the use of TRBs in subsequent processes for the manufacturing of load-aligned structural parts clearly showed the possibility to reduce the weight of the respective final part up to 20 %.

5 Conclusion

The process of flexible rolling has been invented in order to produce semi-finished products with a varying thickness distribution in rolling direction. Thereby, semi-finished products with a load-aligned thickness profile are successfully produced. The further processing of TRBs has been verified in several investigations regarding production steps like deep drawing and stretch forming. A significant weight reduction of the final structural parts in comparison to the use of conventional blanks was achieved. Additionally, TRBs have smooth thickness transitions which are beneficial compared to the welding seams of TWBs. Nowadays, flexible rolling is used on a large industrial scale by the company Mubea. This shows the high demand for enhanced forming processes that can provide semi-finished products with optimized thickness distributions and the accompanying weight reduction.

6 Acknowledgement

The presented research on the development of the flexible rolling process has been funded by the "Deutsche Forschungsgemeinschaft" (DFG) and the German Federal Ministry of Education and Research (BMBF) within several projects. Furthermore, the company Mubea has kindly supported the research, in particularly by the production and supply of tailor rolled blanks. Therefore, the authors would like to acknowledge the received funding and technical support of all project partners.

References

1. Mannesmann, M., 1891, Verfahren und Walzwerk zum Formen und Kalibrieren von stabförmigen Körpern und Platten mit pilgerschrittförmigen Bewegungen des Werkstückes, Patentschrift 59052 Klasse 49 patentiert im Deutschen Reiche.
2. Hachmann, B., Kopp, R., 1992, Walzen belastungsoptimierter Bauteile, 7. Aachener Stahlkolloquium Umformtechnik.

3. Hauger, A., Kopp, R., 1995, Kinematische Erzeugung von belastungsangepassten Langprodukten mit einem über die Länge variablen Querschnitt mittels Walzen, Abschlusskolloquium DFG-SPP Flex. UT Verlag Mainz, Aachen.
4. Allwood, J. M., Utsunomiya, H., 2006, A survey of flexible forming processes in Japan, Int. J. of Machine Tools and Manufacture 46 (15) 1939–1960.
5. Greisert, C.; Ebert, A.; Wiedner, C.; Kopp, R.; Wesemann, J., 2001, Forming Behaviour of Tailor Rolled Blanks, Proc. of the Second Global Symposium in Materials Processing and Manufacturing: Sheet Materials
6. Hauger, A., 2000, Flexibles Walzen als kontinuierlicher Fertigungsprozess für Tailor Rolled Blanks, Dissertation RWTH Aachen University.
7. Dávalos Julca, D., Förster, T., Wietbrock, B., Hirt, G., 2011, Lightweight Design by Flexible Rolling and Strip Profile Rolling, SRI (SE) 10th ICTP 117–122.
8. Kopp, R., Wiedner, C., Meyer, A., 2005, Flexibly Rolled Sheet Metal and Its Use in Sheet Metal Forming, AMR 6-8, 81–92.
9. Meyer, A., Wietbrock, B., Hirt, G., 2007, Increasing of the drawing depth using tailor rolled blanks – Numerical and experimental analysis, Int. J. of Machine Tools and Manufacture 48 (5) 522–531.
10. Urban, M., Krahn, M., Hirt, G., Kopp, R., 2006, Numerical research and optimisation of high pressure sheet metal forming of tailor rolled blanks, JMPT 177 (1–3) 360–363.
11. Van Putten, K., Hirt, G., Thome, M., Romans, T., Dávalos, D., Wietbrock, B., 2009, Cold rolling processes to functionalize semi-finished products, TMS Proc. TMS Aluminum Committee 1209–1214.
12. Hauger, A., Kopp, R., 2001, Verfahren zum flexiblen Walzen eines Metallbandes, EP 1074317A2.
13. Kleiner, M., Chatti, S., Klaus, A., 2006, Metal forming techniques for lightweight construction, JMPT 177 (1–3) 2–7.

Vertical Twin-Roll Strip Casting of Steel

Markus Daamen, Michele Vidoni, and Gerhard Hirt

1 Motivation

New materials such as advanced steel grades pose new challenges for the conventional continuous casting processes. An innovative method for producing these materials is vertical twin-roll strip casting. During this continuous process, the distance from the melt to the solidified hot strip is reduced to a minimum which has several advantages regarding CO_2 emission and energy demand [1]. Characterized by a near-net-shape strip production, it is possible to minimize the need for substantial secondary processing, as shown in Fig. 1a, b. In addition, the extremely rapid solidification of the melt leads to a very fine as-cast structure and, as a consequence, segregation phenomena only occur at wave

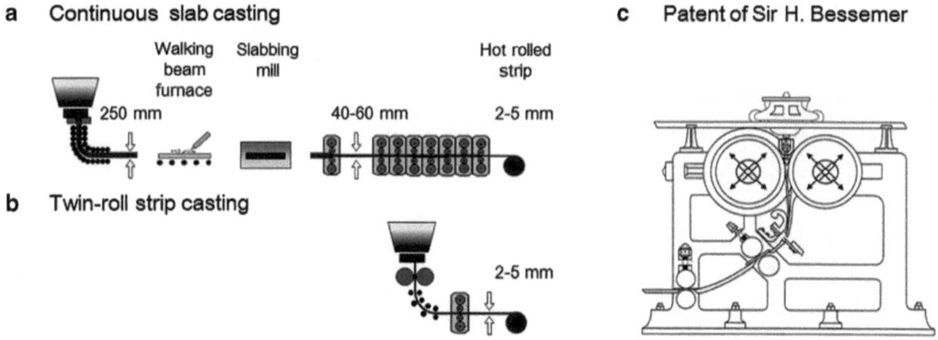

Fig. 1 Typical process layout of continuous slab casting (**a**) and twin-roll strip casting (**b**) for the production of steel hot strip [3] and extract of Bessemers patent [2] (**c**)

Markus Daamen · Michele Vidoni · Gerhard Hirt ✉
RWTH Aachen University, Aachen, Germany
e-mail: hirt@ibf.rwth-aachen.de

A. E. Tekkaya et al. (eds.), *60 Excellent Inventions in Metal Forming*,
DOI 10.1007/978-3-662-46312-3_34

lengths in the range of a few micrometres. In this manner, vertical twin-roll strip casting of steels promises an increased cost effectiveness, energy efficiency and environmental sustainability, especially for the production of highly alloyed steels [1].

2 Process Idea

In 1857, the inventor Sir Henry Bessemer had the vision to gain a thin metal strip directly from the melt and patented his idea of direct strip casting [2]. The basic idea is that the melt comes to the casting gap via a feeding system and solidifies in direct contact with the casting rolls to a solid strip. The two counter-rotating casting rolls form a moving mould without any relative movement between solidifying melt and mould.

In direct contact with the casting rolls, the melt solidifies on the surface of each casting roll to a solid strip shell. These growing strip shells are joined to one solid strip by means of the roll separating force F_{RSF}, as shown in the detailed view of the casting gap in Fig. 2b. In this context, the complexity of twin-roll strip casting becomes evident; casting speed, mass flow, heat transfer and F_{RSF} have to be carefully coordinated in order to enable the joining of the two solidified strip shells at the exit of the casting gap. Therefore, a highly sophisticated process control is of utmost importance. Slightest inaccuracies lead to defects in the thin metal strip such as micro-cracks on the surface, bad strip edges, shrinkage cavities in the strip centre or even an incomplete solidification of the melt.

However, at the time of Bessemers invention, the process could not successfully be established as at least the following prerequisites were missing:

- Materials for the casting rolls which can withstand the adverse conditions in contact with the liquid melt
- Process control systems and highly dynamic electric drives to stabilise the casting process
- Appropriate ceramics and refractory materials allowing to smoothly feed the liquid steel into the roll gap and to laterally seal the melt pool.

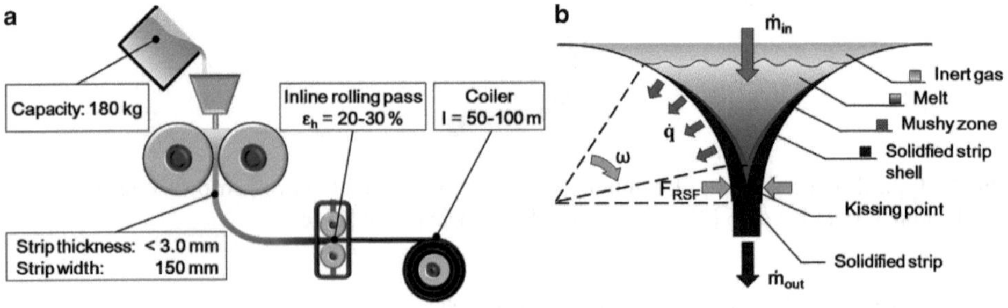

Fig. 2 Principle of the IBF twin-roll caster system (**a**) and detailed view of the solidification in the casting gap (**b**)

Around the 1960 s, these topics appeared solvable and in various places all over the world a lot of efforts and investigations had been undertaken to make use of this sophisticated process. Starting in 1988, ThyssenKrupp first developed a laboratory scale caster in cooperation with the Institute of Metal Forming IBF of RWTH Aachen University. This laboratory scale caster was later upscaled to an industrial caster with a ladle weight of 90 tons and a casting width of 1500 mm at ThyssenKrupp Nirosta GmbH in Krefeld [3]. Figure 2a schematically represents the vertical twin-roll caster at the IBF, which has all of the main units one would find in an industrial caster system. The two water-cooled casting rolls represent the heart of the casting system and are of approximately 580 mm in diameter and are 150 mm wide. The as-cast strip shows the corresponding width and as-cast strip thicknesses of 0.8–3.0 mm can be produced. To close occurring shrinkage cavities and to straighten the strip surface, a subsequent inline hot rolling at a temperature of approximately 1000°C is performed before the strip is coiled.

Recent developments regarding the process technology reveal the opportunities for the production of tailored strips directly from the melt. On the one hand, the alternating closing and opening of the casting gap enables a change in strip thickness along the strip length in flexible casting [4]. One the other hand, profiled hot strip with a defined thickness distribution across the strip width can be produced by using profiled casting rolls [5].

3 Results and Applications

Strip casting at the laboratory twin-roll caster system allows for producing a large variety of steel grades. By the selective setting of the caster system with regard to melting strategy, selection of refractory material, casting additives and an adapted subsequent treatment, it is possible to cast simple carbon steels, commercial stainless steels and even highly alloyed special steels. Currently, strip casting of high manganese TWIP steels with a significant amount of aluminium is under investigation. These highly alloyed steel grades offer superior material behaviour on the one hand but on the other hand cause serious problems in the conventional continuous casting processes which are not solved up to now.

It was demonstrated that strip casting enables the production of TWIP steels in high quality and the produced hot strip was further processed to a recrystallized cold strip with homogeneous element and grain distribution [6]. Tensile tests show the extraordinary mechanical properties with an ultimate tensile strength up to 950 MPa at a uniform strain of 60 % for a recrystallized Fe-17Mn-0.6C-1.5Al TWIP steel. In addition, application-oriented dynamic crash tests reveal the excellent energy absorption capability of these high manganese steels. Due to an intelligent process design using the characteristic strain hardening behaviour, the crash behaviour could be improved from the as-cast strip to a 15 % cold rolled strip, as shown in Fig. 3 [6]. Thus, it is possible to produce crash-boxes with tailored properties in a significantly shortened and energy-saving process line.

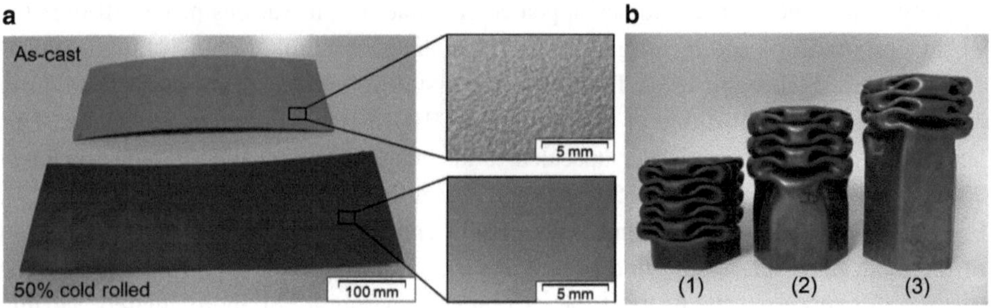

Fig. 3 As-cast strip and 50 % cold rolled strip of a high manganese steel with enlarged view of the strip surfaces (**a**) and hexagonal crash-samples of a Fe-17Mn-0.6C-1.5Al TWIP steel in different material states: (1) as-cast strip; (2) recrystallized strip; (3) 15 % cold rolled strip without heat treatment (**b**)

In the field of process development, the production of tailored strips with a defined thickness distribution directly from the melt has to be mentioned as current innovation. Besides tailored strips with a thickness profile over the strip length, the so called profile casting allows the production of hot strip with thickness transitions over the 150 mm wide strip width, as shown in Fig. 4a. For this purpose, a profile is machined in the surface of the casting rolls. During the casting process, the melt completely fills this profile which gives the final contour to the strip. A numerical model was developed to compute the arising temperature gradients depending on the profile geometry to subsequently choose an appropriate geometry for experimental tests. Using a stainless steel as an example, it was possible to achieve maximum thickness transitions of 1.5 to 2.5 mm over the cross section, corresponding to a difference in strip thickness of 60 % [5]. The tailored strips are then finished by the inline hot rolling pass with profiled hot rolls. Figure 4b presents a

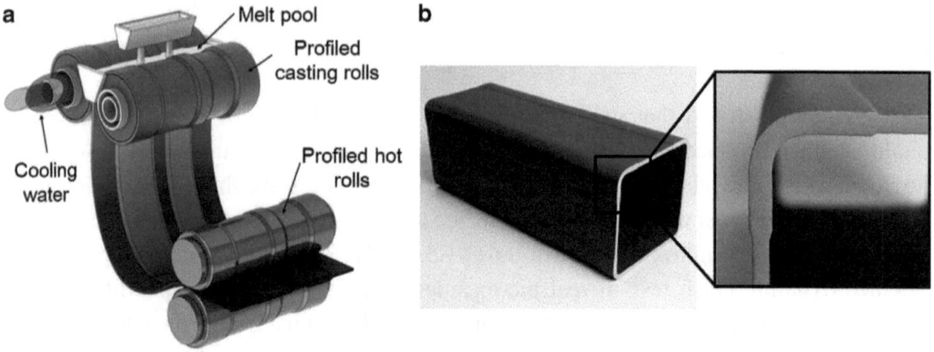

Fig. 4 Principle of profile casting for the production of tailored hot strip (**a**) and tailored profile with thick edges (3 mm) and thin sides (2 mm) produced by laser-welding of two bended tailored strips (**b**)

rectangular profile consisting of two laser-welded tailored strips produced by strip casting. The profile is characterized by thick edges and thin sides. Finally, profile casting enables the production of hot strip with complex strip geometries directly from the melt and thus considerably accelerates the production of tailored strips.

4 Potential

Vertical twin-roll strip casting as near-net-shape casting technology shows great potential for an energy efficient production of steel strip:

1. The production of advanced high strength steels is simplified by small as-cast thicknesses and an inline hot rolling step. Therefore, the expensive post-processing with re-heating and several rolling passes of a conventional production can be minimized and a cost and energy efficient hot strip production may become possible.
2. A fine dendritic as-cast structure is formed by the enormous cooling rates and hence negative segregation effects in highly alloyed steels are significantly reduced.
3. Furthermore, the rapid solidification may allow a nanocrystalline or even an amorphous microstructure, thus enabling the continuous production of amorphous steels or metallic glasses.
4. By an appropriate process design as used in flexible casting and profile casting, tailored strips with defined thickness distribution can be produced in one single process without any subsequent cold rolling or welding operation.
5. Currently, potentials for cladded steel strips by means of vertical strip casting are under investigation. A solid strip is introduced in the melt pool, using the high temperatures of the melt for joining two different materials.

5 Conclusion

Vertical twin-roll strip casting is an innovative process which has great potential for the near-net-shape production of steel. Hot strip is produced directly from the melt in a short process. Various steel grades such as highly alloyed, advanced TWIP steels can be produced in high quality with extraordinary mechanical properties in tensile tests and dynamic crash tests. By using profiled casting rolls, it is even possible to produce tailored strips in one single step with complex geometries and significant differences in strip thickness. In this way, vertical strip casting shows enormous potential for an energy-efficient and environmental friendly production of advanced steels with tailored properties.

6 Acknowledgement

This process has been investigated in several projects. The work has been kindly supported by the German Research Foundation (DFG) under the grant number Hi790/23-1 and within the collaborative research centre (SFB) 761 "Stahl – ab initio" and by ThyssenKrupp Steel Europe AG.

References

1. Ferry M., 2006, Direct strip casting of metals and alloys. The Institute of Materials, Minerals & Mining/Woodhead Pub. and Maney Pub., Cambridge, UK.
2. Bessemer H., 1891, Über die Herstellung von endlosem Blech aus schmiedbarem Eisen und Stahl direkt aus dem flüssigen Metall. Stahl und Eisen 11, 921–926.
3. Ge S., Isac M., Guthrie R.I.W., 2013, Progress in strip casting technologies for steel; Technical developments. ISIJ International 53 (5), 729–742.
4. Hirt G., Klos W., Gräf G., Obieglo A., 2006, Twin-roll casting of strip with tailored thickness variation. Production Engineering. Research and Development 13 (2) 91–94.
5. Vidoni M., Daamen M., Gastreich J., Offermanns N., Hirt G., 2014, Hot rolling of AISI 304 tailored strips produced by twin roll strip casting. Production Engineering. Research and Development 8 (5) 619–626.
6. Daamen M., Güvenç O., Bambach M., Hirt G., 2014, Development of efficient production routes based on strip casting for advanced high strength steels for crash-relevant parts. CIRP Annals – Manufacturing Technology 63 (1) 265–268.

Pair Cross Type Rolling Mill for Hot Rolling

Shunji Omori, Hiroyuki Hino, Kanji Hayashi, and Hideaki Furumoto

1 Motivation

The shape and crown of rolled strips are important not only for the material quality but also for the yield of products and stable rolling. In order to raise the productivity and to save energy, or to achieve the low-volume multi lot production of high quality strips, the development of a new type mill for controlling the thickness distribution along the width direction and rolled strips shape for various size and materials has been strongly required. The thickness distribution along the width direction is a result of elastic deformation of mill and rolls under rolling. Many rolls with initial crown were used for a countermeasure to control thickness distribution and shape. This method is inefficient as many crowned rolls have to be prepared for various rolling conditions. The roll bending method was also developed and used in a practical operation. However, these conventional mill and method were not satisfactory in these respects. And so, the new mill with wide controllability of the rolled strips crown and shape was strongly required to be developed.

2 Process Idea

As shown in Fig. 1, three types are available for crossing of roll axes. The well-known system are the crossing of only the back-up roll axis (Fig. 1a) or only the work roll axis (Fig. 1b). However, these systems cause relative slip in the axis direction between the work roll and the back-up roll due to crossing. The slip in the axis direction causes excessive

Shunji Omori · Hiroyuki Hino
Formerly Mitsubishi Heavy Industries, Ltd., Hiroshima, Japan

Kanji Hayashi · Hideaki Furumoto ✉
Primetals Technologies Japan, Ltd., Hiroshima, Japan
e-mail: hideaki.furumoto@primetals.com

A. E. Tekkaya et al. (eds.), *60 Excellent Inventions in Metal Forming*,
DOI 10.1007/978-3-662-46312-3_35

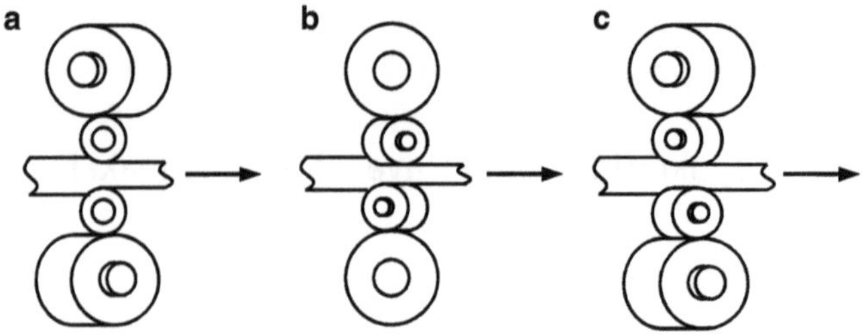

Fig. 1 Roll-crossing systems [5] (Copyright of Photograph: Mitsubishi Heavy Industries, Ltd.). **a** Back-up roll cross, **b** Work roll cross, **c** Pair rolls cross

thrust and power loss due to friction, as well as excessive roll wear. Therefore, the roll cross system has been used for production of paper and plastic only [1, 2] and not used for the steel rolling. As shown Fig. 1c, the PC mill features the pair cross system which crosses the upper and lower roll units with the work roll axis kept in parallel with the back-up roll axis [3, 5]. Therefore, it is a practical system which solves the problems mentioned above. The cross angle is defined as a half skew angle between the upper and

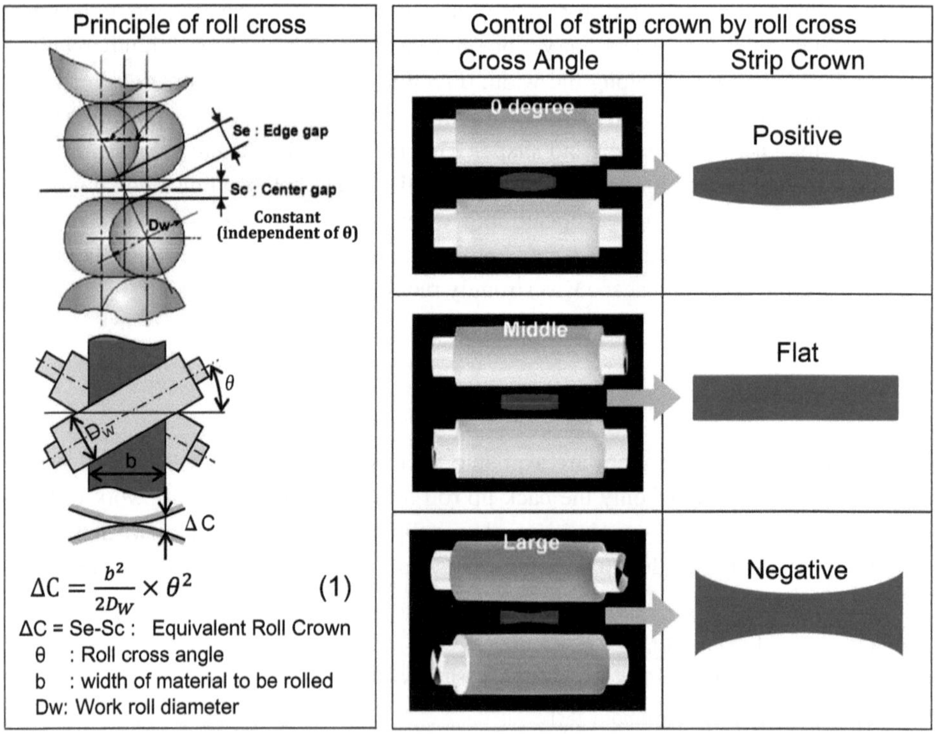

Fig. 2 Principle of control of rolled strips crown by PC mill (Copyright of Photograph: Primetals Technologies Japan, Ltd.)

lower roll units, and could be adjusted by the roll crossing system which skews the upper and lower roll units in a pair.

Figure 2 shows the principle of controllability of rolled strips crown by the PC mill. The roll gap formed by crossing of upper and lower roll units increases in the form of parabola as the distance from the center increases, and it is equivalent to convex roll crowning of the work roll. The equivalent roll crown ΔC is expressed by the Eq. (1) in Fig. 2. Therefore, adjustment of the cross angle enables rolled strips crown control. As compared with the roll bending system, the PC mill can adjust the roll gap without causing forced deflection of the work roll axis and provide a wide range of rolled strips crown control without restriction from the strength of the work roll. The PC mill shows excellent characteristics of controllability of rolled strips crown and shape for various thicknesses, width and materials.

3 Results and Applications

Figure 3 shows the equivalent roll crown ΔC which expresses roll gap change caused by cross angle θ. It is found that the wide roll gap controllability is obtained. Figure 4 shows a relationship between the rolled strips crown and cross angle in case of changing cross

Fig. 3 Equivalent roll crown [6] (Copyright of Photograph: Mitsubishi Heavy Industries, Ltd.)

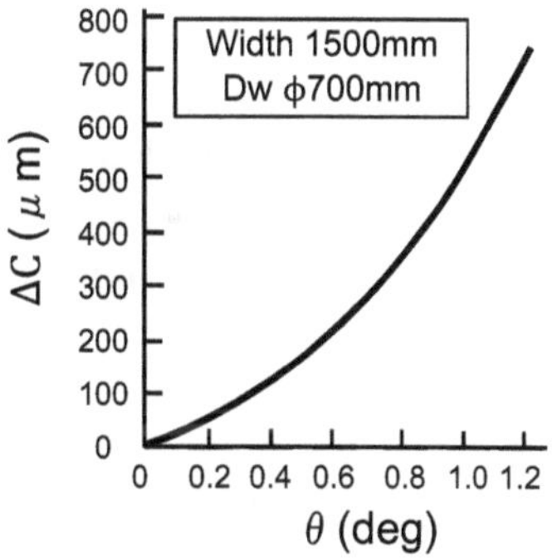

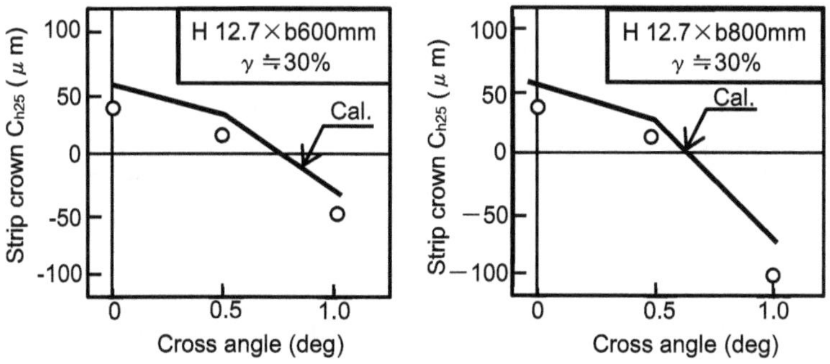

Fig. 4 Effect of cross angle on strip crown (hot single rolling) [5] (Copyright of Photograph: Mitsubishi Heavy Industries, Ltd.)

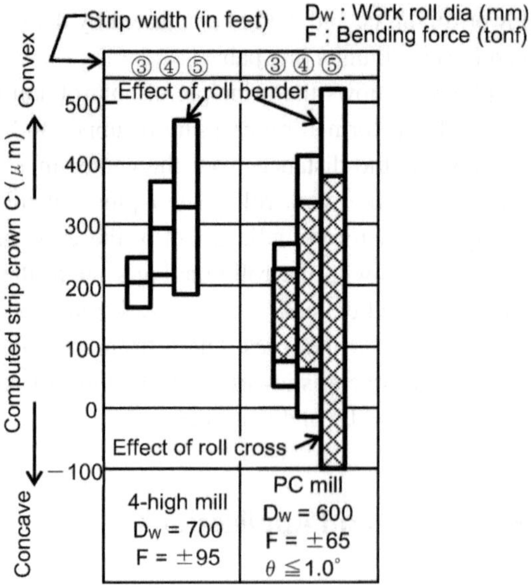

Fig. 5 Comparison of crown control range [5] (Copyright of Photograph: Mitsubishi Heavy Industries, Ltd.)

angle of single-pass rolling for steel strips using a 1/2-scale model mill. It is obvious from the figure that the larger strip crown controllability could be obtained [7]. Moreover, these results agree with the theoretical calculated value. Figure 5 shows the result of theoretical calculation of crown control range for a rolling mill which has an 1820 mm long roll-barrel. As compared with the normal 4-high mill equipped with roll benders only, it is found that the PC mill has much larger control range than that of 4-high mill with bender only. Figure 6 shows the transition of aimed strip crown and actual strip crown in a rolling schedule in which low-crown and high-crown materials are mixed together. It is found that

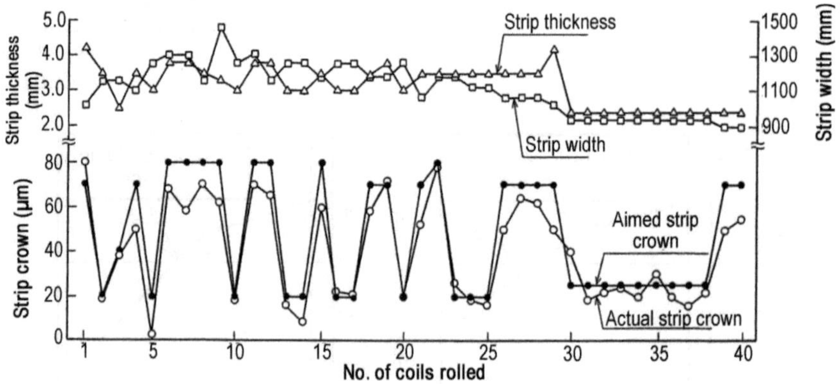

Fig. 6 Result of rolled strip crown in production line [6] (Copyright of Photograph: Mitsubishi Heavy Industries, Ltd.)

Fig. 7 PC mill installed at Nippon Steel Corporation's Hirohata works in 1984 [5] (Copyright of Photograph: Mitsubishi Heavy Industries, Ltd.)

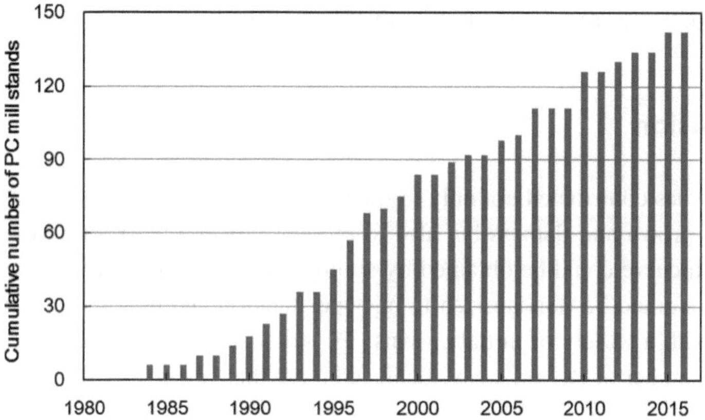

Fig. 8 Cumulative number of PC mill stands (Copyright of Photograph: Primetals Technologies Japan, Ltd.)

changes in actual strip crown precisely follow changes in aimed strip crown and material can be freely rolled to any crown.

Figure 7 shows the first PC mills installed at Nippon Steel Corporation's Hirohata works (now, Nippon Steel & Sumitomo Metal Corporation) in 1984, and Fig. 8 shows a cumulative number of PC mills installed in production lines. A cumulative number of PC mills have been increasing and are widely used in the world. They have achieved more than about 140 stands (including for cold rolling mills).

4 Potential

The PC mill has been developed [8, 9] and it contributes to the improvement of productivity, saving energy and achievement of high accuracy and quality of rolled strips crown and shape·

1. The desirable roll crown could be obtained by simply setting a suitable cross angle. The PC mill could provide a wide range of the rolled strips crown control, thus the schedule of rolling, for example thickness, width or materials, is free from constraints. Therefore, roll curve could be standardized to one specification. There is no need to prepare work rolls with a variety of initial crowns.
2. The PC mill enables the high reduction rolling because of wide controllability of rolled strips crown, and the controllability increases when the rolled strip width becomes larger.
3. The changes in the actual strip crown precisely follow changes in the aimed strip crown and even immediately after roll changing, materials could be freely rolled to any strip crown.
4. The PC mill has been also put into a practical use for plate rolling, and could reduce a number of passes and greatly reduce the plate crown.

5 Conclusion

The PC mill has contributed not only to the achievement of high accuracy and quality of rolled strips crown and flatness but also to productivity and workability of rolling. The desirable roll curve could be obtained by simply setting a suitable cross angle, therefore the roll curve could be standardized to one specification and there is no need to prepare work rolls with a variety of initial crowns. PC mills are widely used in the world and a cumulative number of PC mills have achieved more than about 140 stands.

6 Acknowledgement

The PC mill has been developed in collaboration with Nippon Steel Corporation (now, Nippon Steel & Sumitomo Metal Corporation) and Mitsubishi Heavy Industries, Ltd.

Reference

1. A.R.E. Singer & J.J.O'Brien, JISI (December.1962) p. 1003
2. M.D.Stone & R.Gray, Iron Steel Engr. (August.1965) p. 73
3. Japanese registered utility model No. 1717293 (1978)
4. Hiromi MATSUMOTO, Hidehiko TSUKAMOTO, Seiyo HATAE, Hiroyuki HINO: Development of a Pair Cross Mill for Crown Control of Hot Strips, Proceedings of Advanced Technology of Plasticity Vol. II (1984) p1372–1377
5. Koei Nakajima, Shunji Omori, Toshiharu Kawamoto, Hidehiko Tsukamoto, Seiryo Hatae, Hiroyuki Hino, Hiroshi Aratani: Basic Characteristics of Pair Cross Mill, TECHNICAL REVIEW MITSUBISHI HEAVY INDUSTRIES Vol. 22 N0.2 (1985) p. 143–148

6. Tsuneo Nakano, Ryuichi Ozono, Hiroshi Aratani, Hidehiko Tsukamoto, Kazuo Morimoto: Expansion of PC MILL Application to Plate Mill and Tandem Cold Mill, TECHNICAL REVIEW MITSUBISHI HEAVY INDUSTRIES Vol. 29 N0.3 (1992) p. 159–164
7. Shunji Omori, Hidehiko Tsukamoto, Hiroyuki Hino, Koue Nakajima, Kichi Yoshizawa: Rolling Characteristics of Pair Cross Mill, Journal of the JSTP Vol. 28 no.321 (1987) p. 1067–1074
8. Sunji Omori, Tetsuo Kajiwara, Yoshiharu Taura, Shinji Sadahiro, Hiromi Matsumoto: Analysis of Rolling Load Characteristics of Pair Cross Rolling Mill, Journal of the JSTP Vol. 28 no.323 (1987) p. 1259–1266
9. Shunji Omori, Kanji Hayashi, Hiroshi Aratani, Takao Kawanami, Shigeru Ogawa, Yuji Uebori: Friction and Wear Between Rolls of Pair Cross Rolling Mill, Journal of the JSTP Vol. 29 no.327 (1988) p. 349–355

Endless Hot Strip Rolling

Kanji Hayashi, Hideyuki Nikaido, and Hideaki Furumoto

1 Motivation

The conventional batch rolling process has following problems. (1) Rolled strips thickness and crown or strip quality of head and tail ends region are changed because tension is not applied at these regions. (2) Pincher defects might be occurred when tail ends of rolled strips deliver the last finishing mill stand. This reduces productivity and requires work roll changing due to work roll damage. (3) When thin rolled strips are produced, rolled strips might jump on the hot run-out table in high speed travel. The rolling speed of thin strips is restricted due to this jumping, and the desired productivity or quality of rolled strips could not be obtained. And so, the completely continuous finish rolling has been strongly required.

2 Process Idea

Figure 1 shows the equipment and key technologies for endless rolling [1]. Rolled bars with roughing mills (R_1, R_2 and R_3) are transferred continuously and the sheet bar coiler winds rolled bars. It supplies transfer bars continuously for endless rolling. It has a buffer function which is absorbing the time interval between bars rolled with the roughing mill. The crop shear, installed between sheet bar coiler and joining machine, is used in order to cut the tail end of the preceding bar and head end of the following bar for being uniform along the width direction. The joining of bars is performed at entry side of finishing mills.

Kanji Hayashi · Hideaki Furumoto ✉
Primetals Technologies Japan, Ltd., Hiroshima, Japan
e-mail: hideaki.furumoto@primetals.com

Hideyuki Nikaido
JFE Steel Corporation, Tokyo, Japan

A. E. Tekkaya et al. (eds.), *60 Excellent Inventions in Metal Forming*,
DOI 10.1007/978-3-662-46312-3_36

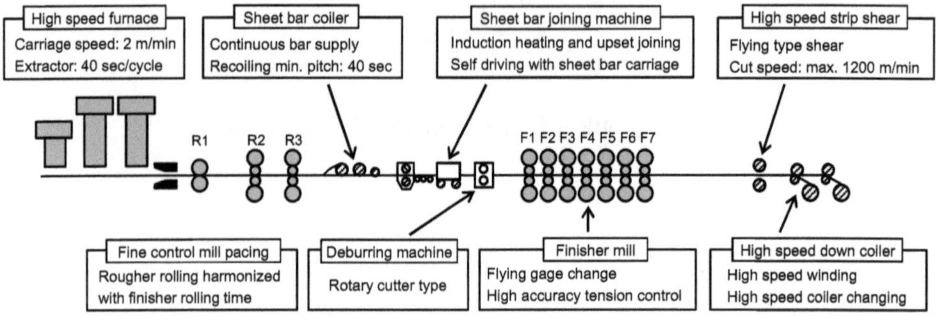

Fig. 1 Endless hot strip rolling equipments and key technologies [1] (Copyright of Photograph: JFE Steel Corporation)

Fig. 2 Principle for induction heating joining (Copyright of Photograph: Primetals Technologies Japan, Ltd.)

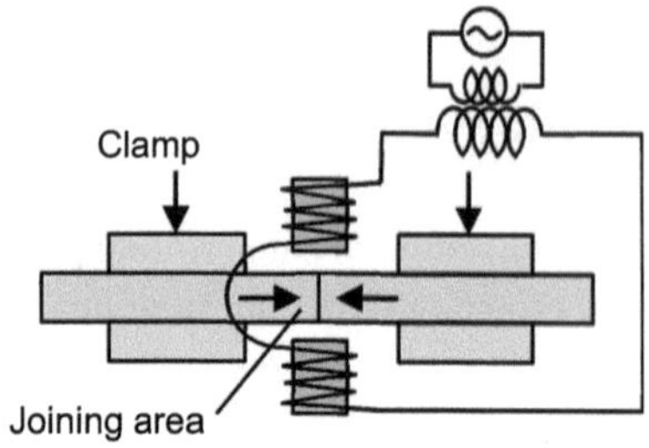

The tail end of preceding bar and head end of the following bar are joined by induction heating joining machine. The joining machine is self-driven and performs heating and upsetting of the bars within a 20 m traveling stroke. The elevator-type table rollers are installed within the traveling range of the joining machine in order to be traveling with the same speed of the bars. The joined bars are rolled with finishing mills continuously and rolled strips are cut and coiled into coils with the high speed strip shear and the high speed down coiler.

Figure 2 shows the principle of joining by induction heating [2, 3]. This is one of key technologies for endless rolling. When an alternating magnetic field is impressed in the thickness direction, an induced current concentrated at the end surfaces of the joint flows through the material and generate Joule heat which raises the temperature of the joint surfaces. The heated surfaces are upset to complete the joining operation. The joining is performed under semi solid condition (solid fraction about 70 %). Figure 3 shows the sheet bar joining machine. It is equipped with an induction heater, clamp frame, upsetting device, drive motor and so on. A centering guide is also provided to ensure that there is no misalignment of the preceding and following sheet bars [4].

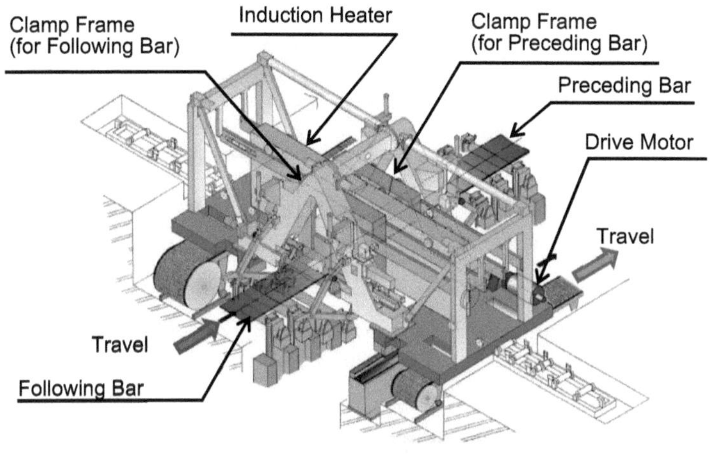

Fig. 3 Sheet bar joining machine (Copyright of Photograph: Mitsubishi-Hitachi Metals Machinery, Inc.)

3 Results and Applications

Figure 4 shows the tensile strength distribution of the joint. It is clear that the joint strength is almost the same as the material strength.

Figure 5 shows the strip thickness accuracy in endless rolling, including 1.26 mm rolled strips. Flying gauge changes were performed at the joints between the first and second and between 9th and 10th coils. A thickness accuracy of $\pm 25\,\mu m$ was achieved continuously

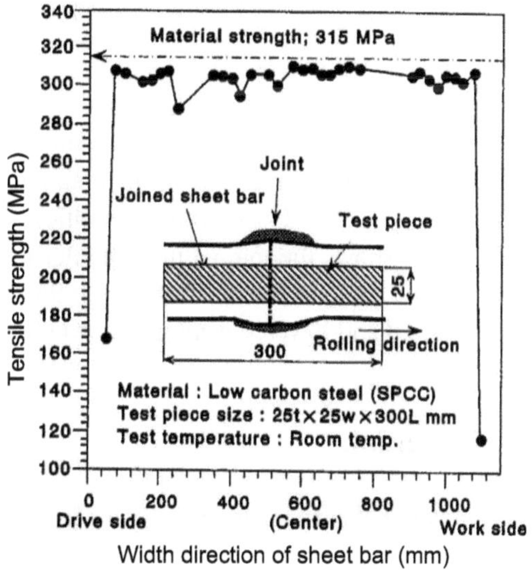

Fig. 4 Tensile strength distribution of the joint [1] (Copyright of Photograph: JFE Steel Corporation)

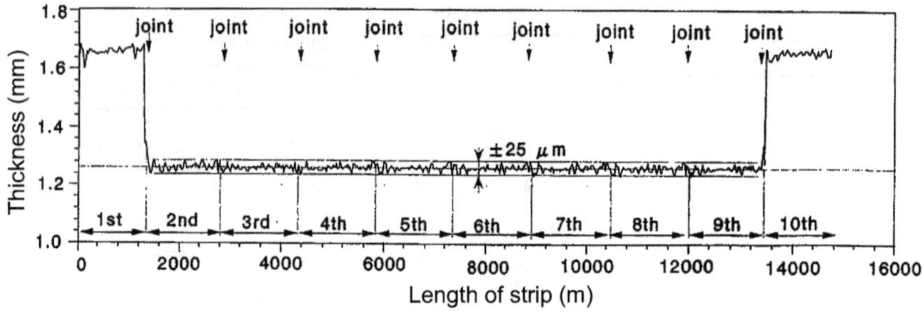

Fig. 5 Strip thickness chart of endless hot strip rolling measured by X-ray thickness meter at F7 delivery side (strip thickness: 1.66 → 1.26 → 1.26 → 1.66 mm)[1] (Copyright of Photograph: JFE Steel Corporation)

Fig. 6 Induction Heating Joining Machine in Operation (Copyright of Photograph: Primetals Technologies Japan, Ltd.)

through rolled coils, and the thickness deviations including flying gauge changes were extremely small. The flatness at the delivery side of the finishing mill for an endless rolling sequence of ten coils was good and no waving occurred in the endless hot-rolled strips during rolling. Good product qualities were obtained [5].

Figure 6 shows the joining machine during joining installed at Kawasaki Steel Chiba works (now, JFE Steel Corporation) in 1996. To realize flying joining, table rollers which can be raised and lowered were installed before and after the joining machine.

4 Potentials

The endless hot strip rolling offers numerous advantages. Because the tail end of the preceding bar and the head end of the following bar are joined, tension could be maintained over almost the entire length of the rolled strip and stable rolling under steady conditions could be achieved at all times. For these reasons, the endless hot strip rolling eliminates the problems of cambers at the head end and pinchers at the tail end. The following advantages are obtained.

1. Extremely slight changes of thickness and width over the entire length of rolled strips.
2. Little fluctuation of the coiling temperature over the entire length of rolled strips.
3. The productivity could be increased and a number of unexpected roll changes could be reduced due to the pinchers.
4. Shape rejects at the head and tail ends and surface defects due to pincher marks could be reduced.
5. New products could be manufactured in hot rolling.
   Ultra-thin strips with less than 1.0 mm thickness could be rolled. With the endless hot strip rolling, it has become possible to perform heavily lubricated and high reduction rolling over the entire length of rolled strips, and so high-formability rolled strips could be obtained.

5 Conclusion

The endless hot rolling process and joining machine have been developed [6, 7]. In this process, transfer bars are joined at the entry side of the finishing mills and then rolled continuously with finishing mills. The thinner, wider and harder steel sheets could be rolled under the completely continuous rolling, and so the productivity and quality of rolled strips could be extremely improved.

6 Acknowledgement

The endless hot rolling process and induction heating joining machine have been developed in collaboration with Kawasaki Steel Corporation (now, JFE Steel Corporation), Mitsubishi Heavy Industries, Ltd. and Mitsubishi Electric Corporation.

References

1. Hideyuki Nikaido, Shigeru Isoyama, Nobuaki Nomura, Kanji Hayashi, Kazuo Morimoto, Hideo Sakamoto: Endless Hot Strip Rolling at No. 3 Hot Strip Mill in Chiba Works, Kawasaki Steel's Technical Review, Vol. 28 No. 4 (1996) p27–32
2. Japanese registered patent No. 2905395 (1994)
3. Japanese registered patent No. 2905400 (1994)
4. Takeshi Takano, Keiji Matsuda, Shoichi Moriya, Nobuhiro Shibatomi, Yoshiki Mito, Kanji Hayashi: Endless hot strip rolling at Kawasaki Steel Chiba Works, Iron and Steel Engineer, February (1997) p. 41–46.

5. Fumiya Yanagishima: Development of Endless Rolling Technology for the Hot Strip Mill, Proceedings of the 7th International Conference on Steel Rolling, (1998) p. 717–726.
6. Hideyuki Nikaido: Journal of JSTP Vol. 40 No. 456 (1999) p. 2–7
7. Hideyuki Nikaido, Yasuo Ichii:The Birth of Endless Finishing Rolling in Hot Strip Mill, Ferrum, ISIJ, Vol. 2 No. 11 (1997) p. 824–829

6-High Type Rolling Mill for Cold Rolling

Toshiyuki Kajiwara, Hidetoshi Nishi, Yasutsugu Yoshimura, and Hideaki Furumoto

1 Motivation

It is required that a variety of materials with different thickness and width could be rolled in flat shape for improving productivity and saving energy. However, work roll deflection changes according to change of the rolling force, and change of the work roll deflection causes the worse shape of rolled strips. And so, the productivity and quality of rolled strips would be reduced. In order to be rolled in good shape, it is desirable to minimize the work roll deflection along the strip width that is caused by the rolling force, and to improve the strip flatness controllability as much as possible.

In case of a conventional 4-High mill, these problems are caused by undesirable contact area between work roll and back-up roll as shown in Fig. 1. Rolling force is applied to work roll at the area of the rolled strip width, and reaction force from back-up roll to work roll is also applied at the contact area between work roll and back-up roll. And so, it is thought that a larger work roll deflection occurs at the area of work roll which extends beyond the strip width.

Furthermore, the effect of work roll bending is diminished by the interference due to the contact between work roll and back-up roll. The effect of work roll bending means that work rolls are bended by the force applied to the end area of work roll and the distribution

Toshiyuki Kajiwara
Formerly Hitachi, Ltd., Hitachi, Japan

Hidetoshi Nishi
Formerly Mitsubishi-Hitachi Metals Machinery, Inc., Tokyo, Japan

Yasutsugu Yoshimura
Primetals Technologies Japan, Ltd., Tokyo, Japan

Hideaki Furumoto ✉
Primetals Technologies Japan, Ltd., Hiroshima, Japan
e-mail: hideaki.furumoto@primetals.com

A. E. Tekkaya et al. (eds.), *60 Excellent Inventions in Metal Forming*,
DOI 10.1007/978-3-662-46312-3_37

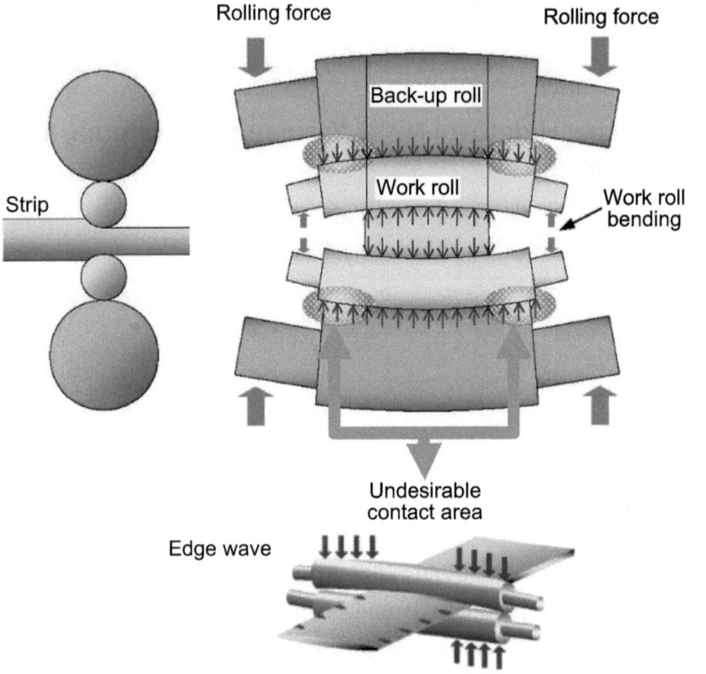

Fig. 1 Conventional 4-High mill and rolled strip shape (Copyright of Photograph: Primetals Technologies Japan, Ltd.)

of top and bottom work roll gap in the width direction could be changed. Accordingly, in actual operation of a 4-High mill, even if the work roll crown and the work roll bending method are used jointly, the range of rolling conditions to flatten the strip shape is greatly narrowed. Even with the same schedule, it is difficult to trace the changes in heat crown and in roll surface conditions which vary with rolling process, so that rolling of strips in insufficient flatness is unavoidable. In additions, to meet strip width or change of rolling force, several rolls with different initial crowns are required, and the frequency of roll changing and a number of spare rolls inevitably increase. And smaller work roll diameter is desirable for reduction of rolling force, but it is difficult because of the undesirable rolled strips shape. The work roll bending for controlling rolled strips shape is ineffective.

Many attempts to overcome these problems have been made, but with little real success.

2 Process Idea

As mentioned above, work roll deflection in a 4-High mill is increased caused by undesirable contact area between work roll and back-up roll. If undesirable contact area can be eliminated, it would be possible to reduce work roll deflection drastically. However, it is

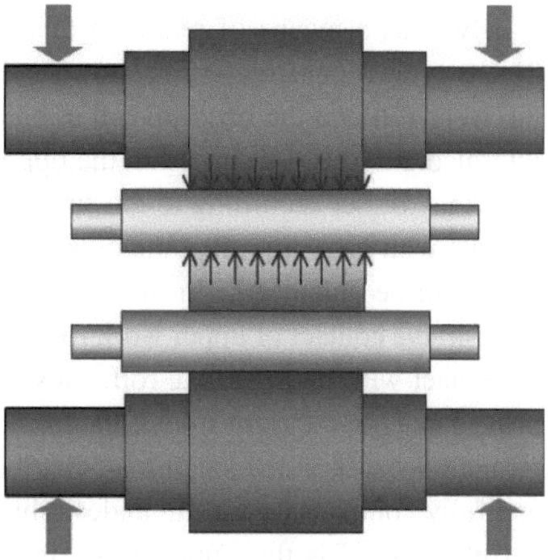

Fig. 2 Ideal 4-High mill (Copyright of Photograph: Primetals Technologies Japan, Ltd.)

not in feasible to adjust the effective contact length between work roll and back-up roll in accordance with strip width as shown in Fig. 2. In a practical operation, it is difficult to frequently exchange the back-up rolls with different barrel length according to different strip width.

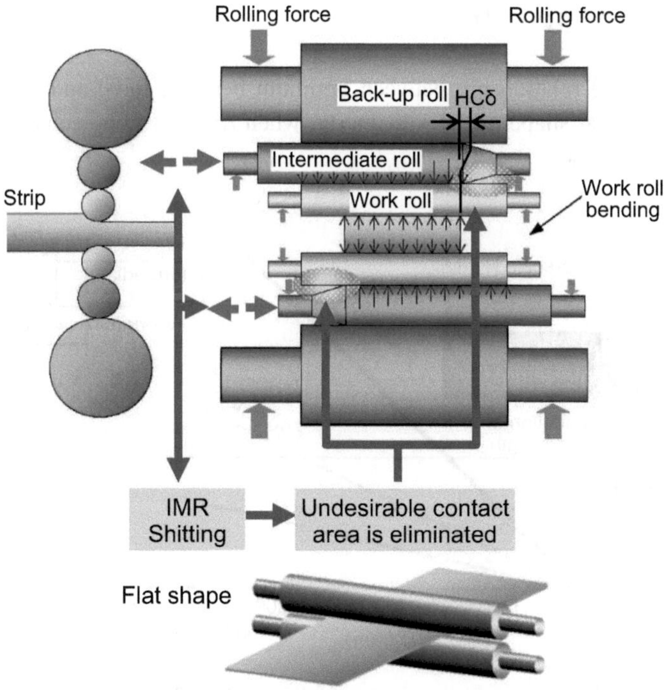

Fig. 3 HC-MILL (Copyright of Photograph: Primetals Technologies Japan, Ltd.)

It was as a means of realizing this adjustment that the HC-MILL was proposed [1, 2]. Its typical construction is shown in Fig. 3. Intermediate rolls are provided between the work roll and the back-up roll at both the top and bottom side. The upper and lower intermediate rolls are shifted in the opposite axial directions so that the intermediate roll ends may be set according to the strip width. In this way, the work roll is freed from the contact with back-up roll at the outside of strip width, and the effect of work roll bending force can be enhanced. In this structure, on the right side of roll barrel in Fig. 3, the undesirable contact of upper work roll can be eliminated, while the lower work roll is kept in contact with intermediate roll. However, it has been theoretically and experimentally proved that, on the right side of the lower roll, the undesirable bending at the contact area is small because of the small load distribution. It has also been shown that the upper rolls can fully compensate for the undesirable bending on the lower roll. The left side of roll barrel in Fig. 3 is the same as the right side, except that the top/bottom relationship is reversed.

From the above, it is found that the HC-MILL is capable of reducing the work roll deflection due to rolling force and increasing the work roll bending effect. The HC-MILL has both a good shape stabilizing capacity and good shape control capacity which have never been seen in a conventional 4-High mill.

3 Results and Applications

Figure 4 shows the relationship between optimum bending force and intermediate roll position HCδ for flat shape of the rolled strip, when rolling force is changed in the pro-

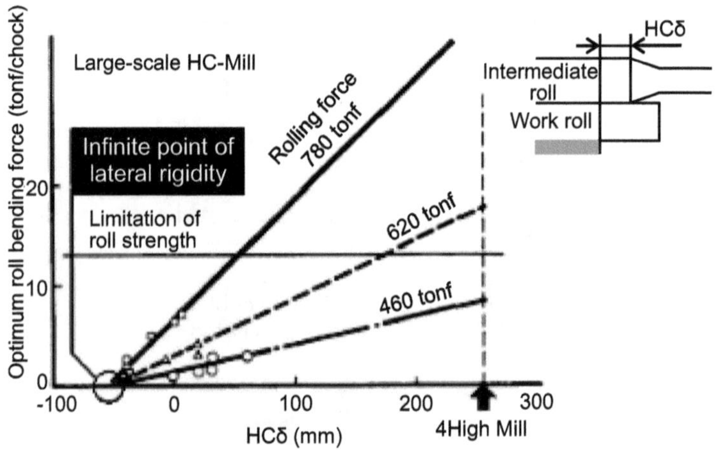

Fig. 4 Relationship between optimum bending force and HCδ [3] (Copyright of Photograph: Hitachi, Ltd)

Fig. 5 Photograph of HC-MILL in production line. (Copyright of Photograph: Primetals Technologies Japan, Ltd.)

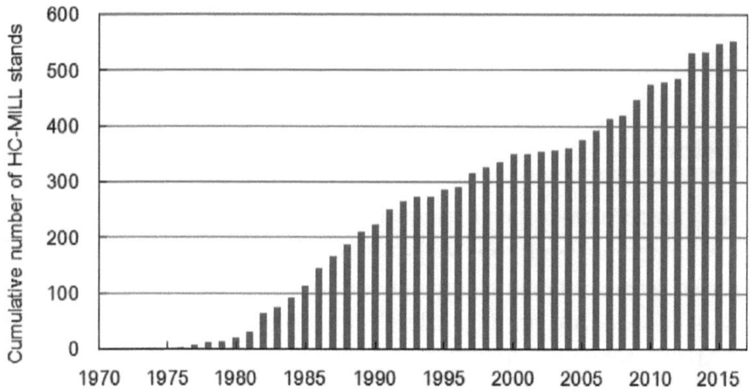

Fig. 6 Cumulative number of HC-MILL stands installed in production line. (Copyright of Photograph: Primetals Technologies Japan, Ltd.)

duction line. HCδ is defined as the distance between intermediate roll end and strip width end. Vertical axis represents optimum bending force which could correct the flatness of rolled strip. Smaller bending force could correct the flatness of rolled strip when HCδ is decreased. When bending force, which corrects the flatness of rolled strip, is more than strength limit of roll and is required, smaller bending force could correct flatness of rolled strip by adjusting intermediate roll position HCδ. It is found that the constant bending force, which corrects flatness of rolled strip, is obtained even with change of rolling force. This is so called in the condition that lateral rigidity is infinite.

Figure 5 shows the photograph of HC-MILL in a production line.

Figure 6 shows a cumulative number of HC-MILLs installed in production lines. A cumulative number of HC-MILL have been increasing and are widely used in the world. They have achieved more than about 550 stands (including for hot rolling mills).

4 Potential

The HC-MILL has been developed in order to improve controlling ability for the rolled strip shape even with changes in rolling conditions and work roll bending effect [4, 5]. The following results are obtained.

1. Undesirable contact area between work roll and back-up roll is eliminated and work roll deflection could be reduced by the HC-MILL.
2. Several rolls with different initial crowns for different width and change of rolling force are not required, and the frequency of roll changing and a number of spare rolls could be reduced.
3. Work roll diameter could be minimized, thus reduction in thickness is increased.
4. Controllability of work roll bending is enhanced because of elimination of undesirable contact area between work roll and back-up roll.
5. The HC-MILL could increase productivity of rolling and quality of rolled strips.

5 Conclusion

The HC-MILL has been developed for improving rolled strip flatness under various rolling conditions. A number of work rolls with different initial crowns and work roll diameter could be reduced because of elimination of undesirable contact area between work roll and back-up roll. And work roll bending effect is enhanced due to intermediate roll shifting system. HC-MILLs are widely used in the world and a cumulative number of HC-MILLs has achieved more than about 550 stands.

6 Acknowledgement

The HC-mill has developed in collaboration with Nippon Steel Corporation (now, Nippon Steel & Sumitomo Metal Corporation) and Hitachi, Ltd.

References

1. Toshiyuki Kajiwara, Nobuhiro Fujino, Hidetoshi Nishi, Shigeru Shida: Properties of Hitachi New Type Mill (HC-MILL), Hitachi Review Vol. 24 No. 8 (1975) p. 341–347
2. Japanese registered patent No. 952086 (1971)
3. Toshiyuki Kajiwara, Nobuhiro Fujino, Hidetoshi Nishi, Yasutsugu Yoshimura,Shigeru Shida: Properties of Large-Scale HC-MILL, THE HITACHI HYORON, Vol. 58 No. 9 (1976) p. 1–6
4. T. Kajiwara, N. Fujino, H. Nishi, S. Shida: The Hitachi HC-Mill – a breakthrough in strip rolling, Iron and Steel International, August (1976) p. 247–255
5. Takashi Furuya, Satoyuki Kitashima, Hideya Sato, Shigeru Hirai, Toshiyuki Kajiwara, Shigeru Shida, Hidetoshi Nishi: New design 6-h cold mill(HC mill)solves shape problems, Iron and Steel Engineer, August (1979) p. 40–45

Riblet Rolling on Ti6Al4V Compressor Blades

Michael Terhorst, Daniel Trauth, and Fritz Klocke

1 Motivation

For the coming years the Advisory Council on Aeronautics Research in Europe (ACARE) predicts an increase of the worldwide air traffic at an annual rate of four to five percent [1]. Therefore, ambitious goals are manifested in its strategic research agenda. The goals include a 50 % reduction of fuel consumption and CO_2 emissions as well as an 80 % reduction of nitrogen oxide emissions per passenger kilometer, both compared to 2008. In order to reach these goals new technologies have to be implemented within the aviation sector.

It has long been known that riblets parallel to the direction of flow can reduce friction drag in turbulent flow [2–4]. Within the field of bionics this effect has been observed when comparing the scales of different shark types. The scales of fast swimming sharks such as white and silky sharks were found to be built up of microscopic riblet structures which could not be proved on slow swimming species [5, 6]. The optimal riblet dimension for maximum drag reduction depends on the fluidic application. It, therefore, results from characteristic fluid properties such as the Reynolds number and the kinematic viscosity of the fluid. So far, riblet structures have either been applied through bonded riblet films or manufactured directly into the material surface by machining, laser structuring or large-area rolling. As for the application of riblets on compressor blades of jet engines the geometrical dimension of optimal riblets lays around the size of a single human hair (40 to 100 µm). Due to high operating temperatures compressor blades are typically made of Ti6Al4 V. The high operating temperatures lead to the non-suitability of riblet films. Besides high temperatures compressor blades are subjected to high dynamic loads such as bending, torsion and edgewise modes [7, 8]. Against this background, riblet structures

Michael Terhorst · Daniel Trauth · Fritz Klocke ✉
RWTH Aachen University, Aachen, Germany
e-mail: F.Klocke@wzl.rwth-aachen.de

A. E. Tekkaya et al. (eds.), *60 Excellent Inventions in Metal Forming*,
DOI 10.1007/978-3-662-46312-3_38

on surfaces are disadvantageous since they cause surface loads to be amplified due to the notch effect. The manufacturing of riblets through forming technologies has the major advantage that strain hardening and compressive residual stresses are induced into the workpiece. This partly compensates the load bearing capacity reduction resulting from the notch effect. Since Ti6Al4 V has a low formability the manufacturing of riblets through large-scale rolling would go along with a too high forming force. This motivates the manufacturing of defined riblet structures through incremental riblet rolling.

2 Process Idea

The process idea is the incremental rolling of riblets path by path as shown in Fig. 1a. The rolling paths are normal to the leading and trailing edges of the compressor blade. In order to avoid asymmetric plastic material flow and, thus, a non-uniform deformation of the blade as it would occur in case of single-sided incremental riblet rolling, a two-sided acting rolling tool was developed at the Laboratory for Machine Tools and Production Engineering (WZL) of the RWTH Aachen University. It allows for the simultaneous forming of riblets on suction and pressure sides of the compressor blade. The rolling force F is induced by a constant hydraulic pressure. The tool can be used in common 3-axis machining centers which allow the positioning of the tool as well as the generation of the rolling velocity v. The prototype is shown in Fig. 1b [9].

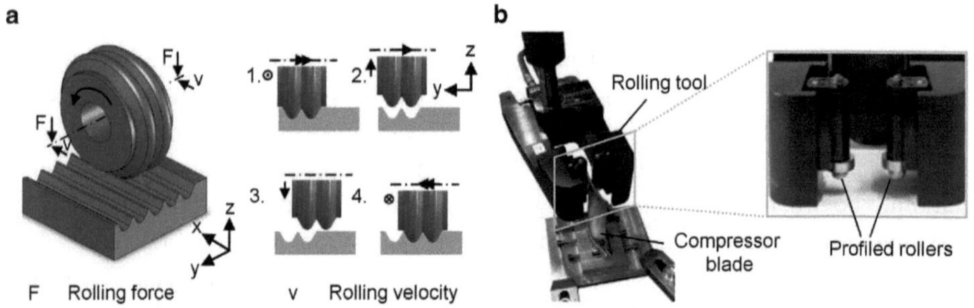

Fig. 1 Process principle (**a**) [10] and prototype tool (**b**) [9] of incremental riblet rolling

3 Results and Applications

Manufacturing of profiled rollers As for the application of riblet structures on compressor blades, a high reduction of drag friction and a low notch effect are demanded. These two requirements, however, constitute a dilemma. For instance, highest reductions of drag friction are achieved with so called blade riblets whose filigree geometry leads to a high notch effect. The optimal riblet geometry for the application of riblets on compressor blades is an aspect ratio $h/s \geq 0.5$, a low riblet angle $\alpha \rightarrow 0$, a high notch radius R and a small riblet spacing $45\,\mu m \leq s \leq 70\,\mu m$ (Fig. 2). The manufacturing of the corresponding riblets is challenging [10].

As for the described process rollers containing the negative riblet profile have to be manufactured. Due to a high notch effect that occurs during riblet rolling on Ti6Al4 V the roller tool material has to be ductile and wear resistant. The high speed steel HS 6-5-2 (1.3343) fulfills these requirements. Based on a design of experiments a turning strategy was derived for the manufacturing of profiled rollers. Figure 2 shows the process chain. For the micro turning, specially ground cemented carbide cutting inserts by Kennametal were used. Prior to riblet rolling the micro turned profiled rollers were polished, vacuum hardened and annealed. For the riblet rolling on Ti6Al4 V, profiled rollers containing the contour of one, two and three riblets were manufactured. As the number of riblets in the roller contour increases, the riblet rolling process can be accelerated. However, the needed forming force increases as well. Due to the limited hydraulic pressure supply three riblets within the roller contour were found to be the maximum number still leading to a satisfying form filling [10].

Long Life Fatigue (LLF) tests on riblet rolled specimens In order to analyze the notch effect of riblets rolled with the previously described process, tool bending fatigue tests were performed. For the tests single paths of riblet rolling was realized on Long Life Fatigue (LLF) specimens. The riblets were rolled at the smallest cross-section area of the specimens since that is where fracture usually occurs in case of too high bending loads. For reference reasons bending fatigue tests with unstructured specimens were executed as

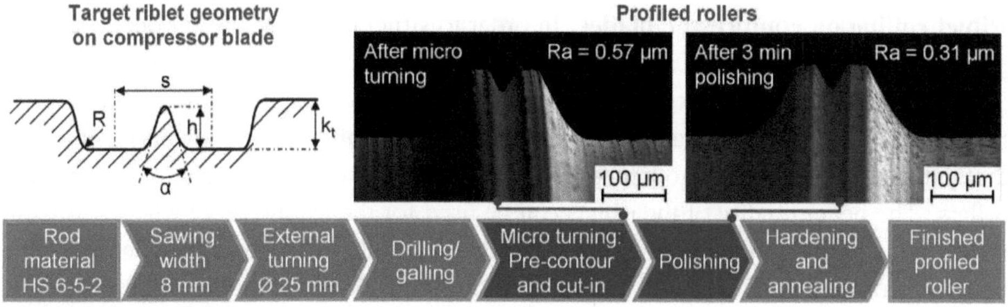

Fig. 2 Target riblet geometry on compressor blade and manufactured profiled rollers [10]

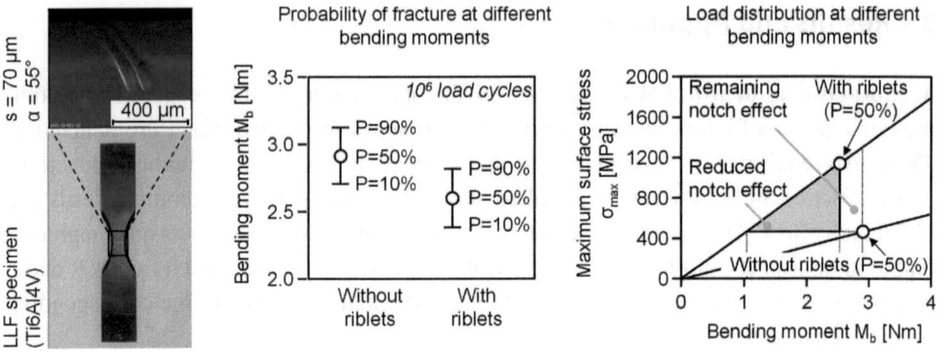

Fig. 3 Riblet rolled LLF specimens and analysis of bending fatigue test [10]

well. Figure 3 shows the test results. The left diagram reveals that specimens with riblets break at lower bending moments (2.57 Nm at fracture probability P = 50 %) than specimens without riblets (2.9 Nm at fracture probability P = 50 %). From FE simulations it has been computed that the tested riblet structure has a notch factor $\alpha_k \approx 2.8$. Keeping this in mind, the specimens with riblets endure maximum surface stresses of $\sigma_{max} = 1167$ MPa (P = 50 %). This value greatly exceeds the flow stress of unformed, thus, not strain hardened Ti6Al4 V. The maximum surface stresses of unstructured specimens at P = 50 % lay at $\sigma_{max} = 475$ MPa. This means that specimens with riblets can endure 2.45 times larger tensile stresses than specimens without riblets. This equals a notch effect reduction of 82 % which can only be attributed to induced strain hardening and compressive residual stresses [10].

Fluid flow analyses In order to quantify the effect of rolled riblets on the reduction of drag friction fluid analyses based on Large Eddy Simulations (LES) were performed at the Institute of Aerodynamics Aachen (AIA). A detailed description of the analyses can be found in Klumpp's dissertation [11]. The analyses reveal that a drag friction reduction of up to 6 % can be reached with the geometry manufactured through incremental riblet rolling [10].

Riblet rolling on compressor blades In order to structure the whole surface of a compressor blade a lateral offset x_{off} has to be implemented. It can either be realized on both leading and trailing edge or at one edge only. In the latter case (Fig. 4a) the profiled rollers move back and forth on the same path prior to the lateral offset. Experiments have shown that the latter strategy leads to better riblet structures at the reversal point of the profiled rollers. The better quality of the riblets can be used for the leading edge where a smooth fluid flow is of main interest. The lateral offset x_{off} results from the actual contour of the inserted profiled rollers which is measured microscopically prior to the forming process. For each profiled roller an optimal lateral offset $x_{off,opt}$ can be determined. Due to manufacturing inaccuracies during micro turning of the rollers, the optimal lateral offset differs

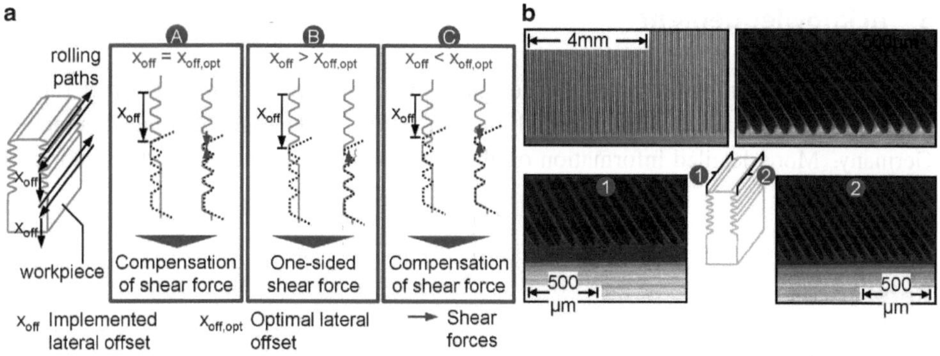

Fig. 4 Shear forces due to lateral offset (**a**) and riblet rolling on Ti6Al4 V specimens (**b**) [9]

from roller to roller. The feeding of the 3-axis machining center, however, only allows a single offset value to be set. As a consequence, the implemented offset can be the optimal one for one profiled roller (Fig. 4a, case A), but differs from the optimal offset of the other roller. In case the implemented offset is larger than the optimal one (Fig. 4a, case B), the material flow exposes a one-sided shear force to one of the negative elevations of the profiled roller, thus, causing its rupture. In case the implemented offset is less than the optimal one (Fig. 4a, case C) the shear force can be compensated and rupture of profiled rollers can be avoided. However, this approach inevitably leads to the partial rollover of a previously produced riblet. Figure 4b shows the experimental outcome of this strategy for the rolling of specimens made of Ti6Al4 V. In Fig. 4b number 2 denotes the side of the specimen where the narrower contoured roller is operating. On this side the implemented lateral offset equals the optimal one. This leads to homogeneous riblets. On the opposite side (number 1) the partial rollover is observed in form of one periodically occurring riblet of smaller height. The period length results from the number of riblets on the profiled roller. For the rolled riblets shown in Fig. 4b profiled rollers with three riblets were used [9].

4 Conclusion

The presented incremental riblet rolling process is able to produce microscopic riblets on Ti6Al4 V material such as compressor blades. The analyses have shown that the riblets reduce the drag friction by 6 %. At the same time the negative notch effect resulting from the notch-like riblets is compensated by 82 % as a consequence of induced strain hardening and compressive residual stresses. Applied for compressor blades of jet engines the process, therefore, promises an effective measure for the reduction of fuel consumption and, thus, emissions of CO_2 as well as nitrogen oxide.

5 Acknowledgement

The research activities linked to the design of the presented incremental riblet rolling process were conducted with financial support of the VolkswagenStiftung, Hannover, Germany. More detailed information on the process can be found in the dissertation of Feldhaus [10].

References

1. Advisory Council on Aeronautics Research in Europe, 2008, Addendum to the Strategic Research Agenda. URL: http://www.acare4europe.com [27.09.2010].
2. Walsh, M.J., Weinstein, L.M., 1978, Drag and heat transfer on surfaces with small longitudinal fins, AIAA, 78–1161.
3. Nitschke, P., 1984, Experimental investigation of the turbulent flow in smooth and longitudinally grooved tubes, Max Planck Institute for Dynamics and Self-Organization, Göttingen, Report 3/1983, NASA TM 77480.
4. Bechert, D.W., Bruse, M., Hage, W., van der Hoeven, J.G.T., Hoppe, G., 1997, Experiments on drag-reducing surfaces and their optimization with adjustable geometry, Journal of fluid mechanics 338, 59–87.
5. Reif, W.-E., Dinkelacker, A., 1982, Hydrodynamics oft he squamation in fast swimming sharks, Neues Jahrbuch für Geologie und Paläontologie 164, Stuttgart, 184–187.
6. Reif, W.-E., 1985, Squamation and ecology of sharks, Courier Forschungsinstitut Senckenberg 78, Frankfurt/Main, 255.
7. Armstrong, E.K., Stevenson, M.A., 1960, Some practical aspects of compressor blade vibration, Fournal of Royal Aeronautical Society 64, 117–130.
8. Cumpsty, N.A., 1989, Compressor aerodynamics, Longman Group Ltd, Pearson Education Ltd, Harlow, Essex, U.K.
9. Hergt, A., Hage, W. Grund, S., Steinert, W., Terhorst, M., Schongen, F., Wilke, Y., 2014, Riblet application in compressors: Towards efficient blade design?, Proceedings of ASME Turbo Expo 2014, GT2014, June 16–20, 2014, Düsseldorf, Germany.
10. Feldhaus, B., 2010, Walzen definierter Ribletstrukturen auf Verdichterschaufeln, Dissertation, RWTH Aachen University.
11. Klumpp, S., 2010, Variation of Friction Drag in Wall-Bounded Flows, Dissertation, RWTH Aachen University.

Part VII

Extrusion and Hot Forging

TR Process for Forging Heavy Crankshafts

Tadeusz Rut, Wojciech Walczyk, Andrzej Milenin, and Maciej Pietrzyk

1 Motivation

Crankshaft is among the largest and the most critically loaded components employed in different types of engines [1]. TR forging process for heavy crankshafts, which allowed manufacturing shafts with continuous grain flow, was a breakthrough in forging technology. This method was developed at the Metal Forming Institute (INOP) in Poznan, Poland (TR follows the name of the inventor Tadeusz Rut). In the first design two connecting rods with joints, which transferred vertical force of the press into two horizontal forces, were applied [2]. This technology gave good results when shape of the crank shoulders was close to an ellipse. New constructions of crankshafts required large juts at the side of journals of main bearings. In consequence, machining of deep grooves in the preform was needed. It was inconvenient and it led to an increase of material waste. To avoid this inconvenience, additional operation was introduced [3, 4]. This was an unsymmetrical pre-upsetting of the stock material, which involved proper flow of the material needed to obtain required shape of the crank shoulders.

2 Process Idea

In Fig. 1a a crankshaft is seen in the press during forging. Idea of the conventional TR process is shown in Fig. 1b. In top figure tools are in the idle positions at the start of a cycle. In the bottom figure the forming cycle was completed and the set of tools was fully

Tadeusz Rut · Wojciech Walczyk
Instytut Obrobki Plastycznej, Poznań, Poland

Andrzej Milenin · Maciej Pietrzyk ✉
AGH University of Science and Technology, Kraków, Poland
e-mail: Maciej.Pietrzyk@agh.edu.pl

A. E. Tekkaya et al. (eds.), *60 Excellent Inventions in Metal Forming*,
DOI 10.1007/978-3-662-46312-3_39

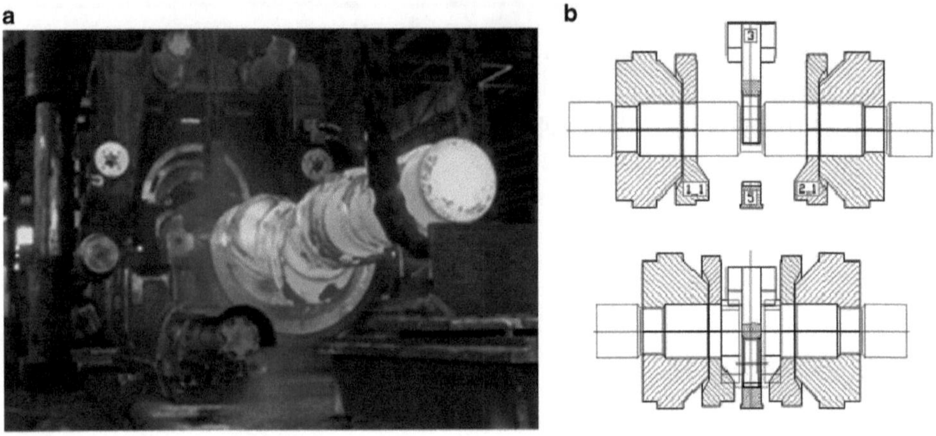

Fig. 1 Crankshaft during forging (**a**) and schematic illustration of the conventional TR process (**b**)

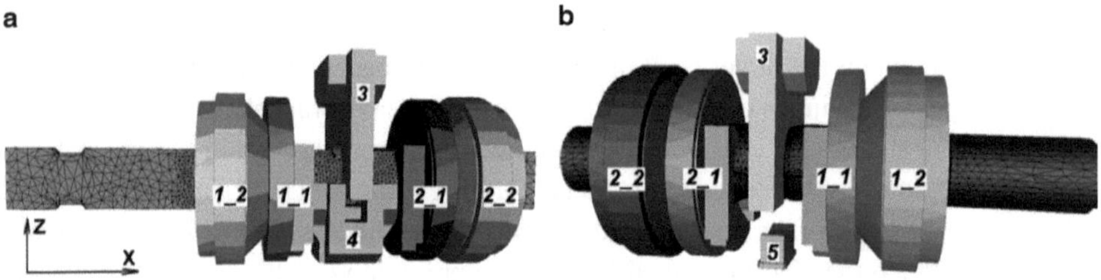

Fig. 2 The process model. **a** unsymmetrical pre-upsetting, **b** forging of the crank throw

closed. A crank throw of a crankshaft was formed. The bending tool (3) and the anvil (5) shaped the pin portion, and the crank webs were formed by upsetting caused by horizontal movement of the face die inserts (1_1; 2_1).

The modification of the TR process is presented in Fig. 2. The set of dies for pre-upsetting, which distinguishes the new process from the conventional one, is shown in Fig. 2a. Beyond face die inserts (1_1; 2_1), clamping die inserts (1_2; 2_2) and bending tool (3), one more tool foreanvil (4) is shown. In forging of the crank throw (Fig. 2b) the foreanvil (4) was replaced with anvil (5).

3 Design of the Best TR Forging Technology

TR forging process shown in Fig. 2 has several variants. Some of the tools may be eliminated and trajectories of the tools can be changed allowing manufacturing of a variety of cranks shapes. On the other hand design of the best technology for a selected crankshaft is a complex and challenging task. Several problems such as excessive material losses due to machining, bending of the shaft at various stages of the process and risk of fracture during heat treatment have to be faced. To avoid costly trial and error method physical and numerical simulations of the TR process were performed. Physical simulations composed forging of the crankshaft made of lead in a scale 1:7. Details of the FE model used in numerical simulation are given in [5]. The objectives of the simulations was prediction of forces acting on the tools, distributions of displacements, strains, stresses and temperatures in the forging as well as microstructure evolution, bending of the shaft and residual stresses after manufacturing.

4 Results

Pilot press installed in the laboratory of the Metal Forming Institute in Poznań (INOP) and selected example of the forging made of lead in the scale 1:7 are presented in Fig. 3. Physical simulations allowed designing the best tool shapes and trajectories for forging of various types of crankshafts.

Numerical simulations were performed for the manufacturing chain composed of heating, forging and heat treatment. The simulations supplied data concerning metal flow, loads and distributions of strains, stresses and temperatures in the forging, see selected results for a single crank shown in Fig. 4 and for multi crank shaft in Fig. 5. Time-temperature history and strains were monitored and used to improve forging technology. Austenite grain size during forging was also calculated, see [6] for details. Since at this stage of research these data have small practical impact, they are not presented in this paper. An attempt of optimization of the TR forging is described in [7]. Due to long computing times metamodelling was applied. The tools positions and trajectories, which gave the correct shape of the crank, were determined.

Fig. 3 Pilot press in the INOP laboratory (**a**) and selected examples of lead forgings in the scale 1:7 (**b**)

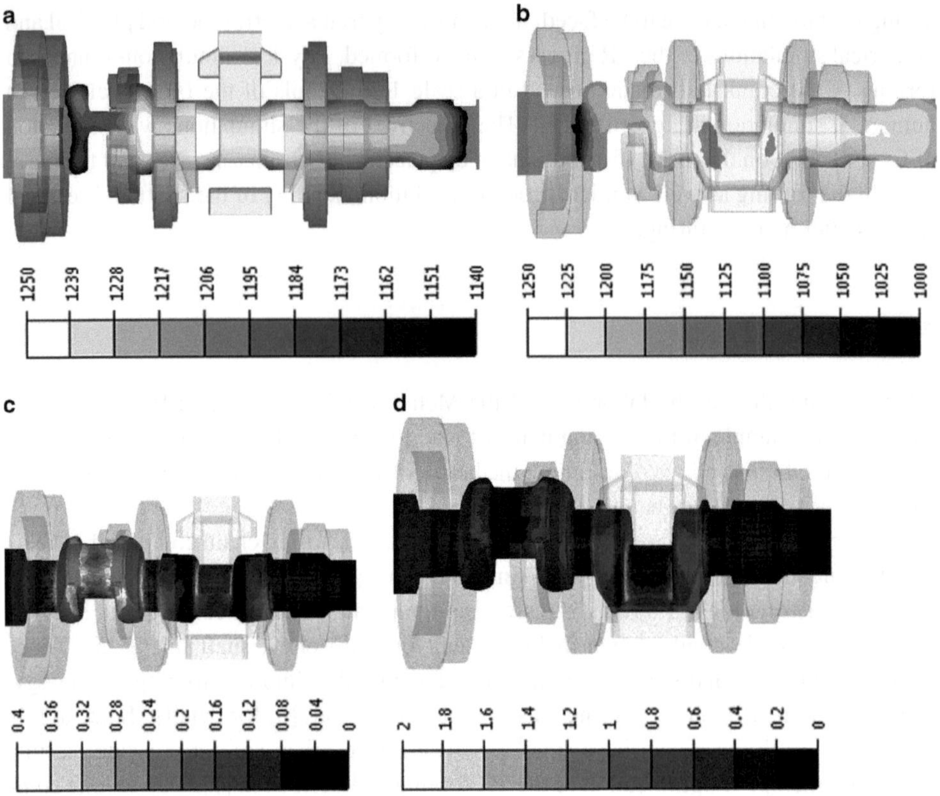

Fig. 4 Selected results of the numerical simulations of the forging process for a single crank. **a**, **b** temperatures, **c**, **d** effective strain, **a**, **c** upsetting, **b**, **d** forging

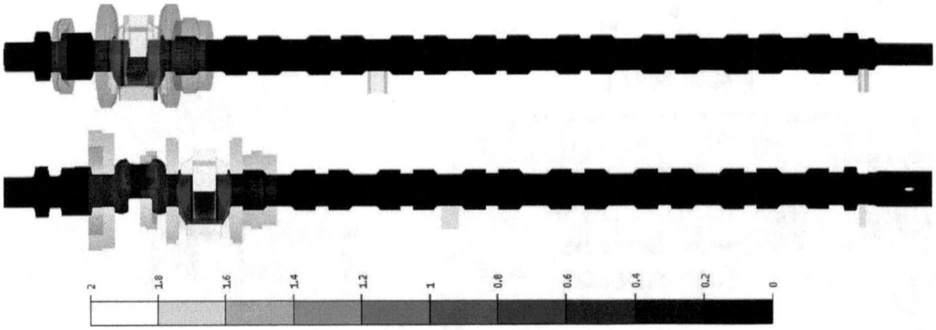

Fig. 5 Selected results of the numerical simulations of the forging process for a shaft with 9 cranks: effective strain distributions during forging of cranks no. 9 and 8

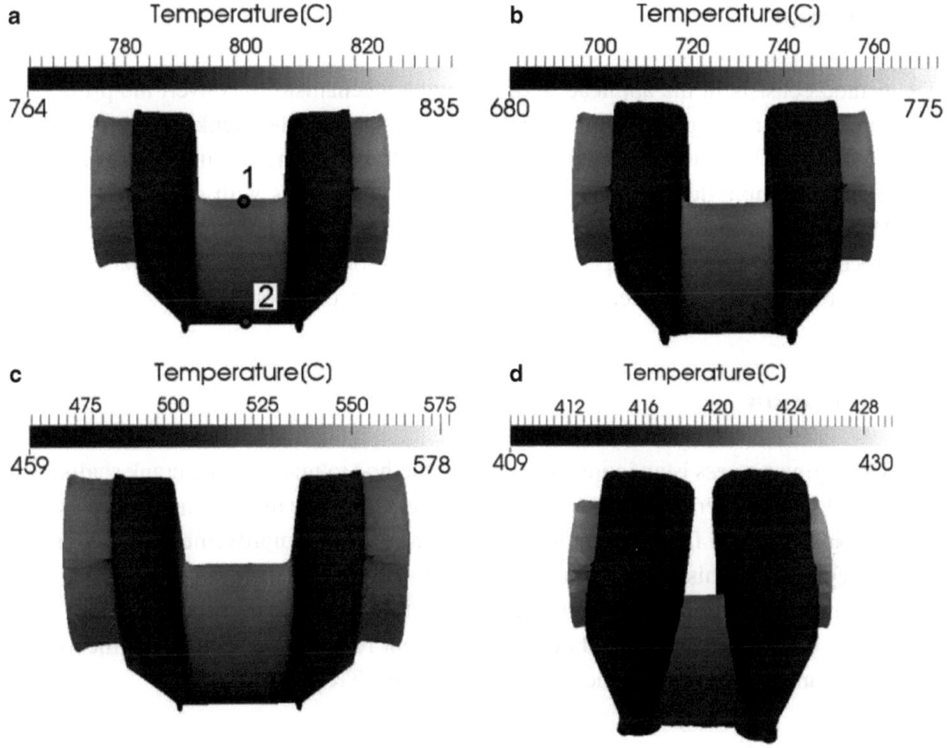

Fig. 6 Temperature distribution and crank deformation during cooling after 4 hours (**a**), 6 hours (**b**), 11 hours (**c**) and 17 hours (**d**): The scale of displacements 200:1

Bending of the crankshaft during heating, forging and heat treatment is one of the difficult problems occurring during production. This problem was investigated by numerical simulations using elasto-plastic FE code coupled with the thermal FE code and phase transformations model. The parameters with the strongest influence on the bending were determined. Results of simulations of bending during cooling, accounting for the dilatometric effect due to phase transformations, are given in [8]. Some of these results are presented in in Fig. 6. Initial stage, when the phase transformations have not yet begun, is shown in Fig. 6a. Bottom part of the crank is cooled intensively and tensile stresses occur in this area. After 6 hours of cooling (Fig. 6b) phase transformations begun, what caused a change of the bending direction. After 11 hours (Fig. 6) the phase transformations were completed, what again changed the direction of the crank deflection. After 17 hours (Fig. 6d) stress relaxation changed again the direction of the crank deflection.

5 Potential

TR method consists in the application of an toggle mechanism to convert the press load into forces self-gripping the forging stock and ones forming the crank throw. It enables forging of crankshafts with complex shapes of crank webs. The forging devices are universal and of simple design. They can be installed on presses with capacities of up to 100 MN. The TR process for heavy crankshafts offers numerous potentialities, qualifying this process for a variety of shapes of crank arms and forming tasks. The advantages are: reduction of material losses, reduction of labour, better product quality.

6 Conclusion

The TR forging process is an innovative forming method to manufacture crank shafts with a variety of shapes of crank arms. This process allows effective manufacturing crankshafts with continuous grain flow. Numerical simulations allowed improvement of the forging technology. Beyond this, simulations have shown that cooling process is accompanied by a three-times changing of stress sign and the direction of bending of the crankshaft. The knowledge on this process gained during subsequent research, as well as the theoretical approach, can be applied to further development of the TR process.

7 Acknowledgement

TR process has been investigated in several projects. The work was supported by the Polish Research Foundation (NCN) under the grant number N 508 391 735. Development of the TR technology in CELSA forging shop in Ostrowiec Świętokrzyski was sponsored by the Applied Research Agency NCBiR under the grant PBS1/B6/3/2012.

References

1. Zoroufi, M., Fatemi, A., 2005, A literature review on durability evaluation of crankshafts including comparisons of competing manufacturing processes and cost analysis, Proc. 26th Forging Industry Technical Conf., Chicago, 1–32.
2. Rut, T., 1986, Forging of long-stroke crankshafts by TR method, in: Steel Forgings, eds, Nisbett E. G., Melilli A.S., American Society for Testing and Materials, Ann Arbor, 504–517.
3. Rut, T., Walczyk, W., 2000, Doskonalenie kucia wałów korbowych metodą TR, Obróbka Plastyczna Metali, 11, 5–8. (in Polish).
4. Rut, T., Walczyk, W., 2002, Rozwój metody TR – urządzenia typu TR do kucia wałów korbowych, Archiwum Technologii i Automatyzacji, 22, Part I: 177–186; Part II: 187–196. (in Polish).

5. Walczyk, W., Milenin, A., Pietrzyk, M., 2011, Computer aided design of new forging technology for crank shafts, Steel Research International, 82: 187–194.
6. Milenin, A., Walczyk, W., Pietrzyk, M., 2012, Numerical modeling of microstructure evolution during forging of crank shafts, Steel Research International, 83; 808–816.
7. Sztangret, Ł., Milenin, A., Sztangret, M., Walczyk, W., Pietrzyk, M., Kusiak, J., 2011, Computer aided design of the best TR forging technology for crank shafts, Computer Methods in Materials Science, 11: 237–242.
8. Milenin, A., Rec, T., Walczyk, W., Pietrzyk, M., 2014, Model of curvature of crankshaft blank during the heat treatment after forging, Procedia Engineering, 81: 498–503.

Chip Extrusion with Integrated Equal Channel Angular Pressing

Matthias Haase and Nooman Ben Khalifa

1 Motivation

The production of primary aluminum, obtained by refining bauxite ore, is one of the most energy intense processes in primary metal production [1]. By remelting aluminum scrap, the energy balance of aluminum production can be significantly improved. However, the remelting of aluminum scrap can lead to a metal loss of 20 % due to oxidation of the molten metal [2]. Machining chips are one of the most difficult to recycle types of scrap, as their high surface to volume ratio intensifies the oxidation of the metal during remelting [3]. The direct recycling of aluminum alloy machining chips into finished or semi-finished products by hot extrusion, first presented and patented by Stern in 1945 [4], is a promising approach to further improve the energy efficiency of aluminum recycling and to overcome the problem of material loss during remelting. In this process, the machining chips are cleaned from remaining lubricants, compacted to chip-based billets and extruded at elevated temperatures to chip-based extrudates [5]. The main challenge for the production of chip-based extrudates with mechanical properties comparable to those of extruded cast material is to break the oxides covering the individual chips and to achieve sound bonding of the pure metal. For the breakage of the oxides, a high amount of (shear) strain is necessary, while a high amount of pressure is required to achieve contact between the surfaces of pure metal [6]. In hot extrusion, the amount of pressure and strain is mainly determined by the extrusion ratio, which is defined as the ratio between the cross sectional area of the upset billet to the cross sectional area of the extruded profile as well as by the design of the hot extrusion die. To achieve high extrusion ratios, high capacity presses with big billet diameters are necessary to produce profiles with cross sections usable in industrial applications. Therefore, the developed process aims on an increase of pressure

Matthias Haase ✉ · Nooman Ben Khalifa
TU Dortmund, Dortmund, Germany
e-mail: Matthias.Haase@iul.tu-dortmund.de

A. E. Tekkaya et al. (eds.), *60 Excellent Inventions in Metal Forming*,
DOI 10.1007/978-3-662-46312-3_40

and strain by altering the design of the hot extrusion die in order to allow the production of high quality chip-based extrudates with conventional size presses.

2 Process Idea

The process idea is to use the combination of hot extrusion with equal channel angular pressing (ECAP) in a single hot extrusion die [7] for the direct recycling of aluminum alloy machining chips. Conventionally, the severe plastic deformation process of ECAP introduces an ultra-large plastic strain into bulk metals by pressing the workpiece through the ECAP die, having an entrance and an exit channel of the same cross section intersected by an angle. By simple shear deformation, ultra-large plastic strain can be introduced into the material without changing the dimension of the workpiece, which results in grain refinement and therefore in an improvement of the mechanical properties of the processed material [8]. For the direct recycling of aluminum alloy machining chips, the concept of ECAP was integrated into a hot extrusion die in order to increase the pressure and strain

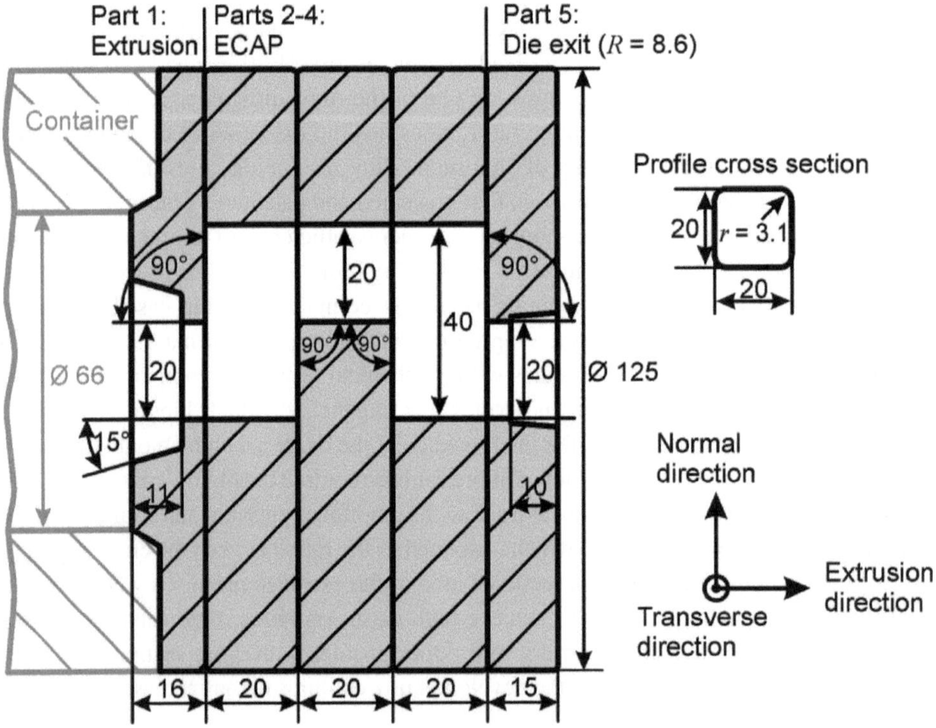

Fig. 1 Design of the developed iECAP die for the direct recycling of aluminum alloy machining chips. All dimensions are in mm [10]

acting on the chips during the hot extrusion process [9]. The design of the resulting hot extrusion die with integrated equal channel angular pressing (iECAP) is shown in Fig. 1.

The iECAP die consists of five parts. Part one contains the conventional extrusion step leading to a rectangular cross section of 20 mm × 20 mm. For the used hot extrusion press with billet diameters of $d = 66$ mm, the size of the profile is resulting in an extrusion ratio of $R = 8.6$. The extrusion step is designed with a prechamber to even the material flow. Parts two to four contain the ECAP steps. Four ECAP steps were integrated into the iECAP die. All channels are intersected by an angle of 90°. The layout of the material flow is equivalent to routes C-A-C in conventional ECAP, where C indicates a sample rotation of 180° and A no rotation between consecutive passes around the extrusion direction axis. This sequence of routes leads to a two-dimensional material flow and to the creation of backpressure after every ECAP turn, except for the last, when the material flows to the die wall. The backpressure is supposed to benefit the quality of the chip-based finished parts, as it can lead to a reduction of residual porosity in the chip-based extrudates [11]. The last part of the iECAP die contains the bearing surfaces and therefore defines the final profile geometry. The die exit of the iECAP die leads to the same 20 mm × 20 mm cross section as the conventional extrusion step in part one of the die. By solely exchanging the last part of the iECAP die, it is possible to produce other profile cross sections [12].

3 Results and Applications

To evaluate the performance of chip-based extrudates fabricated with the iECAP die, the mechanical properties of the processed material were analyzed and compared to the mechanical properties of chip-based extrudates fabricated with state-of-the-art extrusion dies. A conventional flat-face die and a modified four feeder porthole die, having a shortened mandrel for the extrusion of solid sections, were included in this comparison [10, 13]. For all investigated dies, the resulting extrusion ratio was $R = 8.6$. The material extracted from the dies after extrusion are shown in Fig. 2.

Cast billets of the same aluminum alloy were extruded through the three dies with the same process parameters as the chips in order to have a general comparison between the

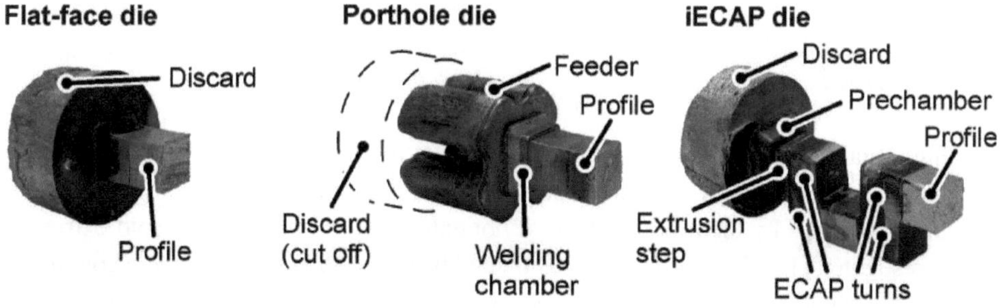

Fig. 2 Material extracted from different types of extrusion dies used for the extrusion of chip-based billets [10]

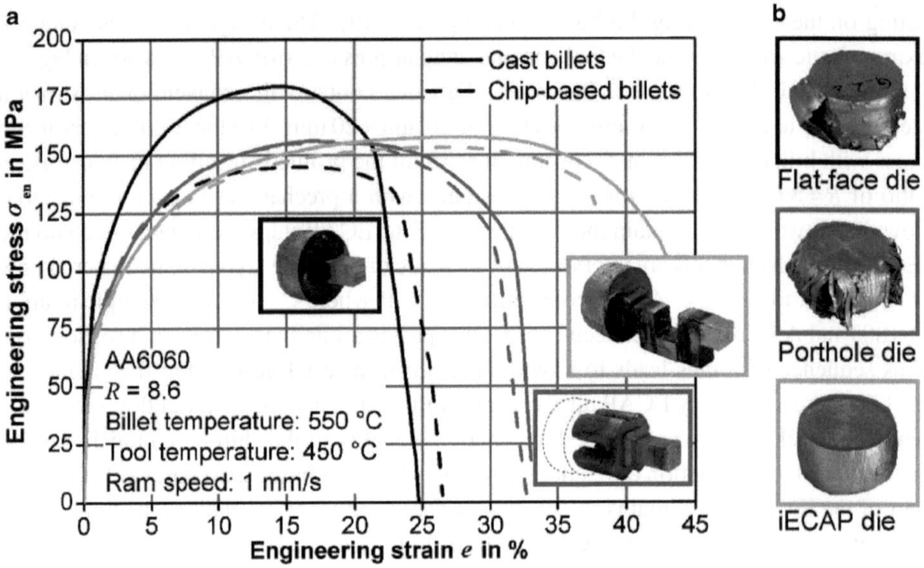

Fig. 3 **a** Tensile test results of chips and cast material extruded through different hot extrusion dies, **b** Results of compression tests of chip-based extrudates fabricated with different dies [12, 14, 15]

performance of chip-based extrudates and extruded cast material. The results of tensile and compression tests are shown in Fig. 3a and b, respectively.

The tensile test results show that the tensile strength of hot extruded chips and hot extruded cast material is only comparable for the porthole die and the iECAP die processed material. The tensile strength of the chip-based extrudates fabricated with the flat-face die is significantly lower compared to the tensile strength of cast material extruded through the same die. With increasing complexity of the die design, not only the tensile strength, but also the uniform elongation of the chip-based extrudates increases. Although the tensile test results indicate sound chip-bonding for the porthole die processed extrudates, compression test results reveal inferior bonding between the individual chips when using the porthole die or the flat-face die. Only by using the iECAP die, the chip-based extrudates remain defect-free after compression tests.

4 Potential

The high quality of the chip-based extrudates offers the possibility of further processing the material in subsequent forming processes. To investigate this possibility, recycling of aluminum chips by the process chain of hot extrusion with subsequent cold extrusion was investigated [16]. Forward rod extrusion and backward can extrusion were conducted at room temperature with preforms based on chip-based extrudates fabricated with the iECAP die. For both processes, chip-based finished parts with a quality comparable to similarly processed cast material were successfully produced [17].

5 Conclusion

The direct recycling of aluminum alloy machining chips by hot extrusion with integrated equal channel angular pressing is an innovative forming process, which allows the fabrication of high quality chip-based products. This process combination leads to a significant improvement of the mechanical properties of the chip-based extrudates compared to conventional hot extrusion. This is related to the additional amount of (shear) strain and pressure acting on the material during the extrusion process, which is leading to an improved breakage of the oxide layers covering the individual chips and to an improved bonding of the pure metal. The combination with subsequent forming processes offers numerous potentialities for the production of chip-based finished parts. This process can be a possible alternative for the conventional recycling of aluminum alloy machining chips by remelting, as it offers the benefit of increased energy and resource efficiency.

6 Acknowledgement

This process has been kindly supported by Research School of Energy Efficient Production and Logistics at TU Dortmund University.

References

1. Schmitz, C., 2006. Handbook of Aluminum Recycling, Vulkan Verlag, Essen, 27–30.
2. Gronostajski, J., Matuszak, A., 1999. The recycling of metals by plastic deformation: an example of recycling of aluminium and its alloys chips. Journal of Materials Processing Technology 92–93, 35–41.
3. Sharma, C. S., Nakagawa, T., Takenaka, N., 1977. Recent Development in the Recycling of Machining Swarfs by Sintering and Powder Forging. CIRP Annals Manufacturing Technology 26 (1), 121–125.
4. Stern, M., 1945. U.S. Patent 2,391,752.
5. Tekkaya, A. E., Schikorra, M., Becker, D., Biermann, D., Hammer, N., Pantke, K., 2009. Hot profile extrusion of AA-6060 aluminum chips. Journal of Materials Processing Technology 209, 3343–3350.
6. Cooper, D.R., Allwood, J.M., 2014. The influence of deformation conditions in solid-state aluminium welding processes on the resulting weld strength. ournal of Materials Processing Technology 214, 2576–2592.

7. Estrin, Y., Ferkel, H., Hellmig, R. J., Lamark, T., Popov, M. V., 2008. German Patent DE102005049369.
8. Valiev, R. Z., Langdon, T. G., 2006. Principles of equal-channel angular pressing as a processing tool for grain refinement. Progress in Materials Science 51, 881–981.
9. Haase, M., 2014. Mechanical Properties Improvement in Chip Extrusion with Integrated Equal Channel Angular Pressing, Aachen, Shaker (Dortmunder Umformtechnik, 77).
10. Haase, M., Ben Khalifa, N., Tekkaya, A.E., Misiolek, W.Z., 2012. Improving mechanical properties of chip-based aluminum extrudates by integrated extrusion and equal channel angular pressing (iECAP). Materials Science and Engineering A 539, 194–204.
11. Lapovok, R., Tomus, D., Bettles, C., 2008. Shear deformation with imposed hydrostatic pressure for enhanced compaction of powder. Scripta Materialia 58, 898–901.
12. Haase, M., Jäger, A., Tekkaya, A. E., 2012. Influence of extrusion ratio on AA6060 aluminum alloy chips solid state recycled by integrated extrusion and equal channel angular pressing. In: Mori, K., Pietrzyk, M., Kusiak, J., Majta, J., Hartley, P., Lin, J. (Eds.), Steel Research International, Special Edition: 14th International Conference Metal Forming, Wiley, Weinheim, Germany, 479–482.
13. Güley, V., Ben Khalifa, N., Tekkaya, A. E., 2011. The Effect of Extrusion Ratio and Material Flow on the Mechanical Properties of Aluminum Profiles Solid State Recycled from 6060 Aluminum Alloy Chips. In: Proceedings of the 14th International ESAFORM Conference on Material Forming, Belfast, 1609–1614.
14. Misiolek, W. Z., Haase, M., Ben Khalifa, N., Tekkaya, A. E., Kleiner, M., 2012. High quality extrudates from aluminum chips by new billet compaction and deformation routes. CIRP Annals Manufacturing Technology 61, 239–242.
15. Tekkaya, A. E., Güley, V., Haase, M., Jäger, A., 2012. Hot extrusion of aluminum chips. In: Proceedings of the 13th International Conference on Aluminum Alloys (ICAA13), Pittsburg, 1559–1573.
16. Haase, M., Tekkaya, A. E., 2014. Recycling of aluminum chips by hot extrusion with subsequent cold extrusion. Procedia Engineering 81, 652–657.
17. Haase, M., Tekkaya, A. E., 2015. Cold extrusion of hot extruded aluminum chips. Journal of Materials Processing Technology 217, 356–367.

Non-graphite Water Soluble Lubricant for Hot Forging

Nobuhiro Ikeda

1 Motivation and Process of Development

Graphite type lubricants have been commonly used in the hot forging factories because the graphite provides good anti-galling performance due to the excellent heat resistance and low friction at high temperatures. However, in recent years, it is required to be aware of the environmental issues, such as keeping the working environment to be compatible to other manufacturing processes and reduction of the amount of waste liquid. In the late 1980's, Toyota Motor Co. and Daido-chemical Industry started to develop non-graphite type lubricant to substitute the graphite type lubricants, and in 1986 Toyota began to use the developed non-graphite lubricant. In 2003 all the hot and warm forging lines of Toyota were changed to use the non-graphite lubricants. From around 2000, many of the hot forging factories in Japan and in other countries increasingly changed to non-graphite lubricants.

2 Function of Water Soluble Hot Forging Lubricant

Since heated work piece slides over the die surface under a high contacting pressure in hot forging, the hardness of the die surface is reduced and rapid wear proceeds. In addition, the direct contact between the die and the work piece leads to galling and surface defects of product. The requested performance of hot forging lubricant is to create effective film to protect the die surface during forging, and to provide good lubricity in forging and in releasing the product from the die.

Figure 1 shows the process of water soluble lubricant experiencing in a cycle of hot forging. When the lubricant is sprayed on the die surface the die is cooled and lubricant

Nobuhiro Ikeda ✉
Daido-chemical Industry co. Ltd., Yamato-kouriyama-shi Nara, Japan
e-mail: giken3-5@daido-chemical.co.jp

A. E. Tekkaya et al. (eds.), *60 Excellent Inventions in Metal Forming*,
DOI 10.1007/978-3-662-46312-3_41

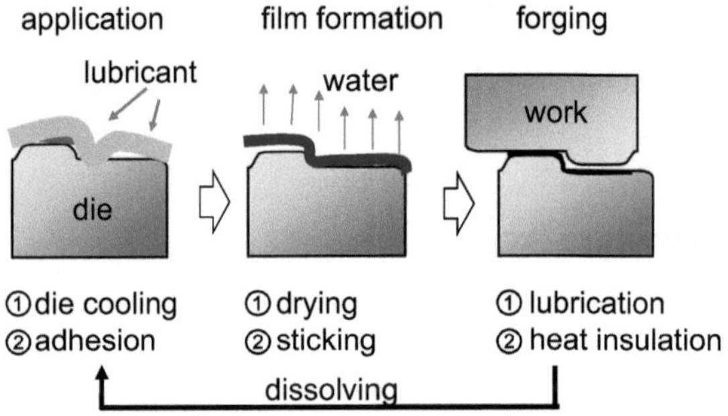

Fig. 1 States and functions of water soluble lubricant in hot forging

film sticks onto the die surface. Then the film is dried by the heated die and dried film sticks to the die surface after evaporation of water. The solid film functions as a good lubricant during forging. The residue and surplus caused by burning act as insulator of heat, and are resolved into the liquid lubricant splayed in the next spray [1].

3 Graphite Type Lubricant

As shown in Fig. 2, graphite has a layered crystal structure with a weak bonding strength between the layers, and thus sliding occurs easily between the layers when shear stress is applied. This property of graphite offers effective lubrication. Since graphite is a heat resistant material as well, it is often used as hot forging lubricant.

When graphite is dispersed in water to be used as hot forging lubricant, the following problems occur.

(1) Degradation of working environment due to sticking of graphite to building and machines.
(2) Problems related to electric conductivity of graphite such as short circuit and erosion.
(3) Difficulty of separation of graphite from oil leading to large amount of oil sludge.

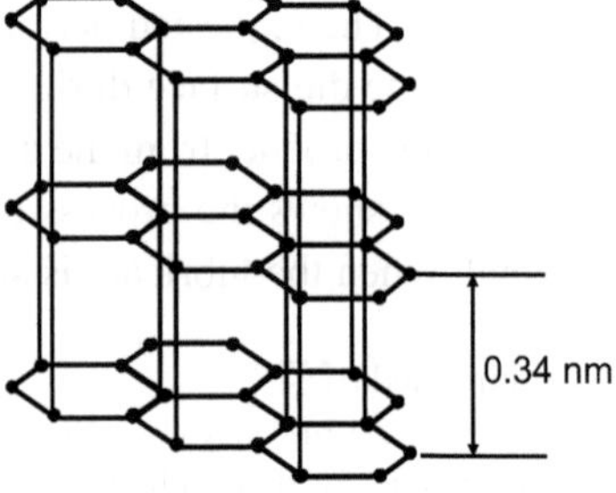

Fig. 2 Crystal Structure of Graphite

4 Development of Non-Graphite Type Lubricant

To overcome the problems of the graphite type lubricant, non-graphite lubricants were examined using various polymers. Since carboxylate showed the best result, it was decided to be used as the base component and water soluble high polymer, inorganic acid and solid lubricants were mixed. The obtained lubricants are transparent or white when they are dried, and thus they provide good working environment. Since oil-water separation performance of the lubricants is superior, waste disposal in case of recirculation is reduced to 1/4 of the graphite type lubricant [2].

The roles of the water-soluble high polymer compound are as follows:

(1) Forming viscous film with good sticking characteristics

The viscosity of the water-soluble high polymer compound increases sharply as the degree of concentration increases when the contained water decreases. This means that the viscosity of the lubricant increases sharply as the contained water is evaporated when it is applied to the hot die surface. By the increased viscosity, the lubricant sticks well to the die surface and finally forms solid lubricant film when the water is completely evaporated.

(2) Formation of decomposed residue during contacting with heated work-piece

The created lubricant film has a function of reducing friction during forging and it is chemically decomposed by contacting with the high temperature work piece. The decomposed residue including alkali salt helps to protect the die surface by insulating the heat from the work piece and to prevent galling during ejection of the forged product from the die cavity.

5 Comparison of Graphite Type and Non-Graphite Type Lubricants

Because various properties, such as viscosity, cooling performance, heat resistance, etc., are requested to the lubricant depending on the forging condition, various non-graphite type lubricants with different molecular weight were developed. The performance of the standard type lubricant is as follows:

(1) Adhesion performance and ability of heat resistance of coating film

Adhesion performance of lubricant is evaluated by the weight of the lubricant coated on the heated test piece after splaying by air mixed spray. Figure 3 shows a comparison of volume of lubricant adhering to the test piece between a general purpose commercial water-soluble non-graphite type lubricant and a graphite type lubricant. The result shows that larger volume of the non-graphite type lubricant is left on the surface than the graphite type lubricant when the temperature is higher than 300°C.

Figure 4 shows the remaining ratio of the lubricant films after heating to a temperature up to 700 °C. The result highlights that the non-graphite type lubricant begins to be decomposed at around 450 °C About 40 % of the non-graphite lubricant remains at 700 °C,

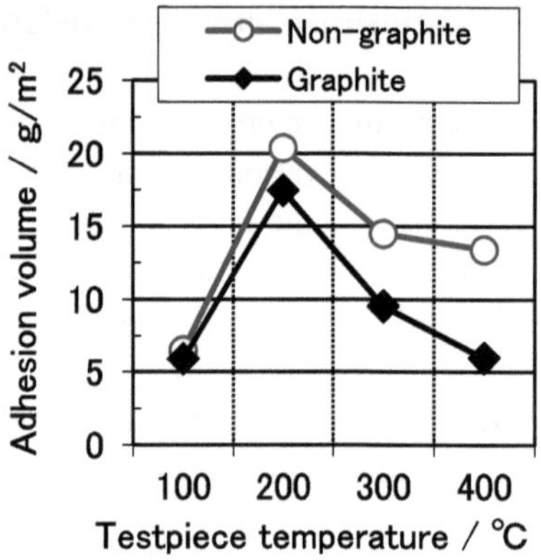

Fig. 3 Adheasion volume on each testpiece temperature

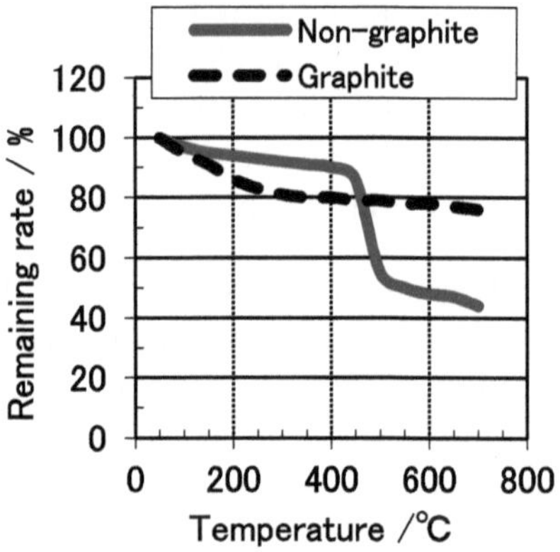

Fig. 4 Remainig ratio after heating to various temperatures

while 80 % of the graphite type remains at temperatures at 300 °C, and almost no change is observed at higher temperatures. This result of the non-graphite type lubricant is attributed to the fact that the main component of the organic substances begins to be decomposed at around 450 °C, while the dispersant and the binder components of the graphite type lubricant evaporates at low temperatures and then graphite itself remains without decomposition up to high temperatures. The remaining volume of residue of the graphite type lubricant is twice as large as that of the non-graphite type. This result suggests that two folds of non-graphite type lubricant should remain on the die surface is the lubricity of the two lubricants is similar.

(2) Effect of film thickness on lubricity

In the early stage of development of the non-graphite type lubricant, there were many cases in which the non-graphite type resulted in inferior performance to the graphite type lubricant in terms of lubricity and die wear. To clarify the reason, ring compression tests were carried out to measure the friction coefficient. Figure 5 shows the friction coefficient plotted against reduction in height when the thickness of the lubricating film is approximately 3.0 μm. The non-graphite type lubricant showed a little higher friction coefficient

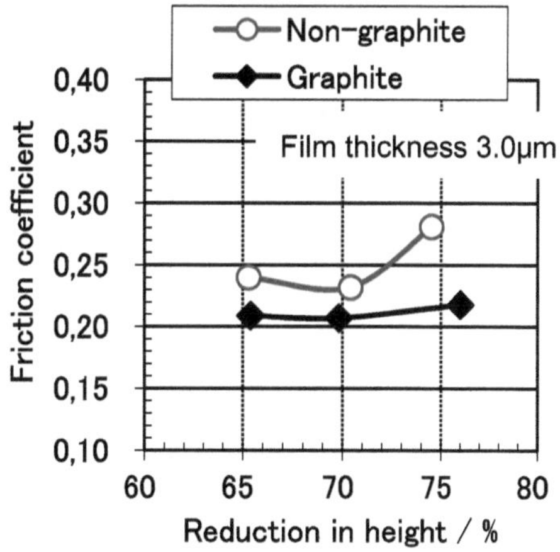

Fig. 5 Effect of reduction in height on friction coefficient at 3.0 μm film thickness

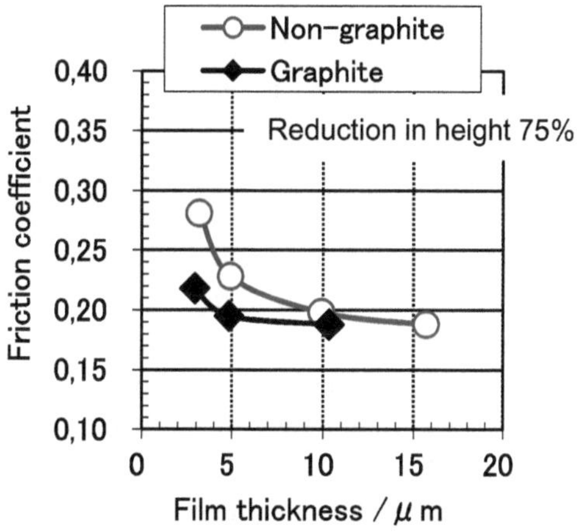

Fig. 6 Effect of film thickness on friction coefficient at 75 % reduction

than the graphite type. Figure 6 shows the effect of film thickness on the coefficient friction at a reduction in height of 75 %. It is observed that the non-graphite type lubricant provides almost the same friction coefficient as the graphite type lubricant when the film thickness is double as thick as the graphite type. This result is consistent with Fig. 4 which suggests that the residual volume in the non-graphite type lubricant is about 1/2 of the graphite type. From this result together with the result of Fig. 3, thicker film is formed by keeping the die temperature around 300 °C.

6 Lubrication Under Wet Condition

(1) Problem under wet condition

The results shown above are obtained when completely dry films are created on the die surface. However, in actual forging operations, there are undesirable cases when the cycle time is not long enough or the die temperature is kept low and lubricant films are not dried completely.

(2) Increase of viscosity by forming pseudo-giant high molecular compound

When the lubricant film is semi-dried, it tends to be squeezed out from the interface between the die and work piece. To solve this problem of squeezing out of lubricant under the wet condition, an idea to change the lubricant to be extremely viscous during forging not to be squeezed out was put forward.

Figure 7 shows the sticking stage of the water soluble lubricant on the die surface without being dried well. Bonding substance is added to the water soluble high polymer compound of ordinary viscosity. When the semi-dried lubricant film contacts the hot work piece, water is evaporated and the high molecule polymer and the bonding substances are concentrated, and then the bounding substance connects the high molecular compound making pseudo-giant high molecular compound. When the pseudo-giant high molecular compound is formed, the viscosity increases extremely and the lubricant is hardly squeezed out from the working surface.

It is found that metal soaps act as good bounding substance to make extremely viscous pseudo-giant molecules. Figure 8 shows the laboratory simulation of pseudo-giant high molecular compound for liquid concentration with a bounding substance of metal soap. The liquid with high polymer compound and the bonding substances turns to a very viscous syrup state when the amount of water is reduced. The extremely viscous pseudo-giant high molecular compound is hardly squeezed out from the work piece surface and can work as good lubricant during forging. Although the most of the pseudo-giant

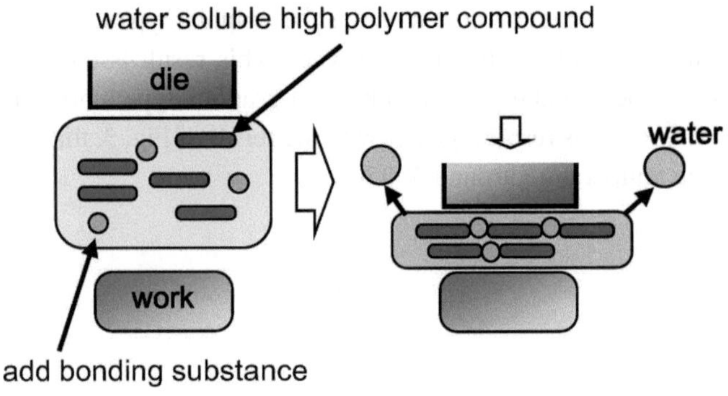

Fig. 7 Model of producing pseudo-giant high molecular compound

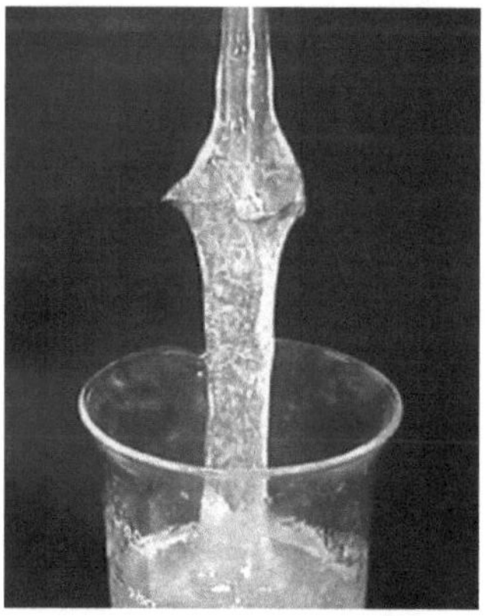

Fig. 8 Laboratory simulation of pseudo-giant high molecular compound

molecule is decomposed during forging, some part remains on the die surface without being decomposed.

It is found that the very viscous pseudo-giant molecular compound shown in Fig. 8 rapidly changes to low viscos material when water is added. This means that the remaining pseudo-giant molecule in the die is dissolved into the water soluble lubricant applied for the next strike and gives no harm to the forging operation [3].

7 Conclusion

To solve the problems of the long-used graphite base lubricants for hot forging, new polymer based lubricants were developed. The water-soluble non-graphite type lubricants offer comparable performances as the graphite lubricants in hot forging when they are dried by contacting with the heated dies. But when they are not dried completely they are squeezed out from the gap between the die and the work piece, and thus careful control of the die temperature was needed. This problem was solved by adding bonding substance which changes the semi-dried water soluble lubricants to an extremely viscous pseudo-giant high molecular compound when it contacts the hot work piece. The pseudo-giant molecular compound is dissolved into the newly applied lubricant for the next strike.

References

1. Ikeda, N: Seminar Text of Materials Process Technology, The Materials Process Technology Center, (2011), 17.
2. Uda, K,Tatsumi, K, Kuroda, M, Ikeda, N: Proceedings of JAST Tribology Conference, Japanese Society of Tribologists, 2009-5(2009), 325.
3. Ikeda, N: Symposium Text of Plasticity Technology 264th, Japan Society for Technology of Plasticity, (2008), 49–56.

References

Composite Extrusion

Christoph Dahnke, Thomas Kloppenborg, Martin Schwane,
Marco Schikorra, Daniel Pietzka, Matthias Kleiner, and
Michael Schomäcker

1 Motivation

Due to economic and environmental reasons, especially in the automotive and aerospace industry as well as in the field of architecture, lightweight strategies have become an important aspect in the design of structural components [1]. Thereby two partially conflicting challenges have to be considered. In the automotive sector for instance, on the one hand a lower overall weight of the vehicle should lead to a reduction of fuel consumption and CO_2 emissions. On the other hand, the components require consistent or improved mechanical properties due to safety aspects. The development of hybrid structural components is a promising approach to cover this demand, as a multi-material design allows the combination of selected mechanical, physical or chemical material properties in one component. With regard to the weight reduction, aluminum is known as an ideal lightweight material, due to its low density associated with good mechanical properties. To meet the mentioned requirements regarding higher strength and stiffness associated with a lower weight, various innovative approaches for the manufacturing of hybrid components have been developed and investigated in the field of extrusion. One of these approaches is the continuous composite extrusion process, which allows the manufacturing of endless reinforced profiles. The process was developed and investigated within

Christoph Dahnke ✉ · Thomas Kloppenborg · Martin Schwane · Matthias Kleiner
TU Dortmund, Dortmund, Germany
e-mail: Christoph.Dahnke@iul.tu-dortmund.de

Marco Schikorra
Vallourec Deutschland GmbH, Mülheim an der Ruhr, Germany

Daniel Pietzka
Schuler Pressen GmbH, Göppingen, Germany

Michael Schomäcker
TU Dresden, Dresden, Germany

A. E. Tekkaya et al. (eds.), *60 Excellent Inventions in Metal Forming*,
DOI 10.1007/978-3-662-46312-3_42

the Transregional Collaborative Research Center/Transregio 10. In comparison to other processes like discontinuous composite extrusion [2], co-extrusion [3] or extrusion of fiber-reinforced billets [4], continuous composite extrusion has several process-related advantages. Composite extrusion allows a continuous reinforcement of the profile at defined positions in the cross section, whereby a wide range of profile geometries can be realized. Furthermore the process is less cost-intensive, due to the usage of conventional billets, which also offers a higher wear resistance of the used dies.

2 Process Idea

The general idea of the process is to embed high strength material into an aluminum matrix by using modified porthole dies (see Fig. 1). In conventional extrusion, porthole dies are used for the manufacture of hollow profiles. There, the material flow is separated by the bridges of the mandrel in at least two strands. Within the welding chamber, the strands are rejoined before the material leaves the orifice of the die. Thereby the orifice of the die defines the geometric shape of the profile. In the continuous composite extrusion process, the reinforcing or functional elements are guided from outside the extrusion press through the bridges and the mandrel. Within the channels of these feeding mandrels, the elements are redirected by 90° and orientated in extrusion direction before they enter the welding chamber. Inside the welding chamber, the elements are embedded in the plane of the longitudinal weld. The elements and the aluminum are extruded through the die, which results in the composite profile.

With regard to a reinforcement of the profile, high strength stainless steel wires, ropes or flat ribbons can be used. Furthermore, the usage of non-metallic reinforcing elements, like ceramic oxide fibers, as well as the usage of functional elements, like isolated electric

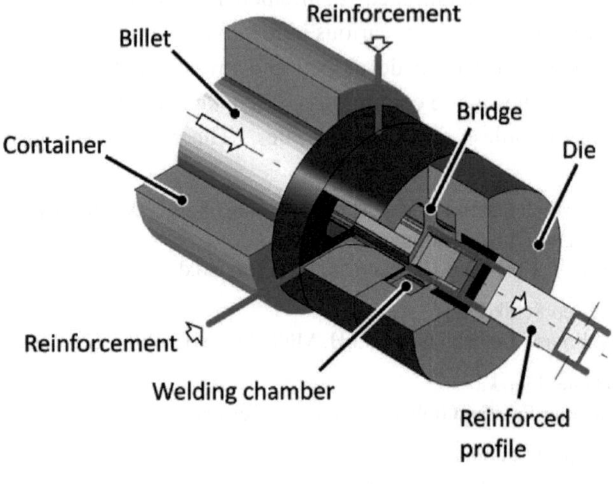

Fig. 1 Process principle of the composite extrusion process

conductors, is possible. Due to the thermal and mechanical loads, which are acting on the elements inside the welding chamber, every die design has to be adapted to the mechanical properties of the elements. Furthermore, the process parameters, like the extrusion ratio, the temperature of the billet or the extrusion velocity have to be considered to achieve a stable and reliable process, without any failure.

3 Results and Applications

In twelve years of continuous development, the composite extrusion process was advanced in various numerical and experimental investigations. In the initial stage of the project, the general feasibility of the process has been proven by the manufacture of rods and simple rectangular profiles with only one or two reinforcing elements. It could be shown, that the positioning of the elements within the profile is depending on the extrusion velocity, the temperature, the position of the feeding and the extruded length [5]. The fundamental investigations lead to the development of basic guidelines for the die design. However, the production of more complex profile geometries with a higher number of embedded elements required an adaption of the particular die. Especially the material flow and the positioning of the longitudinal weld seam had to be determined by the use of 2D [6] and 3D [7] simulations. Due to the numerical optimization methods [8], the manufacture of hollow profiles with up to 14 reinforcing elements or reinforced double-tee profiles could be achieved.

The characterization of the manufactured profiles has shown that in dependence of the reinforcing volume, which is defined as the ration between the cross section of the reinforcing elements and the cross section of the base material, the reinforcement leads to a significant increase of the tensile strength under quasistatic conditions [9]. Furthermore the notched flexural impact strength [10] as well as the bending stiffness [7] can be increased. Beside the dependence on the reinforcing volume, the improvement of the mechanical properties of the reinforced profile is depending on the element material. Primarily high strength stainless steel (X10CrNi18-8) wires with a diameter of 1 mm or 1.5 mm have been used for the manufacturing of continuously reinforced profiles. It could be shown that the maximum reinforcing volume for the embedding of steel wires is limited by the maximum extrusion force, the thickness of the profile, the distance between the single elements as well as the number of the embedded elements, which is depending on the layout of the extrusion press [11]. The theoretical maximum reinforcing volume was calculated to 31 %. In experimental trials a profile with a maximum reinforcing volume of 13.5 % could be manufactured in a stable and reliable process [11], whereby the specific tensile strength could be increased by nearly 50 % (see Fig. 2). A further increase could be realized by the embedding of non-metallic reinforcements like ceramic oxide fibers, which offer a higher strength and a lower density in comparison to metallic elements. The general feasibility could be shown in experimental investigations [11].

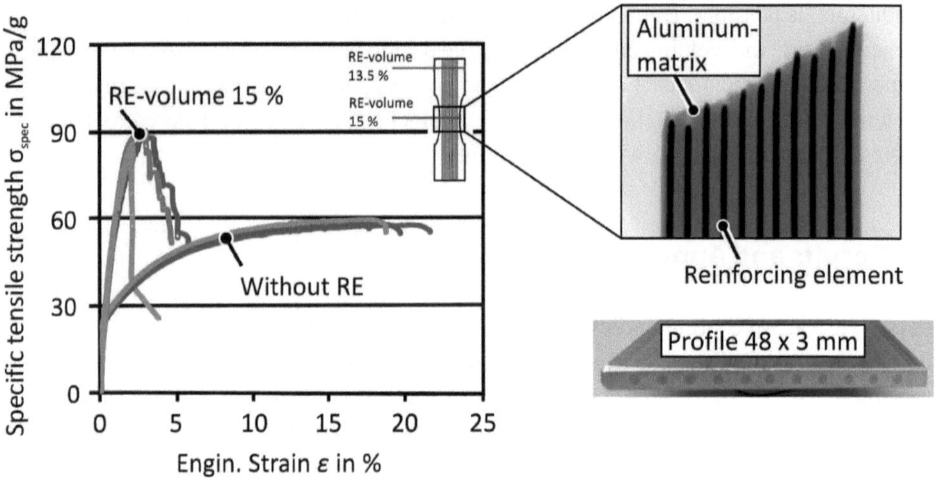

Fig. 2 Specific tensile strength of reinforced profile with a reinforcing volume of 13.5 % [11]

4 Potential

The composite extrusion process offers numerous possibilities for potential industrial applications:

1. Increase of tensile strength: The embedding of reinforcing elements leads to an increase of the tensile strength in dependence of the reinforcing volume as well as the reinforcing material associated with a reduction of the ductility.
2. Increase of bending stiffness: According to an increase of the tensile strength, the bending stiffness can be increased by the integration of reinforcing elements, whereby the increase of the stiffness is disproportional with regard to an increase of the weight [7].
3. Load adjusted components: Due to the possibility of local positioning of the elements as well as the possibility of discontinuous variation of the geometric shape of the reinforcements, load adjusted profiles can be manufactured.
4. High flexibility: The process offers a high flexibility according to the profile geometry, the material of the elements and the matrix material.
5. Increase of functionality: Functional elements, like electrical conductors or data cables, can be embedded in the aluminum matrix (see Fig. 3). This offers a high potential for the transmission of signals or data through the profile as well as a component monitoring.

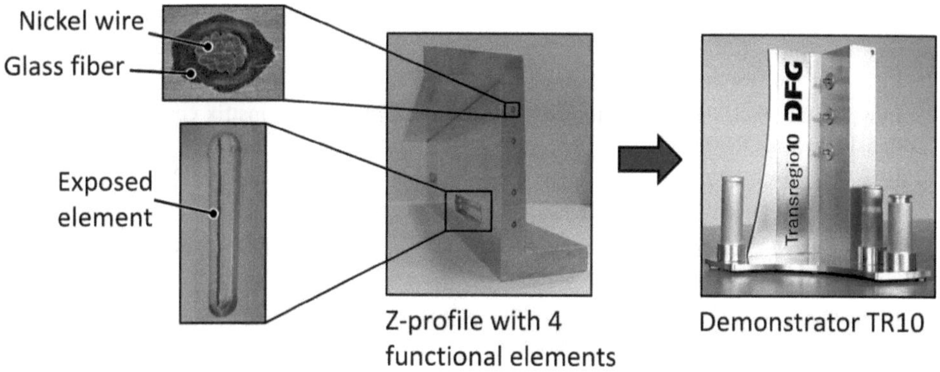

Fig. 3 Z-Profile with functional elements and application in Transregio 10 demonstrator [12]

5 Conclusion

Composite extrusion is an innovative forming process to embed reinforcing or functional elements into a metallic matrix. Various analytical, numerical and experimental investigations have shown that the process is influenced by many parameters, e. g. extrusion velocity, temperature or position of the feeding. The fundamental research as well as the continuous development of the die design in dependence of the mechanical and thermal properties of the elements led to a stable and reliable process in the last couple of years. Lightweight profiles with embedded reinforcing elements possess improved mechanical properties, like higher strength and stiffness. Furthermore, functional elements like isolated electric conductors extend the functionality of the profile. Due to the high flexibility regarding the matrix material, the material of the elements and the profile geometry, composite extrusion offers a high potential for future industrial applications.

6 Acknowledgement

The work is based on the investigations of the subproject A2 – "Composite Extrusion" as well as B1 – "Integral design, simulation and optimization of extrusion dies" of the Transregional Collaborative Research Center/Transregio 10, which is kindly supported by the German Research Foundation (DFG).

References

1. Kleiner M., Geiger M., Klaus A., 2003, Manufacturing of lightweight components by metal forming. CIRP Annals – Manufacturing Technology 52 (2), 521–542.
2. Foydl A., Haase M., Ben Khalifa N., Tekkaya A.E., 2011, Co-extrusion of Discontinuously, Non-centric Steel-reinforced Aluminum, AIP Conference Proceedings 1353 (1), 443.

3. Güzel A., Jäger A., Ben Khalifa N., Tekkaya A. E., 2011, Influence of initial billet geometry on plastic flow during composite rod extrusion, Proceedings of the International Conference on Extrusion and Benchmark, Bologna, Italy.
4. Lloyd D. J., 1994, Particle reinforced aluminium and magnesium matrix composites. International Materials Reviews, 39, 1–23.
5. Schomäker M., 2007, Verbundstrangpressen von Aluminiumprofilen mit endlosen metallischen Verstärkungselementen, Dissertation, Universität Dortmund.
6. Schikorra, M., 2006, Modellierung und simulationsgestützte Analyse des Verbundstrangpressens, Dissertation, Universität Dortmund.
7. Kloppenborg T., 2012, Analyse- und Optimierungsmethoden für das Verbundstrangpressen, Dissertation, Technische Universität Dortmund.
8. Schwane M., Citrea T., Dahnke C., Haase M., Ben Khalifa N., Tekkaya A.E., 2014, Simulation of Composite Hot Extrusion with High Reinforcing Volumes, 11th International Conference on Technology of Plasticity, ICTP 2014, 81, 1265–1270.
9. Merzkirch M., Weidenmann K. A., Schulze V., 2011, Werkstoffkundliche Charakterisierung verbundstranggepresster Leichtmetallmatrix-Verbundwerkstoffe. In: Fortschritt-Berichte VDI, Tekkaya, A. E., Baier, H., Biermann, Fleischer, J., Schulze, V., Zäh, M. F., Pietzka, D. (Hrsg.). Integration von Umformen, Trennen und Fügen für die flexible Fertigung von leichten Tragwerkstrukturen. Reihe 2. Nr. 678. VDI Verlag, Düsseldorf. 49–72.
10. Weidenmann K. A., Hammers T., Merzkirch M., Kerscher E., 2009, Charakterisierung des mechanischen Verhaltens verbundstranggepresster Leichtbauprofile unterschlagartiger Beanspruchung. In: Verbundwerkstoffe, 17. Symposium Verbundwerkstoffe und Werkstoffverbunde. Krenkel W. (Hrsg.). Wiley-VCH Verlag, Weinheim. 168–173.
11. Pietzka D., 2013, Erweiterung des Verbundstrangpressens zu höheren Verstärkungsanteilen und funktionalen Verbunden, Dissertation, Technische Universität Dortmund.
12. Dahnke C., Pietzka D., Haase M., Tekkaya A.E., 2014, Extending the Flexibility in the Composite Extrusion Process, Proceedings of the International Conference on Manufacturing of Lightweight Components – ManuLight 2014, 18, 33–38.

Novel Billet Design for Co-extrusion of Bi-metallic Shapes and Tubes

Mario E. Epler and Wojciech Z. Misiolek

1 Motivation

Co-extrusion of solid shapes and tubes is a manufacturing technique where an initial extrusion billet is composed of multiple materials that are concurrently extruded into a bonded product. Two or more different materials are used to form a final composite shape. Production of composites in this manner maximizes the benefits of the individual component materials. Often, one alloy or material is selected because of a specific property, such as corrosion resistance. This is then paired with another material that may either be of lower density or less expensive, or has other desirable physical properties for the final product. One alloy may be used in the form of a cladding in the final product with the additional material present to provide support and strength. Among the major benefits of using extrusion to produce composite parts is that there is no need for any additional post deformation bonding process to manufacture the final part. As the materials are deformed into a net or near net shape, they are concurrently metallurgically bonded while formed into a final shape. This eliminates the need for processes such as fusion welding (both overlays and joining) or mechanical joining (through the use of fasteners or adhesives). However, the process is limited to geometries that can be formed in extrusion. Extrusion of bi-materials presents problems that are not encountered when extruding a single material. Dimensional tolerances of the materials used and the bond location become an important consideration, along with the strength of the bond, and the potential for greater amounts of scrap due to uneven flow of the two materials.

Mario E. Epler · Wojciech Z. Misiolek ✉
Lehigh University, Bethlehem, USA
e-mail: wzm2@lehigh.edu

A. E. Tekkaya et al. (eds.), *60 Excellent Inventions in Metal Forming*,
DOI 10.1007/978-3-662-46312-3_43

2 Process Idea

The presented invention is focused on predicting and controlling the final extruded geometries in reference to the bond location (i. e. thickness and location of the component layers). Growth of applications utilizing bi-material extruded products inspired many theoretical and experimental works that have been performed in recent years to understand the bond development in bi-material extrudates [1–6]. Many factors controlling bi-material solid rod and tube manufacturing processes were established and optimized. As shown before in literature [7], a new successful method was developed for improving the amount of material that is "within tolerance" after co-extrusion of solid shapes made from different aluminum alloys. This method was initially developed using round-to-round extrusion, where the core material was not the full billet length as it is presented in Fig. 1b.

The presented innovation was applied and then confirmed in the case of extruded tubes composed of a stainless steel (UNS S30400, referred to as 304) core and a plain carbon steel sleeve (UNSG10211, referred to as 1020). Testing was performed on multiple billet geometries in reference to initial core and sleeve thickness as well as two different extrusion ratios. Initially the FEM simulation was performed using Deform 2D™ software to predict the influence of process parameters as well as billet geometry on the final product geometry for extrusion ratios of 3.3 and 10.6. The FEM simulations (see Fig. 2) show that the core in a traditional billet with its full length is "ballooning" in the front end of extrudate. In the case of a shorten core length, the modified billet designs produce im-

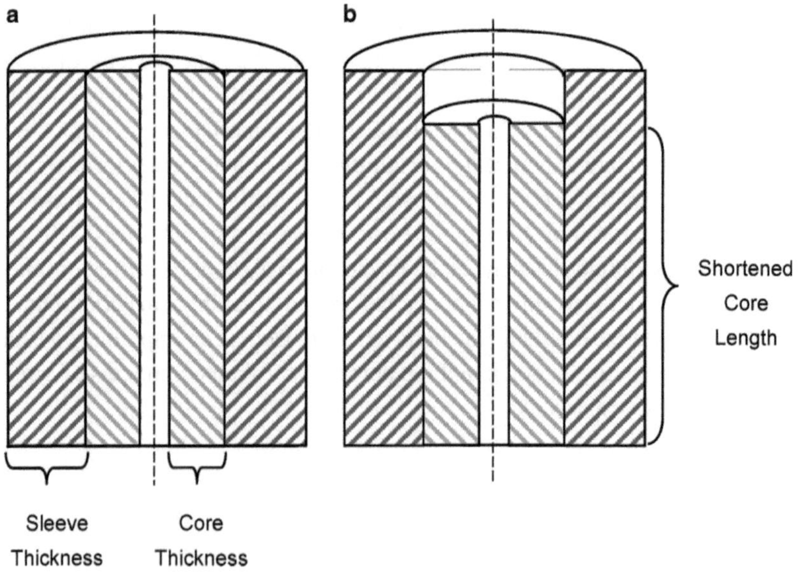

Fig. 1 Schematic cross section of unextruded tubular billet with the extrusion direction upward. **a** traditional design, **b** core shortened geometry

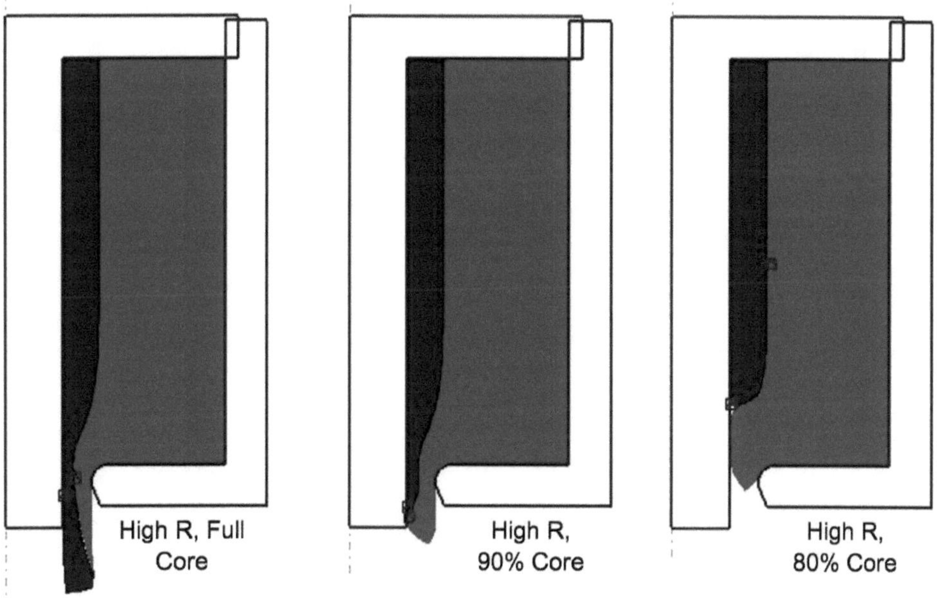

Fig. 2 FEM predictions at early stage of extrusion and front-end geometry development for higher extrusion ratio (R = 10.6) extrudates

proved extrusion results. The sleeve material upsets and fills the gap left by the missing core material and extrude ahead of the core material. This way a limited scrap material is formed and it is followed by extrudate with controlled location of the interface bond. By reducing the core by only 10 %, concurrent flow of both materials is still promoted. However, a reduction of 20 % results in better material flow (depending on the extrusion ratio) and lower material loses.

This has been confirmed for a range of extrusion ratios in both modeling and actual industrial extrusion experiments. The detailed description of numerical simulation and experimental extrusion verification are available in literature [6, 7, 9].

3 Proof of Concept

For verification purpose the core thickness data points were extracted from the FEM simulations and were compared to the core thicknesses measured in the actual extrudates. For the extrudates, the average core thickness value was calculated for each slice from four measurements equally spaced around the tube circumference. Core thickness was plotted as a function of tube length from the front of the extrudate as presented in Fig. 3 showing very good agreement between predicted and actual values.

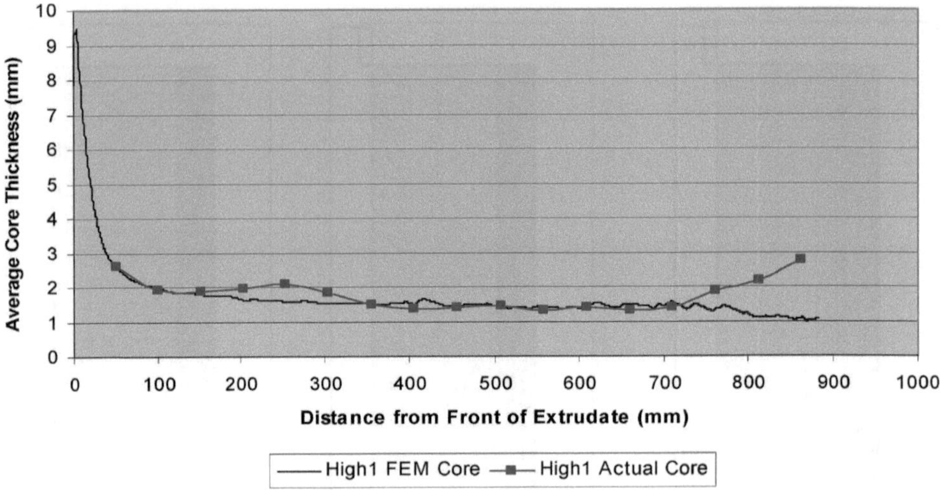

Fig. 3 FEM predictions vs Actual Average Core Thickness as a Function of Distance for a traditional billet design and extrusion ratio, R = 10.6

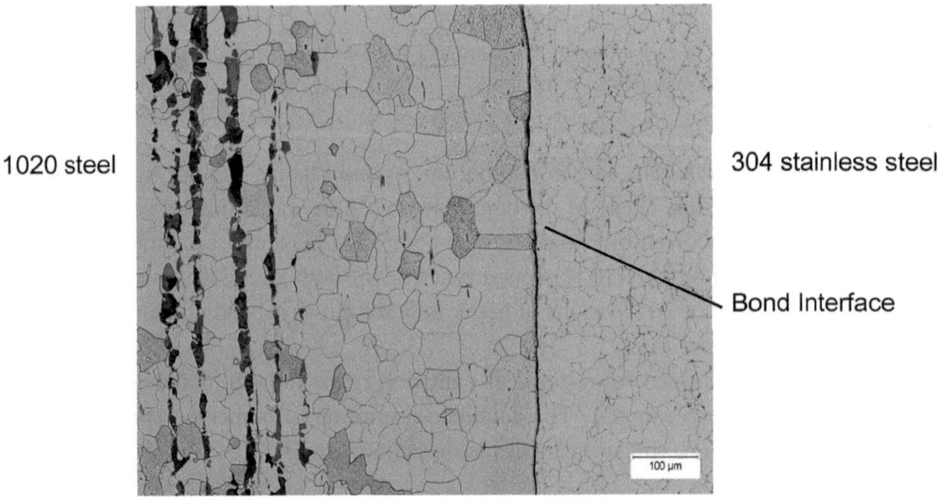

Fig. 4 Interface between 1020 steel (**a**) and 304 stainless steel (**b**) after 2 hours at 1100°C and co-extruded followed by air cool to room temperature extracted from the middle of the extrudate length. Etched with agitated 60 mL HCl, 60 mL HNO_3, and 60 mL distilled H_2O for 30 seconds

It is extremely important to prove that the proposed method does not negatively influence integrity of the weld between two materials along the extrudate and allows reaching steady state sooner. Metallographic analysis was performed and continuous, porosity free metallurgical bond is presented in Fig. 4.

4 Summary

The proposed innovative design of bi-material extrusion billet allows for significant reduction of the product "out of the desired tolerances" (measured by the interface bond location) through the improvement of metal flow during the co-extrusion process. The core shortening method controls material flow at the front of the extrudate during the initial part of the extrusion process. In addition to minimizing the out of tolerance material in the front of the extrudate, which directly corresponds to scrap, a material savings is also realized in the initial billet by shortening the amount of core material. Shortening the core material length up to 20 % can be effective in managing sound material flow while still extruding the same amount of in-tolerance material. Core shortening acts to cause the sleeve material to upset into the recessed region of the billet so that concurrent flow is achieved. Less material is lost to early flow out of the die and thus steady-state core thickness flow conditions are reached sooner. The actual amount of shortening is governed by the specific extrusion process parameters for the intended solid shape or tube geometry. During steady state material flow, there is little or no difference in core material distribution between extrudates produced from a billet with a full length or shortened core. Maximum savings can be achieved with the amount of upsetting being in balance with the flow of the core material into the die orifice, which is also influenced by the die angle. The performed numerical simulations and industrial experimental verification showed sound metallurgical bond between two different materials.

5 Acknowledgement

The authors would like to acknowledge financial support from U.S. Department of Energy – grant No. DE-FC07-01ID14254. Partial support of Wojciech Z. Misiolek is provided by the Loewy Family Foundation through the Loewy Professorship at Lehigh University. The authors would like to acknowledge fruitful discussions with Dr. Steven Claves and Dr. Pawel Kazanowski.

References

1. Sliwa, R., J. Mat. Proc. Techno. Vol. 67, pp. 29, 1996
2. Bandar, A. R., Misiolek, W.Z., Kloske, K.E., Jeong, T.H., Improving Flow in Soft-Core Bi-material Billets, Proceedings of the 7th International Aluminum Extrusion Technology Seminar ET2000, Chicago, Illinois, Aluminum Extruders Council and Aluminum Association, Vol. 2, p. 223, May 16–19, 2000
3. Erman, E., Semiatin, S.L, Physical Modeling of Metalworking Processes, The Metallurgical Society, Inc., Warrendale, PA, p. 1, 1987
4. Gomez, X., Echeberria, J., Mater. Sci. Eng. A, Vol. 348, pp. 180, 2003

5. Alcaraz, J.L, Gil-Sevillano, J., Inter. J. Mech. Sci., Vol. 38, pp. 157, 1996
6. Kazanowski, P., Misiolek, W.Z., Sikka, V.K., Materials Science Forum, Vols. 426–432, pp. 3795–3800, 2003
7. Kazanowski, P, Epler, M.E., Misiolek, W. Z., Journal of Materials Science & Engineering A, Vol. 369, no. 1-2, pp. 170–180. 25 Mar. 2004
8. Coextrusion welding, ASM International, ASM Handbook Vol. 6: Welding, Brazing, and Soldering (USA), pp. 311, 1993
9. Epler, M. E., Misiolek, W.Z., Materials Science and Engineering A 429 (2006) 43–49

Curved Profile Extrusion

Alessandro Selvaggio, Dirk Becker, Alexander Klaus, Dieter Arendes, and Matthias Kleiner

1 Motivation

In space frame structures of the modern transport industry, the application of straight and curved profiles is common use. In low volume production, the space frame construction in the automotive industry focuses on the application of extruded aluminum profiles. In particular, three-dimensionally curved profiles are used to avoid multiple profiles joined, for example, with cast nodes with regard to the lightweight properties. Today, the manufacture of such profiles is very complex and often requires expensive tooling. In the initial stage of the traditional production of curved profiles, a straight semi-finished product is extruded and formed into the curved shape in a subsequent bending process. However, certain problems of the bending processes have to be taken into account, in particular the springback, the cross-section deformation with wall thinning, the tooling costs, etc [1]. The challenge is to obtain a profile with an acceptable accuracy by using a cost efficient manufacturing technology, also for small quantities. In the following, a technology and its enhancement for the manufacturing of curved profiles is presented to avoid the mentioned negative aspects of the conventional production process.

2 Process Idea

The curved profile extrusion (CPE) is a substitute process for the conventional process chain for the manufacturing of curved profiles by extrusion and bending. In the new process, which was initially introduced by Kleiner [2], curved profiles are produced by influencing the material flow during the conventional extrusion process. The basis of this

Alessandro Selvaggio ✉ · Dirk Becker · Alexander Klaus · Dieter Arendes · Matthias Kleiner
TU Dortmund, Dortmund, Germany
e-mail: Alessandro.Selvaggio@iul.tu-dortmund.de

A. E. Tekkaya et al. (eds.), *60 Excellent Inventions in Metal Forming*,
DOI 10.1007/978-3-662-46312-3_44

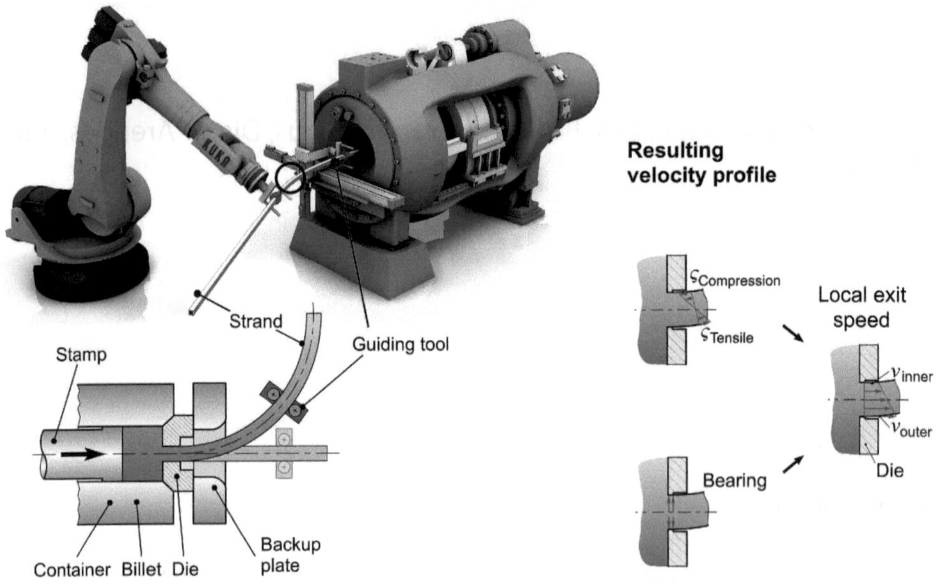

Fig. 1 Principle of the curved profile extrusion process

method is the kinematic shape generation. The exiting strand of an extrusion process is fed into a guiding tool, which can be moved by a linear axis system in at least one dimension (see Fig. 1).

The deflection of the profile influences the material flow within the die. Due to the high temperature and large press forces, the condition within the extrusion die can be assumed as visco-plastic. Therefore, only small additional forces are necessary to move the guiding tool. By moving the guiding tool and keeping the equation of continuity in mind, the material in the die is decelerated at the inner side and accelerated at the outer side of the profile radius. Due to the differences in the local exit velocities over the cross-section of the profile at the die exit, curvature is generated (see Fig. 1).

Different geometries of curved profiles are produced by moving the guiding tool on a path synchronously to the profile velocity defined by the press conditions. Due to the exploitation of the viscous material behavior within the die, this forming technology causes no cross section deformation, no springback, and nearly no decrease in formability [3, 4].

3 Results and Applications

To show the difference between CPE and warm bending, the forces for the shape generation were measured in two experiments. The first experiment was CPE with altering the material flow in the die during extrusion, to produce a plane curved profile with a radius of 1,500 mm. In a second experiment, the warm bending test was performed subsequently

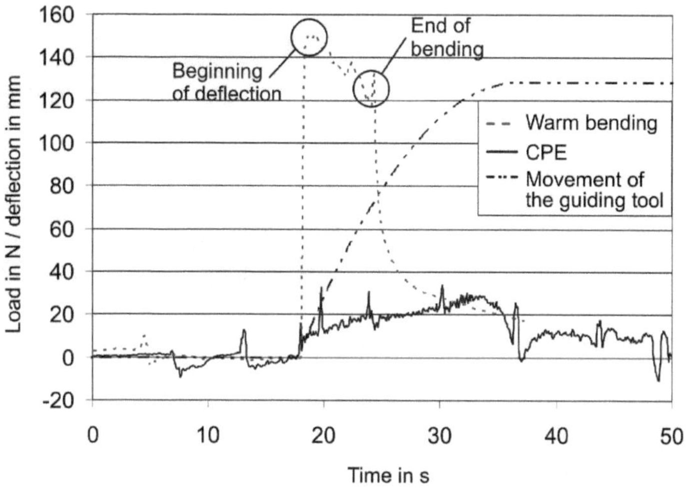

Fig. 2 Load results of warm bending and CPE

to the extrusion process. Here, the extrusion process was started and run until the steady state was reached, which is the case when the target speed is attained and the press force slowly decreases. Then, the press stops and the same deflection of the guiding tool starts to introduce the bending process. The results of both force measurements are shown in Fig. 2.

The diagram indicates that CPE requires a maximum force of only 34 N to deflect the strand for the curvature generation. In comparison to CPE, the warm bending process needs initial forces which are up to five times higher (150 N) to deflect the profile. This result makes the difference between CPE and warm bending obvious [5]. Due to the low necessary forces for the CPE, it is important that no unintended forces, like the forces related to the dead weight of the extrudate, act on the profile, as this would result in an altered final shape of the profile. To manufacture a large three-dimensionally curved profile, an additional supporting tool is required to avoid unintentional bending of the profile by dead weight. The maximum possible flexibility concerning the movement and the acting range is given by a 6-axis-robot system (e. g. Kuka KR150-2) [6]. Here the guiding tool and the support tool have to be synchronized with the profile speed very accurate in order to have small deviation of the realized 3D-contour of the profile. But the PLC of the extrusion press only provides processed values of the ram speed, where the output of the profile speed is different compared to the real measured one. For this reason, a high-resolution displacement transducer for the ram with direct coupling with the control computer was used for the following experiments [5]. During the involvement of the incremental measuring system into the existing extrusion press, it was considered that no thermal and mechanical influences falsify the measurement result. By using the

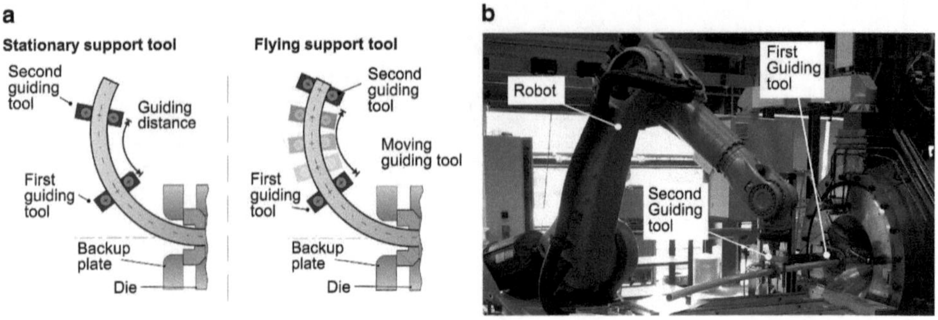

Fig. 3 **a** Support strategies, **b** experiments with profile support

new displacement transducer, the repeatability of the profile lengths could be improved from a deviation of ±8 mm to a deviation of ±3.8 mm [7, 8].

In CPE even small forces have an influence on the resulting radius. Therefore it is necessary to support the profile in order to reduce unwanted deformations caused by the weight, which would affect the material flow and therefore result in an unintended profile contour. In experimental investigations, the accuracy of different support strategies regarding the actual and desired contour of the profile were analyzed (see also [9]). The following alternatives were considered as support strategies (see Fig. 3): 1. The profile is guided on a fixed distance behind the guide tool, so that the guiding tool moves on a hemisphere with a constant radius 2. The profile is always supported at the center of gravity, so that the bending forces are reduced to a minimum

The results of the experimental investigation are quantified by a comparison between the actual and the desired contour. The curved profile is digitized using a 3D geometry acquisition system and then compared to a CAD drawing. Deviations could be reduced by 25 % by the usage of the flying support tool (see Fig. 4). Without supporting tool,

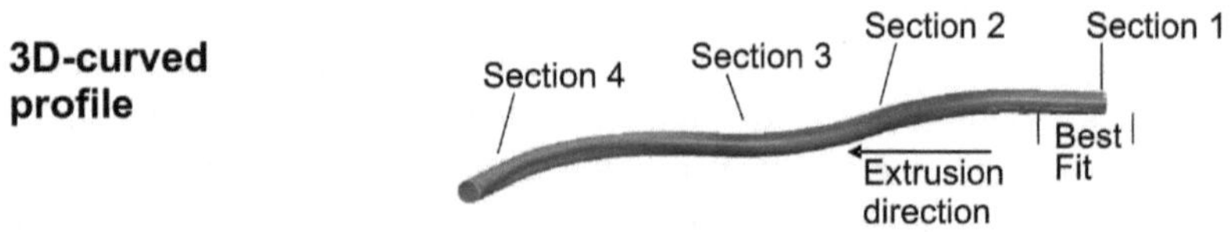

Stationary support tool: **2295 mm**
Profile velocity: **17.5 mm/s**

Section / Deviation in mm	1	2	3	4
Min. Deviation	0.6	0.9	0.9	0.9
Max. Deviation	1.5	1.3	1,6	2.0
Avg. Deviation	1.0	1.1	1.4	1.6

Flying support tool: **center of gravity**
Profile velocity: **17.5 mm/s**

Section / Deviation in mm	1	2	3	4
Min. Deviation	0.5	1,0	0,9	0.6
Max. Deviation	0.7	1.1	0,9	2.0
Avg. Deviation	0.6	1.0	0.9	1.3

Fig. 4 **a** Support strategies, **b** experiments with profile support

the CPE is not useful because of the high temperatures, since the profiles are strongly deformed. Regarding the accuracy of the profiles, the influence of gravity has to be considered during CPE. With the usage of a second guiding tool handled flexibly by a robot and the choice of the optimal supporting strategy, the influence of the gravity could be successfully compensated.

4 Potentials

The process of CPE offers the possibility for the production of 3D-curved profiles and numerous potentialities. With the CPE, even profiles with complex cross sections can be curved in 3D without deformation and reduction of the formability. This means, that a multitude of profiles types can be considered by the designer of new components (e. g. space frame constructions). Also, the absence of springback is leading to a high accuracy of the produced profiles. Therefore, the CPE is highly suitable for the small batch production, due to the flexibility of the process and due to the lack of additional costs for the bending tool.

5 Conclusion

The curved profile extrusion represents an alternative technology for the production of curved profiles with better properties compared to bent profiles concerning the cross-section deformation and the residual stress. In addition, the tooling costs are significantly reduced. The investigations show that CPE, in comparison to a conventional warm bending process, requires considerably less forming force. And in addition, due to the deflection of the strand during the extrusion, a moment is superimposed on the already plasticized material in the die and influences the material flow. Due to the exploitation of the viscous material behavior within the die, this innovative forming technology causes no cross section deformation, no springback, and nearly no decrease in formability. The process of CPE allows the production of complex shaped structure components with high accuracy.

6 Acknowledgement

This paper is based on investigations of the subproject A1 – "Multi-Axis Curved Profile Extrusion" – of the Transregional Collaborative Research Center/Transregio 10, which is kindly supported by the German Research Foundation (DFG).

References

1. Vollertsen, F.; Sprenger, A.; Kraus, J.; Arnet, H.: Extrusion, channel, and profile bending: a review. Journal of Materials Processing Technology 87 (1999), p. 1 27.
2. Kleiner, M.: Patent EP 0 706 843 B1: Verfahren und Vorrichtung zur Herstellung von gekrümmten Werkstücken. 1999.
3. Arendes, D.: Direkte Fertigung gerundeter Aluminiumprofile beim Strangpressen. Dr.-Ing. Dissertation, Universität Dortmund, Shaker Verlag Aachen, 1999
4. Klaus, A.: Verbesserung der Fertigungsgenauigkeit und der Prozesssicherheit des Rundens beim Strangpressen. Dr.-Ing. Dissertation, Universität Dortmund, Shaker Verlag Aachen, 2002
5. Becker, D.; Schikorra, M.; Tekkaya, A. E.: Manufacture of 3D Curved Profiles for Structure Components, In: Advanced Materials Research: Flexible Manufacture of Lightweight Frame Structures – Phase II: Integration, Volumes 43 (2008), pp. 1–8, ISBN 0-87849-385-9
6. Kleiner, M.; Becker, D.; Klaus, A.: 3D-Enhancement of Curved Profile Extrusion by Means of a Robot Support, In: Annals of the German Academic Society for Production Engineering, WGP, Vol. XIII/1 (2006) S. 35–38, ISSN 0944-6524
7. Schneider, M.; Koehler, G.; Becker, D.; Selvaggio, A.; Tekkaya, A. Erman.; Munzinger, C.; Schulze, V; Kleiner, M.: Towards the flexible and near-net-shape production of three-dimensionally curved extrusion profiles, In: Production Engineering, 2010, Springer Verlag, Berlin/Heidelberg, ISBN/ISSN 1863-7353, DOI: 10.1007/s11740-010-0246-1
8. Selvaggio, A.; Tekkaya, A. E.: Mehrachsiges Runden beim Strangpressen, In: VDI Fortschritt-Berichte, Integration von Umformen, Trennen und Fügen für die flexible Fertigung von leichten Tragwerkstrukturen, Reihe 2, Nr. 678, VDI Verlag Düsseldorf, 2011, S. 3–25
9. Becker, D.; Selvaggio, A.; Marré, M.; Tekkaya, A. E.: Industrielle Anwendungen für das neue Fertigungsverfahren Runden beim Strangpressen, In: Lightweightdesign, Ausg. 4/5 2009, S. 78–86

Part VIII

Cold Forging

Joining of a Shaft-Hub Connection by Lateral Extrusion

Florian Dörr and Mathias Liewald

1 Motivation

Joining by plastic deformation in the field of bulk metal forming is currently experiencing a renaissance. The main reason for these developments is that there are current requirements regarding a lightweight material design of bulk metal-formed components and function integration. The second reason for these developments is that powerful machining processes are available with growing automation, fewer staff and hence lower unit cost at the same or higher quality. In addition, the number of derivatives due to more individual products is increasing at a constant total number of units so that more flexible manufacturing processes are required. To enhance the attractiveness of manufacturing processes in the field of bulk metal forming, new and more complex products as well as the integration of additional production steps to reduce production costs will be necessary in the near future. Combining the two basic manufacturing processes of forming and joining allows metal forming processes to remain competitive. In contrast to conventional joining methods, joining by forming allows the connection of parts in a wide range, showing less distortion, high process reliability and a simple quality control. However, such methods disclose numerous disadvantages or challenges, such as the geometrical unevenness of the joining, lack of standardization or other difficulties in terms of correction and repairing [2].

Florian Dörr · Mathias Liewald ✉
University of Stuttgart, Stuttgart, Germany
e-mail: Mathias.Liewald@ifu.uni-stuttgart.de

A. E. Tekkaya et al. (eds.), *60 Excellent Inventions in Metal Forming*,
DOI 10.1007/978-3-662-46312-3_45

2 Process Idea

In the field of bulk metal forming, numerous joining methods through plastic deformation are known, and some are extensively researched in science and already used in industrial applications. Such joining methods through plastic deformation are associated with intermediate joining methods due to the fact that no auxiliary part is necessary. According to Lange [3] and DIN 8593-5 [4], two or more raw parts stacked upon each other or stuck into each other can be joined by cold forging of both parts. For this purpose, Fig. 1 shows an overview of possible joining methods through plastic deformation in the field of cold forging. The structure of this scheme corresponds to the classification of basic cold forging processes. In addition, it is distinguished how the joint members are deformed. Thus, both components can be deformed in an elastic-plastic manner, or just one of the joining parts is deformed elastic-plastically and the second joining part only suffers an elastic deformation. This overview does not claim completeness because many other raw part combinations are possible just by a further subdivision of the two joining partners shown here.

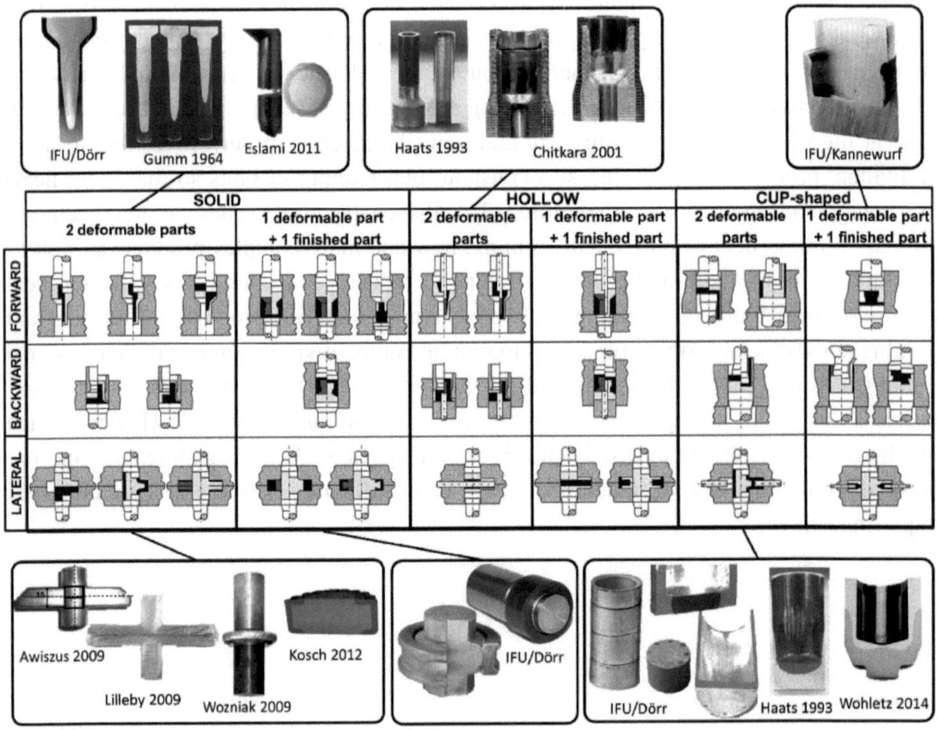

Fig. 1 Scheme of common cold forging methods [5]

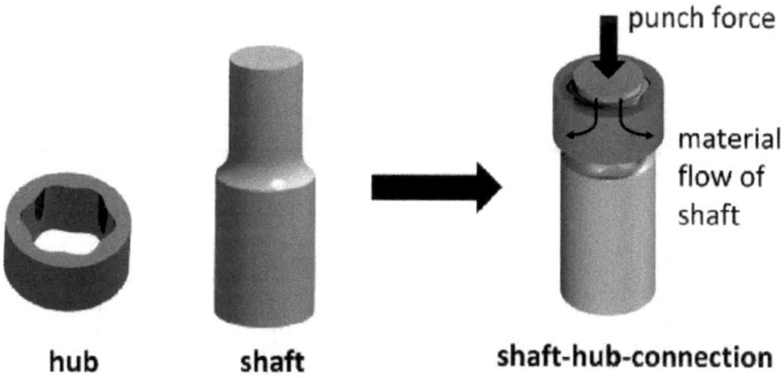

Fig. 2 Joining of a shaft-hub connection by lateral extrusion

The manufacturing of a shaft-hub connection by lateral extrusion can also be found within this scheme of well-known cold forging methods. In this joining by forming process, the shaft is deformed in an elastic-plastic manner whereas the hub is deformed purely elastically. As a result, a positive and force-closed connection of both parts is generated (see Fig. 2)

3 Results and Applications

Two important factors influencing the connection strength of such components are the tribological conditions within the contact area of both parts on the one hand and the usage of a suitable internal hub profile. By using cylindrical, thin-walled hubs and measuring the radial deformation of the hub after the forming and joining process, it is possible to determine the contact pressure and to calculate a normal force applied to the contact surface of the shaft and the hub. If the hub and the shaft are separated in a destructive manner, an extensive axial force is applied and measured. According to Coulomb's friction law, specific friction coefficients are calculated depending on tribological parameters, such as shaft and hub contact surface topographies and lubricant. The static and sliding friction coefficients determined in this way are shown in Fig. 3.

During the joining process, the hub is necessarily expanded in an elastic manner to realize a non-positive joint. If non-cylindrical internal hub profiles are used and the lateral material flow of the shaft starts, the first contact between the shaft and the hub occurs at the minimum turning point of the internal hub profile. Therefore, the radial deformation shows a positive value in this area at this process time, and the roundness deviation increases (see Fig. 4). At the maximum turning point, the two joint partners do not come into contact and the hub sinks in, which causes negative values for the radial deformation. This effect significantly depends on the profile eccentricity because a higher profile eccentricity causes a greater difference between minimum and maximum turning points, wall thicknesses and the structural stiffness of the hub. A further increase in filling degree leads to an increase in contact area and thus to an elastic expansion of the hub at the maximum turning point as well. A local minimum occurs at different filling degrees, depending on

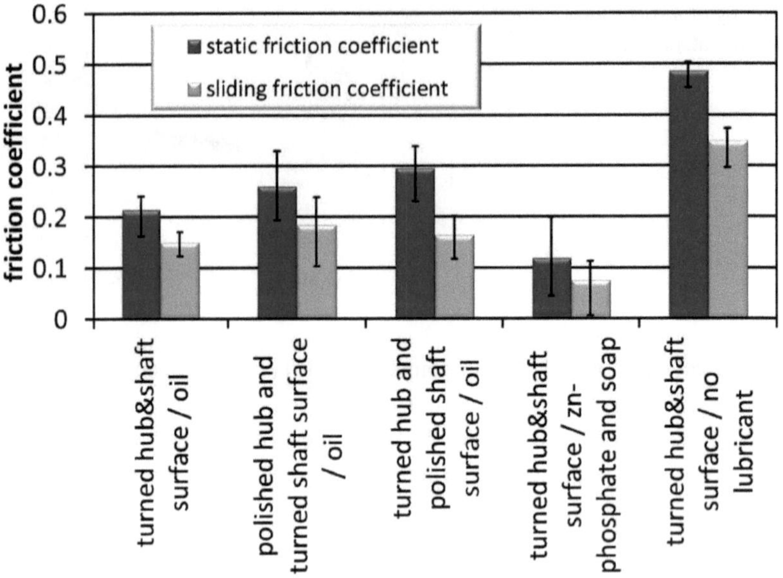

Fig. 3 Influence of tribology conditions on static and sliding friction coefficients [6]

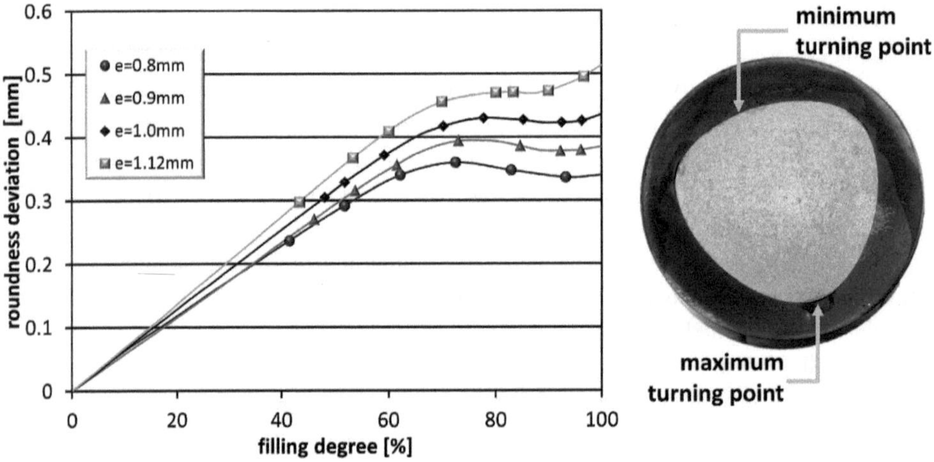

Fig. 4 Influence of profile eccentricity and filling degree on the roundness deviation of joined hubs [7]

the profile eccentricity. To realize a low radial deformation and also a high filling degree, different filling degrees are recommended for different profile eccentricities.

Apart from the process investigations, the performance characteristics of these components, such as fatigue strength under torsional load, are also determined. Therefore, endurance tests have been performed at a purely pulsating torsional load of 30 Hz using a polygonal internal hub profile and a profile eccentricity of 0.8 mm. At the beginning of the fatigue test, a training effect or rather a hardening of the connection strength occurs, which causes a change in the rotation angle. Hence, a reduction in the rotation angle by about 3.5 times is measured within the approximately first 200 load cycles, when applying a purely pulsating torsional load of 450 Nm. A shaft-hub connection is regarded as having high fatigue strength if it shows no functional failure, for example such as a hub or shaft fracture above $2 \cdot 10^7$ load cycles. The tested polygonal shaft-hub connections have high fatigue strength for a purely pulsating torsional load of 500 Nm. If such thin-walled shaft-hub connections are used, the stress levels within the shaft and the hub are very similar. This is why even small deviations of the surface quality or microstructures of the quenched and tempered hub can reduce the fatigue life of such connections. An SEM (scanning electron microscope) study on a hub failing due to a hub fracture has shown that even an unfavourable coincidence of turning grooves and manganese sulphide accumulations within the microstructure at the internal surface of the hub can cause a fracture at a very low load level. The investigations of fatigue strength also show that the determined fatigue strength of such shaft-hub connections with a nominal joining diameter of 25 mm and a purely pulsating torsional load of 500 Nm is higher than the plasticization limit of the undeformed shaft with a diameter of 30 mm (about 400 Nm). This increase in connection strength is mainly caused by the cold work hardening of the shaft during the joining and forming process [8].

4 Conclusion

Manufacturing of shaft-hub connection by lateral extrusion of the shaft stands for an innovative process combination which allows producing heavy duty components. For that reason in future there is no need to produce both partners with tight tolerances and to overcome hurdles of combinations of tolerances to ensure a proper connection between both joint partners or to heat external joint partner. If optimal process parameters are used for the internal hub profile or the tribological conditions, high-strength torsional loads can be transmitted by such components. Such a joining process also allows using highly complex internal hub profiles, from which a further increase in the connection strength is to be expected.

5 Acknowledgement

The presented research is part of the project "Manufacturing of a shaft-hub connection by lateral extrusion", funded by the German Research Foundation (LI 1556/24-1 and BI 746/10-1). The authors would like to thank the German Research Foundation (DFG) for their financial support.

References

1. Funk, M.; Dörr, F.; Binz, H.; Liewald, M.: Investigations of transmission capacity of polygonal shaft-hub-connection manufactured by lateral extrusion, VDI-Report 2176, Nürtingen, 2012, 83–93; ISBN 978-3-18-092176-1
2. Mori, K.; Bay, N.; Fratini, L.; Micari, F.; Tekkaya, A. E.: Joining by plastic deformation. CIRP Annals – Manufacturing Technology 2013, Volume 62, Issue 2, 2013, pages 673–694
3. K. Lange: Umformtechnik – Handbuch für Industrie und Wissenschaft, Bd. 4: Sonderverfahren, Prozesssimulation, Werkzeugtechnik, Produktion; 2. Auflage, Springer-Verlag Berlin, Heidelberg, 1993; ISBN: 3-540-55939-6
4. DIN 8593-5: Fertigungsverfahren Fügen Teil 5: Fügen durch Umformen Einordnung, Unterteilung, Begriffe. Normenausschuss Technische Grundlagen (NATG) – Geometrische Produktspezifikation und -prüfung – im DIN Deutsches Institut für Normung e. V., September 2003
5. Liewald, M.: Current Research Activities along the process flow for Manufacturing of Hollow Lightweight components at University of Stuttgart. Proceedings of 34th SENAFOR, Porto Alegre/RS, 2014
6. Dörr, F., Funk, M., Liewald, M., Binz, H., Savino, M.: Determination of friction coefficients of interstice of a shaft-hub-connection manufactured by lateral extrusion. Advanced Materials Research, Volumes 966–967, 2014, 659–670
7. Dörr, F., Funk, M., Liewald, M., Binz, H., Köstlmeier, R.: Influence of Internal Hub Profile on Joining Process of Shaft-hub-connection by Lateral Extrusion. Procedia Engineering, Volume 81, 2014, 1988–1993
8. Funk, M., Dörr, F., Binz, H., Liewald, M.: Übertragungsfähigkeit mittels Quer-Fließpressen hergestellter Welle-Nabe-Verbindungen mit polygonaler Nabeninnengeometrie bei zyklischer Torsionsbeanspruchung. In: VDI-Bericht 2238 (6. VDI-Fachtagung Welle-Nabe-Verbindungen Dimensionierung – Fertigung – Anwendungen). Karlsruhe, 2014, 201–212; ISBN 978-3-18-092238-6

Divided Flow Method

Kazuyoshi Kondo

1 Motivation

The benefit of high productivity of cold forging is often deteriorated by the necessity of successive finishing operation. Especially, almost all of the precision machine components must be produced by machining, grinding and polishing. If the net shape forging which does not need successive finishing operation is achieved, production cost of industrial components will be reduced remarkably. In this research, development of the net shape forging process is aimed.

2 Process Idea

When net shape forging is aimed at conventional cold die forging, tool failure occurs easily because of high working pressure. "Enclosed die forging" which utilizes enclosed pressure of upper die and lower die as safety guard will be a remedy. But the application range of this process is limited to relatively simple contour shape of the product. From the basic consideration of the components of working pressure in forging two basic concepts are obtained. 1. To reduce pure deformation resistance completely enclosed condition must be avoided by utilizing relief axis or relief hole. 2. To reduce frictional resistance the ratio of the contacting tool width w to material thickness t, w/t must be reduced. In order to realize these concepts in the actual production process, the idea of "Divided Flow Method" shown in Fig. 1 is proposed at first. Basic research shows that filling up state of gear teeth utilized relief hole is better than that utilized relief axis. But these processes are too simple to achieve net gear shape. To achieve real net shape forging "Divided Flow Method" must

Kazuyoshi Kondo ✉
Toyota Technological Institute, Nagoya, Japan
e-mail: kondoh@toyota-ti.ac.jp

A. E. Tekkaya et al. (eds.), *60 Excellent Inventions in Metal Forming*,
DOI 10.1007/978-3-662-46312-3_46

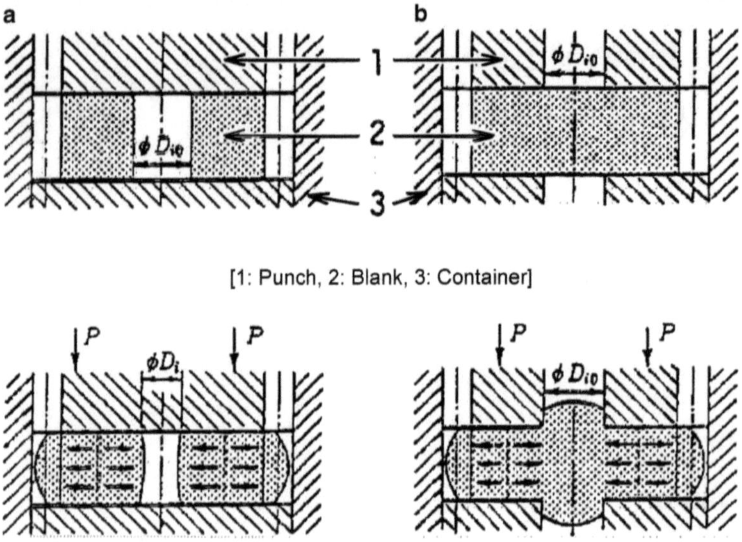

Fig. 1 Forging process utilizing divided flow. **a** Flow relief hole principle, **b** Flow relief axis principle

be constructed as two steps process as shown in Fig. 2. Figure shows the case when a ring specimen is used. The first step is a conventional forging in which mandrels are inserted in the hole of specimen. Punch stroke must be stopped within allowable working pressure for tools. At this stage, the top face of the product gear does not reach to the top face of the die. At the second step, mandrels are withdrawn by the specified amount to open relief hole and punch is depressed. Appropriate selection of the size of relief hole

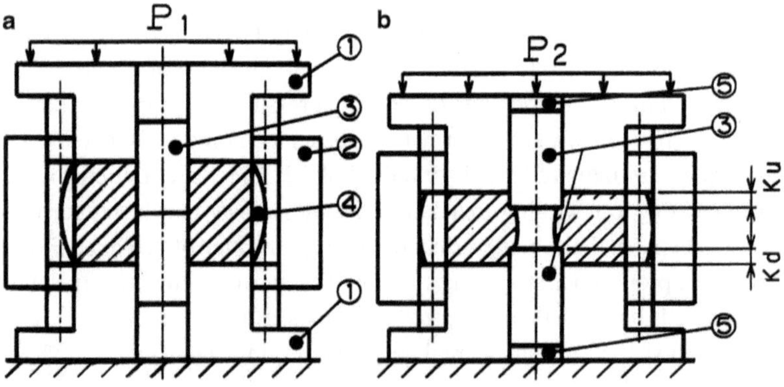

Fig. 2 Divided flow method for a spur gear (from ring specimen). **a** Conventional closed die forging, **b** Divided flow process

is important. In this step, gear teeth are completely filled up within allowable working pressure by the divided flows towards gear teeth and the relief hole. The divided flow behavior is the same as that of ring compression test. Relieved portion of material must be removed afterwards [1, 2].

3 Applications

Divided flow method was successfully applied to the net shape forging of sun gear at Aishin AW Co. Ltd. in 2002. Figure 3 shows working process of sun gear. At the initial stage, inner punch which is supported by back up pressure is protruded by a specified amount from outer punch. Both punches descend simultaneously. When the working pressure of outer punch exceeds the back up pressure, inner punch goes back relatively from outer punch until the protrusion becomes almost zero. At this stage divided flow utilized flow relief axis principle occurs. By the appropriate selection of the back up pressure gear teeth are filled up completely. Obtained accuracy of the product is high and necessary successive operation is only shaving to give crowning to teeth. Material utilization is very high because flow relief axis portion is almost disappeared [3]. By the improvement of tool material and polishing method, tool life is prolonged by almost 50 % afterwards [4].

The most complicated product which has been developed until now is external helical gear with boss, internal spline and two annular grooves shown in Fig. 4. This product

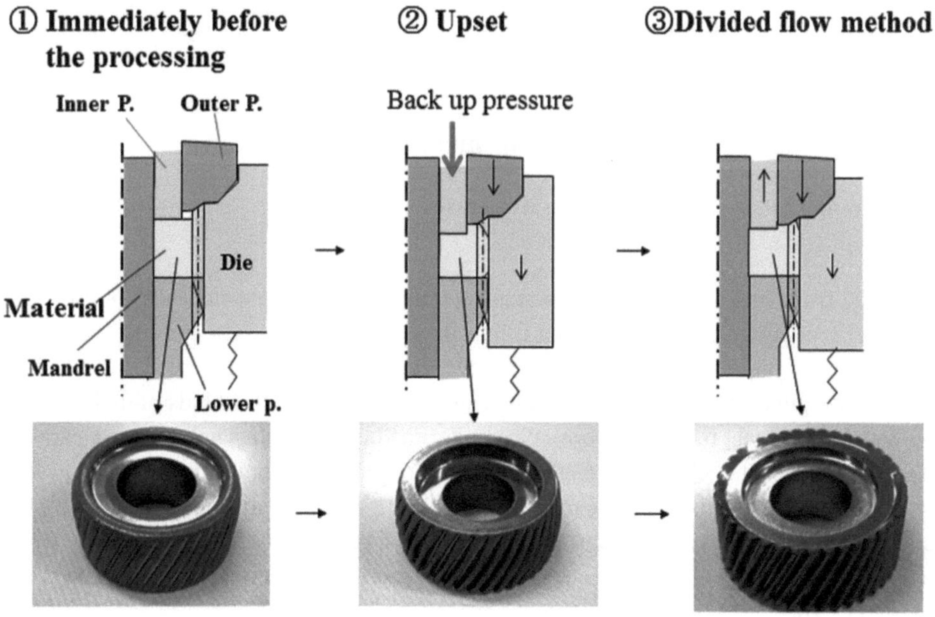

Fig. 3 Cold forging process of sun gear

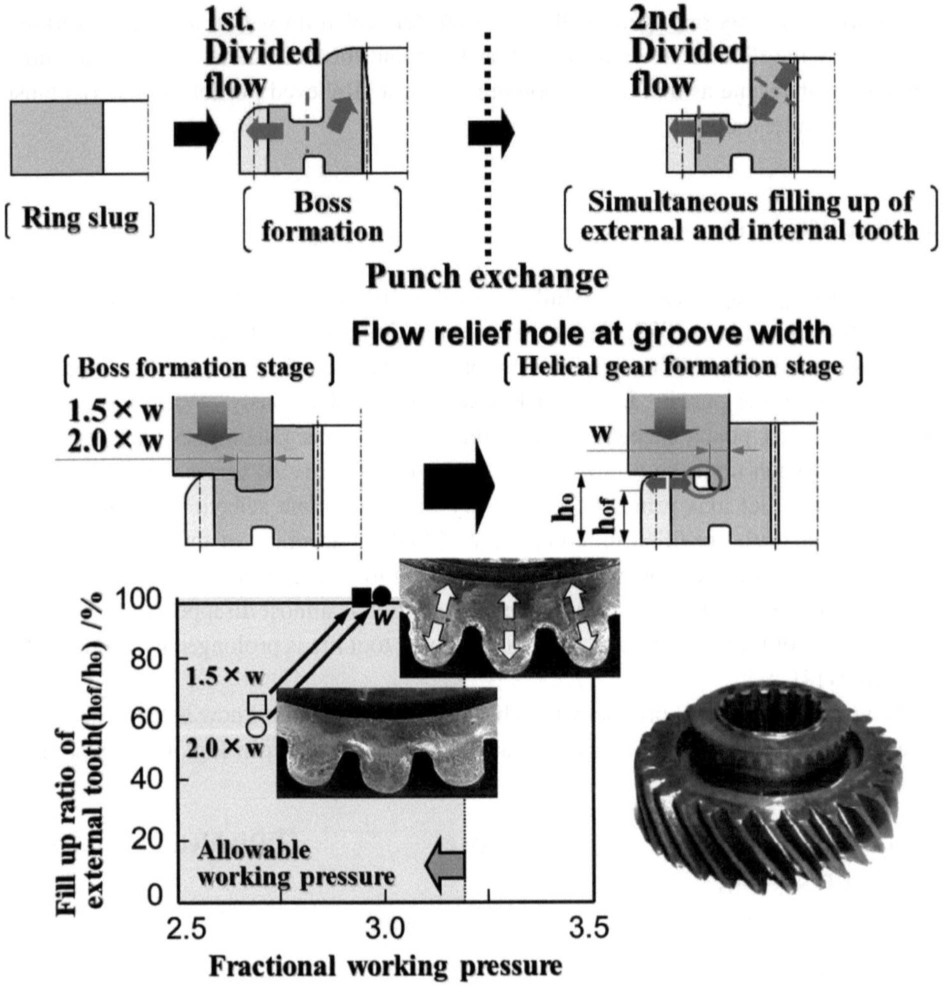

Fig. 4 Working Process with helical gear with boss

is forged from ring specimen by two steps operation. At the first step boss portion and oversize groove are formed. But the filling up state of helical gear portion and spline portion is insufficient. At the second step oversize groove punch is replaced by desired size groove punch and the punch depresses again. In this stage two kinds of divided flow between helical gear teeth and groove clearance, and between spline and groove clearance occur. As the result, helical gear teeth and spline are filled up completely within allowable working pressure. This process can be practiced as a single continuous operation by the adoption of one die two blows system. Distortion of helical gear portion after heat treatment is comparable with that of machined gear because the material at gear portion is not twisted [5, 6].

4 Potential

Divided flow method has the following features.

1. Application range of net shape forging is considerably wide. Intimate cooperation with product design division is inevitable.
2. Remarkable reduction of working pressure hence prolongation of tool life can be attained also in hot or warm forging.
3. In combined forging of hot or warm forging and cold sizing, acceptable accuracy of hot or warm forged product becomes considerably wide hence tool life of hot or warm can be prolonged remarkably.

5 Conclusion

Net shape forging process named "Divided Flow Method" is newly proposed. This process utilizes flow relief axis principle or flow relief hole principle to reduce working pressure. This process is successfully practiced in the actual production of gear toothed components for automobile. Because of the simplicity of the idea, wide range of practical applicability is expected.

References

1. Kondo, K., 1984, Development of New Cold Die Forging Processes, Proceedings of 1st International Conference on Technology of Plasticity, Tokyo, 878–887.
2. Kondo, K., 1999, Improvement of Product accuracy in Cold Die Forging, Proceedings of 6th International Conference on Technology of Plasticity, Nuremberg, 41–48.
3. Kondo, K., 2005, Profitable Net Shape Forging of Automotive Components, International Conference on New Developments in Forging Technology, Stuttgart, 35–43
4. Kondo, K., 2007, Net Shape Forging of Automotive Components by Divided Flow Method, Proceedings of International Conference on New Developments in Forging Technology, Stuttgart, 105–11
5. Kondo, K. et al., 2002, Net Shape Forging of, an External Helical Gear with Boss and Internal Spline, Proceedings of 7th International Conference on Technology of Plasticity, Yokohama, 49–54.
6. Kondo, K. et al., 2003, Net Shape Forging of an External Helical Gear with Boss and Internal Spline for Manual Transmission, Proceedings of International Conference on Accuracy in Forming Technology, Fraunhofer, 193–202

Potential

Enclosed Die Forging Using Die Set

Yoshihiro Ishihara and Kozo Osakada

1 Motivation

Enclosed die forging is a sort of side extrusion in which the metal flow is either perpendicular or inclined at an angle to the direction of the action of the machine as shown in Fig. 1 [1]. To make the die cavity, it is necessary to divide the die to close the die cavity during forging and open when the product is taken out, and thus the name of "enclosed die forging" was given.

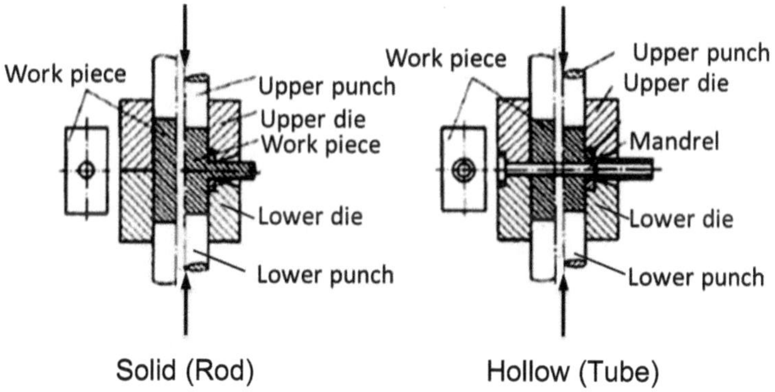

Fig. 1 Side extrusion [1]

Yoshihiro Ishihara
Osaka University, Osaka, Japan

Kozo Osakada ✉
Osaka University, Toyonaka, Osaka, Japan
e-mail: k.osakada@hm.h555.net

A. E. Tekkaya et al. (eds.), *60 Excellent Inventions in Metal Forming*,
DOI 10.1007/978-3-662-46312-3_47

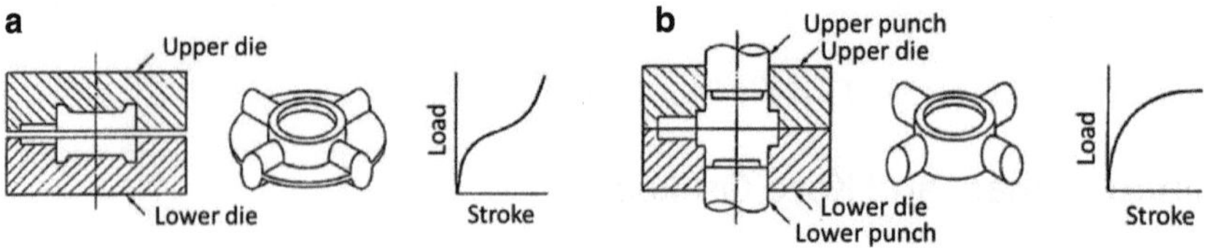

Fig. 2 Comparison of semi-closed die forging and enclosed die forging of cross pin. **a** Semi-closed die forging, **b** Enclosed die forging

Figure 2 illustrates a comparison of forming processes for producing a cross pin with conventionally used semi-closed die forging and enclosed die forging [2]. In semi-closed die forging, the maximum working load is extremely high because the contacting area between the dies and the work piece increases as deformation proceeds and the large thin burr is compressed. Further additional processes of cutting off the burr and machining of the extruded rods are needed.

In the case of enclosed die forging, the contacting area of the punches and the work piece is limited to the punch ends during the whole process, and the forming pressure and the forming load are kept low. Thus enclosed die forging is advantageous for cold forging of dimensionally accurate products with high yield ratio of the work piece material [2].

Figure 3 shows two types of enclosed forging. Bevel gear and inner race (CVJ component) are formed by moving one punch. Cross pin (steering component) and tripod (CVJ component) necessitate two moving punches and the die closing mechanism, and usually three moving axes are requested.

Fig. 3 Enclosed die forging methods and products. **a** One moving punch, **b** Two moving punches

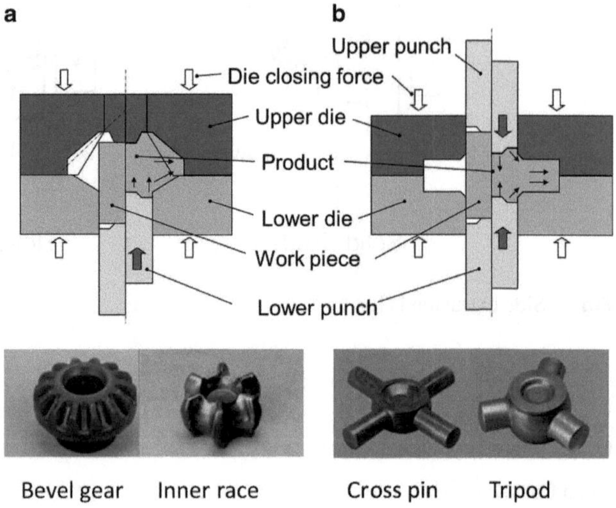

Fig. 4 Hydraulic 1100 tonf 3-axis press

Fig. 5 Hydraulic 800 tonf 3-axis press

To realize three moving axes for enclosed die forging, Mitsubishi Heavy Industries constructed 3-axis numerically controlled hydraulic presses (Fig. 4) up to 1100 tonf capacities in around 1975, and Kurimoto Ltd. and Nichidai Corp. constructed 800 tonf 3-axis hydraulic press (Fig. 5) in 1986.

2 Die Set with Pantograph Mechanism

Let us consider the case of Fig. 6a in which the upper punch moves downwards with a velocity 1, and the lower punch moves upwards with the same velocity while the divided die does not move. To realize this relation of the velocities on a conventional press, Nichidai Corp. proposed a die set with a pantograph link as Fig. 6b [3]. The upper hinge of the pantograph moves with the same velocity as the press slide, and the lower punch and the lower hinge are fixed to the lower plate and do not move. The divided die is fixed to the middle plate supported by the middle hinges of the pantograph. When the upper punch moves with a velocity of 2, the die moves down with a velocity 1, and the lower punch does not move, and thus the same relative velocities are achieved as Fig. 6a.

Figure 7 illustrates a die set with the pantograph link. The lower die is fixed to the middle plate which is pushed up by the lower hydraulic cylinder. The upper die is moved by the upper hydraulic cylinder and provides the force to close the die. Figure 8 shows the pantograph links in the die set.

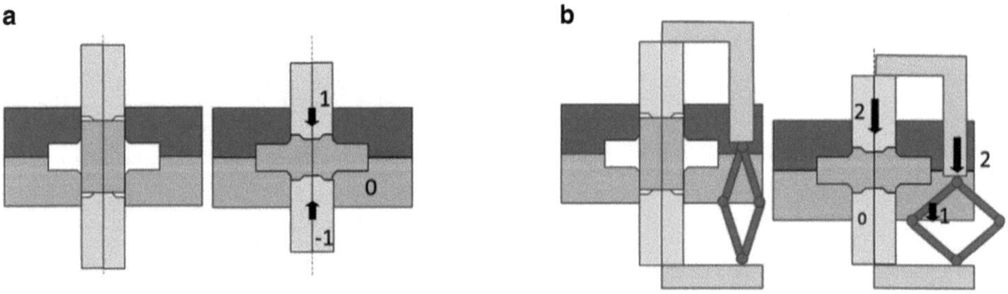

Fig. 6 Velocity relations in enclosed die forging methods. **a** Upper and lower symmetric structure, **b** Die set with pantograph link

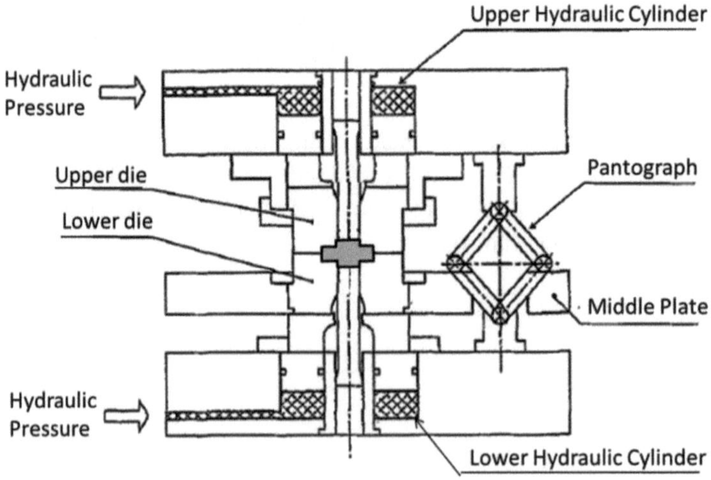

Fig. 7 Die set with pantograph link

Fig. 8 Photograph of die set with pantograph

3 Vertical Division of Die

Since the commonly used die for enclosed die forging is divided horizontally as Fig. 9a, partition line or small burr are formed on the surface of the extruded rods of a cross pin as shown in Fig. 9b. Further, the circularity of the extruded rod is deteriorated as Fig. 9c because the shapes of the initially circular die exits are squashed elastically by the closing force applied to the perpendicular direction to the axes of the rods. Thus machining of the extruded rods is necessary to eliminate the burr and the non-circularity, but it is desirable to eliminate machining process by extruding accurate rods.

Nichidai Corp. invented a vertically divided die for enclosed forging of cross pins and tripods as shown in Fig. 10a [4, 5]. In this die structure, the die closing force is applied horizontally via the cams that change the direction of the applied closing force from the vertical direction to the horizontal direction, i.e. parallel to the axes of the rods. It is confirmed that very good circularity of the extruded rods is attained by optimizing the die

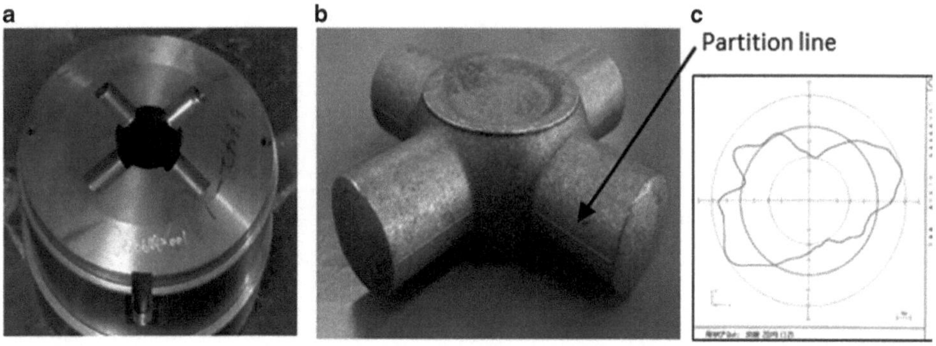

Fig. 9 Horizontal division of die and forged cross pin. **a** Horizontally divided die, **b** Cold forged cross pin, **c** Circularity

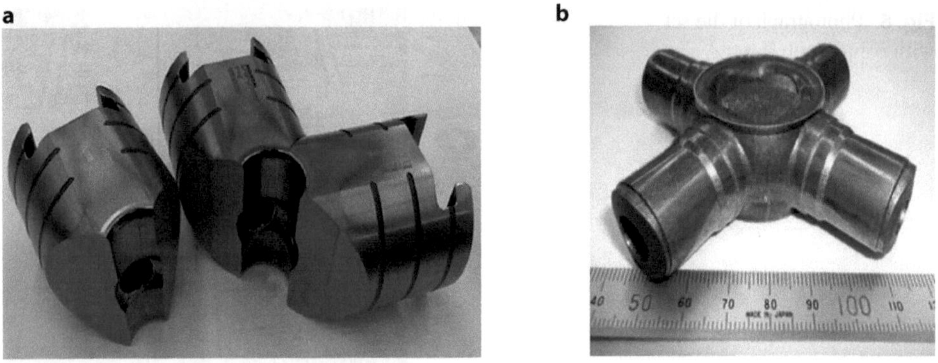

Fig. 10 Vertically divided die and cold forged cross pin. **a** Vertically divided die for cross pin, **b** Cross pin made with vertically divided die

closing force. As shown in Fig. 10b, no parting line appears on the extruded rods and thus machining after cold forging is eliminated.

References

1. Lange, K., 1972, Handbook of Metal Forming, McGraw-Hill Book Company, 13.2–13.3.
2. Shinozaki, K., Yoshimura, H., Ando, H., 1999, Enclosed die forging, Its prosperous aspect as a precision forging process (in Japanese), 40(459), 316–321.
3. Japanese Patent, 1992, No. 1676368.
4. Japanese Patent, 2014, No. 330204
5. Ishihara, Y., 2011, Development of precision forging method with die structure controlling elastic deformation of die (in Japanese), Die and Mould Technology, 26 (5), 34–37.

Joining of Serrated Shaft with Holed Disk by Indentation

Kazuhiko Kitamura, Kenji Hirota, Yoshihiko Ukai, and Kei-ichi Matsunaga

1 Motivation

An axle shaft with disk-like parts such as gears or coupling drums is an important component for the automotive industry/technology. A large axle-shaft is generally produced by hot forging and machining if high strength is required. This process, however, wastes energy and material because it necessitates heating the work-piece to a high temperature and a large quantity of machining scrap is caused.

Another process to make an axle-shaft is joining of a shaft and a disk after they are made separately. Joining of two parts has advantages if large differences in shape, diameter, and material of the two parts exist. Welding is most commonly used to join two parts with a high strength. The welding method, however, has problems of low productivity due to long welding time and of environmental pollution by noxious fumes.

Another joining process using plastic deformation is characterized by high efficiency and low environmental pollution. Some inventions of this type of joining process have been used in industries in Japan. Nippon Piston Ring Corp. has applied shaving joining [2] to a camshaft [1]. Nichidai Corp. has made a disk with plural pins using indentation joining proposed by Matsumoto et al. [3]. Hitachi Corp. has used upsetting joining invented by Kanamaru [4] for large power electric parts with high pulling-out resistance. It is con-

Kazuhiko Kitamura ✉
Nagoya Institute of Technology, Nagoya, Japan
e-mail: kitamura.kazuhiko@nitech.ac.jp

Kenji Hirota
Kyushu Institute of Technology, Kitakyushu, Fukuoka, Japan

Yoshihiko Ukai
Toyota Motor Corp., Toyota, Japan

Kei-ichi Matsunaga
MEG Inc., Anjo, Aichi, Japan

A. E. Tekkaya et al. (eds.), *60 Excellent Inventions in Metal Forming*,
DOI 10.1007/978-3-662-46312-3_48

sidered that economic and ecological joining processes for automotive parts with high torsional strength could be developed with joining using plastic deformation.

2 Serration Joining of Shaft and Disk

To attain more accurate axle shafts with high torsional strength, serration joining [6, 7] has been developed. The following procedure of serration joining is shown in Fig. 1:

1. The shaft is partially serrated by cold extrusion using a serrated die at the 1st step.
2. This serrated shaft is hardened by carburizing and subsequent quenching at the 2nd step.
3. The serrated shaft is pushed into the small hole of the soft disk at room temperature at the 3rd step.

After this process, the disk material fills the spaces between the teeth of the shaft as shown in Fig. 1. The disk material near the hole is work-hardened by plastic deformation and the high residual compression stress fastens the shaft.

Figure 2 illustrates the shaft and the disk before joining at the third step. A lead angle α and an overlap ratio δ/h are the most important parameters to attain a high filling ratio and a high torsional strength. Lead angle α is the taper angle of the front end of the shaft, and the overlap ratio δ/h of the shaft and the disk is defined by $(d_s - d_h)/(2h) = \delta/h$. The diameter of the shaft is described by d_s, d_h is the diameter of the disk hole, and h is the height of the teeth serrated on the shaft by cold extrusion.

The teeth shape of the serration shaft is specially designed as shown in Fig. 3 to enhance the following effects:

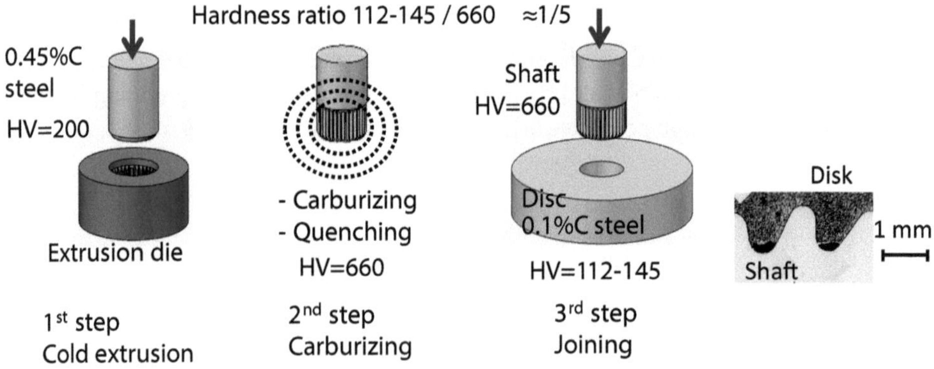

Fig. 1 Procedure of serration joining of shaft and disk, and cross-sectional view of joining region

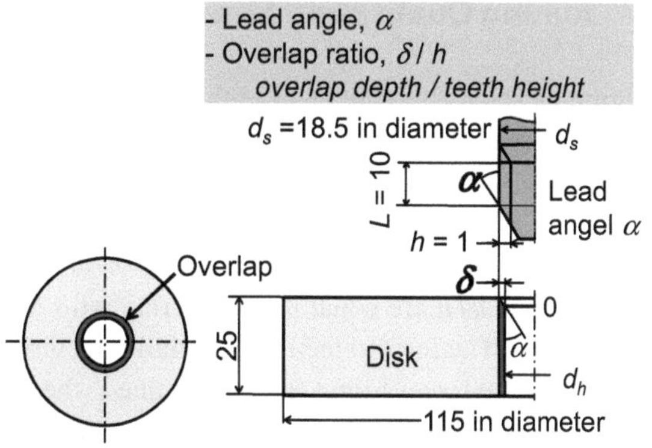

Fig. 2 Important parameters for good joining

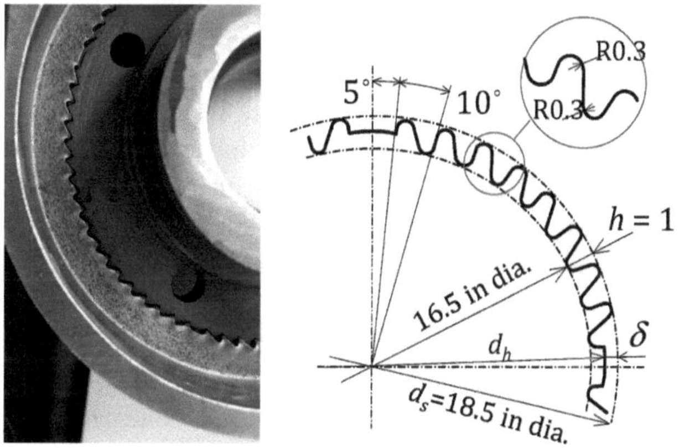

Fig. 3 Specially-designed teeth shape

1. Each tooth is not symmetric: one side is straight in the radial direction and another side is inclined, and the top of the tooth is rounded. This shape gives efficiently high work-hardening to the disk material without causing high load if the shaft is pushed into the hole of the disk.
2. The inclined directions of the teeth are altered at every 90° to avoid slipping both in clockwise and anti-clockwise directions if alternating torque is applied during practical use.
3. At every 90° the serration has four toothless parts with a diameter that corresponds to the diameter of the hole. These toothless parts guide the shaft to be inserted smoothly into the hole of the disk by keeping concentricity between the shaft and the hole of the disk.

3 Joining Quality and Strength

Figure 4a shows the influence of the lead angle α and the overlap ratio δ / h on the filling ratio d / h when the overlap ratio $\delta / h = 0.5$. The largest filling ratio of $d / h = 0.7$ is obtained when α is 30°. The disk material does not flow into the open spaces between the teeth if α is greater than 45° because larger α causes mainly shaving without filling. Figure 4b shows the influence of the overlap ratio δ / h on filling ratio d / h when $\alpha = 30°$ and 90°. The filling ratios d / h are equal to the overlap ratio δ / h when $\alpha = 90°$, meaning that shaving takes place. The formed teeth are partially fractured when $\alpha = 90°$ and δ / h is not less than 0.5. Thus, the larger filling ratio is obtained when $\alpha = 30°$ and $\delta / h \geqq 0.5$.

After indenting with a stroke of 10 mm, the joined product is subjected to static torsion using a torsion tester. Fracture occurred in the shaft, not in the joined portion. To measure the joining strength, the joined part is shortened by machining to 6 mm after indenting by 10 mm. The yielding torque and the maximum torque increase with increasing d / h as shown in Fig. 4c.

Figure 5 shows the results of the torsion tests of the joints made by serration joining and mechanical assembly, in which plastic deformation is not caused. In mechanical assembly, a serrated shaft is only fit in a serrated hole. Both serrated shaft and the serrated hole were made with a wire electric discharge machine before fitting. The joint made by serration joining is stronger than the mechanical assembly. The yield torque of the joint made by serration joining is 1.5 times as high as that of the joint made by mechanical assembly. Due to the strong joining strength, Toyota Motor Corp. applied practically the serration joining process to manufacturing of automobile parts.

The serration joining process has attained a high torsional strength, but it may not be so strong when the shaft is subjected to pulling out force from the disk. If a high resis-

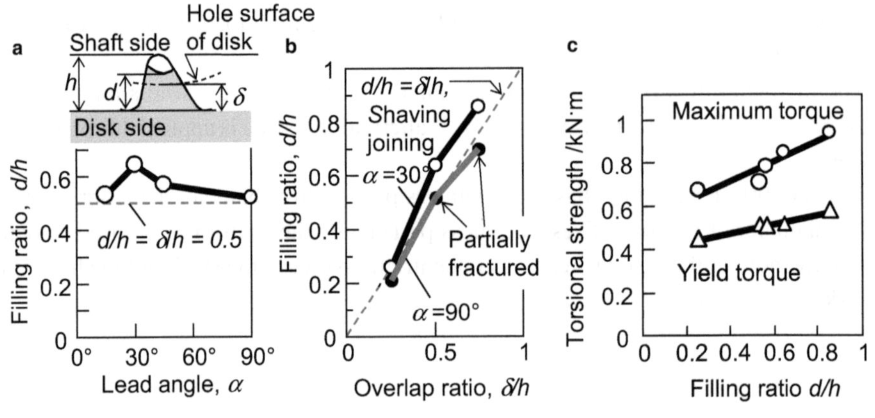

Fig. 4 Influence of lead angle α and overlap ratio δ/h on filling ratio d/h to enhance torsional strength. **a** Influence of α on d / h, **b** Influence of δ / h on d / h, **c** Torsional strength for d / h

Fig. 5 Comparison of torsional strength between serration joining and mechanical assembly

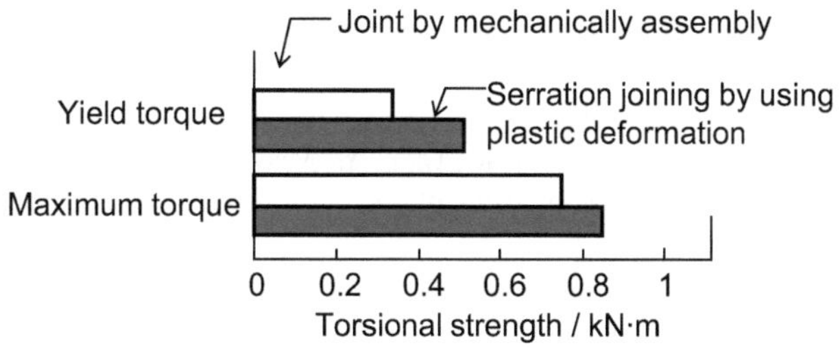

Fig. 6 Improvement of pulling-out resistance by additional caulking after serration joining

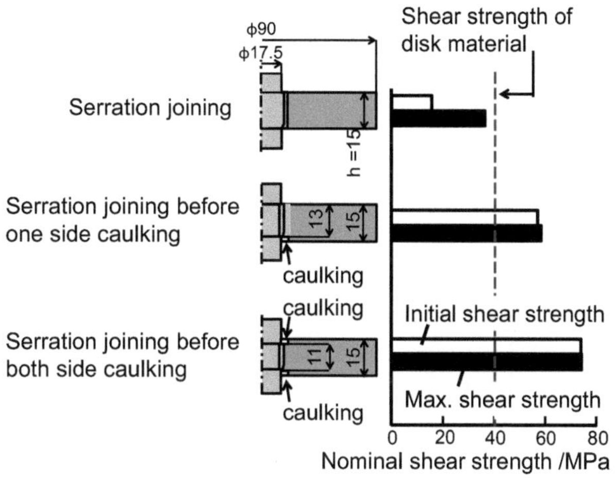

tance force against pulling-out in the axial direction is required, caulking process is added by indenting narrow area of the disk near the hole after serration joining to cause small plastic deformation. As shown in Fig. 6, the caulking process is effective in improving the resistance against pulling-out force, because the additional caulking increased in filling ratio and elastic compression stress between the serration shaft and the disk [8].

References

1. Egami, Y., Abe, Y., 2011, Joining technology of composite camshaft (in Japanese), Journal of JSTP 52 (603), 439–442.
2. Machida, T., 1987, Shave-joining process of dissimilar materials and its application for mild steel (in Japanese), Journal of JSTP 28 (322), 1158–1165.
3. Matsumoto, R., Hanami, S., Ogura, A., Yoshimura, H., Osakada, K., 2008, New plastic joining method using indentation of cold bar to hot forged part, Annals of the CIRP 57, 279–281.
4. Kanamaru, H., Tsuruoka, K., Oku, M., Tatsumi H., Development of metal flow (combination with plastic flow) and application to automotive Parts, Society of Automotive Engineers of Japan, Review, Jul, 1984, 83–89.
5. Monaghan, J., Naughton, P., 1991, The production of permanent joints by plastic flow. International Journal of Machine Tools and Manufacture, 31(3):283–293.

6. Hirota, K., Matsuura, T., Kitamura, K., Ukai, Y., Matsunaga, K., 2007, Plastic flow joining of quenched shaft and flange with emphasis on torsional strength (in Japanese), Journal of JSTP, 48 (522), 66–70.
7. Kitamura, K., Hirota, K., Ukai, Y., Matsunaga, K., Osakada, K., 2012, Cold Joining of rotor shaft with flange by using plastic deformation, CIRP Annals – Manufacturing Technology, 61, 275–278.
8. Kitamura, K., Inishi, H., Hirota, K., Ukai, Y., Matsunaga, K., 2014, Combination of plastic upsetting joining and plastic serration joining of disk and shaft, Advanced Materials Research, 966–967, 21–28.

Development of Orbital Forging Processes by Using Marciniak Rocking-Die Solutions

Andrzej Kocańda

1 Motivation

It has always been crucial for the metal forming processes to reduce a deformation force as much as possible by controlling the microstructure of a workpiece, friction effect, lubrication conditions, temperature, etc. One of the natural ways used by our ancestors and craftsmen was to reduce the instantaneous area of contact between the hammer and the workpiece. Thus, the local deformation repeated successively by hammering in neighbouring areas led to the required shape change with substantial reduction of the force level. This idea of using incremental deformation has been successfully applied to the production of lighter and cheaper forging machines as comparing with those where the whole workpiece is deformed at once. Another benefit has been associated with the possibility of forging materials difficult to deform and obtaining high degree of deformation. The incremental method has been used in rotary and orbital forging machines for almost 100 years. In orbital forging, the workpiece placed in the lower die moves towards the upper die with orbital motion and the force reduction could be up to 90 percent as comparing to conventional forging. At present, there are also incremental deformation machine systems available, which operate with multi movement tooling under microprocessor control and offer flexible manufacturing. Unfortunately, the high cost of such systems and relatively longer time of the forming process than in conventional forming, limit the opportunity for economic advantage to high value and complex geometry products. Under such circumstances, further development and applications of orbital and rotary forging machines has been still of interest to industry and research centres.

Andrzej Kocańda ✉
Warsaw University of Technology, Warsaw, Poland
e-mail: a.kocanda@wip.pw.edu.pl

A. E. Tekkaya et al. (eds.), *60 Excellent Inventions in Metal Forming*,
DOI 10.1007/978-3-662-46312-3_49

2 Past Development of Orbital Forging Process

It is believed that the first modern rotary forging machine was invented by E.E. Slick in 1918 and built in metallurgical plant [1]. The machine known rather as Slick Wheel Mill was used to produce railway wheels from heated ingots. Two dies were placed horizontally in the mill. They were rotated with permanent angle of 10.7 degrees between the axes. In 1920, E.E. Slick [2] had also designed the first rotary forging press for heading a shaft end. This horizontal press was equipped with a stationary die in which the shaft was clamped and the rotating die, which was moved towards the headed shaft by hydraulic system. The rotating die was inclined to the axis of the stationary die with a constant angle and its face was slightly conical.

In 1929 M.F. Massey patented a design of vertically operated orbital forging press [3]. The upper and lower dies did not rotate at all. However, the upper die had a constant angle circular rocking motion about a central point between the dies. The upper die movement was limited to one direction of circular motion only. The prototype of machine called "Rotaform" was manufactured many years later in 1969. In 1972, the first "Rotaform" press was built for production purposes [4]. The upper die was inclined at a constant angle of 4 degrees.

In the early 1960's, a number of independent works on development of orbital forging machines had been conducted. However, these machines were equipped only with circular motion of die as it was in the case of "Rotaform" press. The solution proposed by Z. Marciniak from Warsaw University of Technology made a significant breakthrough in this area. He had been working on development of vertical mechanism with one [5] and two eccentric sleeves for driving the housing of upper die [6]. The second mechanism proved to be revolutionary. Cross sectional view of this mechanism is shown in Fig. 1a. The axis

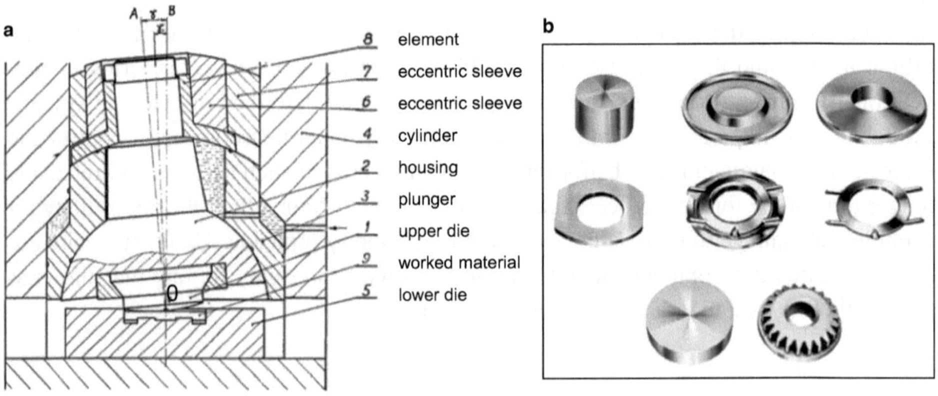

Fig. 1 Cross sectional view of a mechanism for any rocking motion of the upper die **a** – the original drawing from Z. Marciniak paper [6], **b** three examples of the first parts made by using press PXW-100

of movable die (1) mounted in the housing (2) may be inclined from the vertical position at an angle γ through spherical bearing of the housing. Upper pivot of the housing is subjected to oscillations of the axis around the point 0 what results from simultaneous rotation of the eccentric sleeves (6) and (7). Having an independent rotation of the two eccentric sleeves, performed with different speeds, enabled any rocking motion of the upper die ensuring the best condition for plastic flow of material [6, 7]. Generally, four types of the complex upper die motions have become possible: circular or orbital, spiral, planetary or daisy and straight-line or rocking. A choice of motion type depends on the geometric features of the final product. Circular motion is suitable for deformation of axisymmetric products, straight-line motion for elongated products, planetary or daisy motion for products with radial gears and ribs, see Fig. 1b. This revolutionary idea was applied for patents in 1967 [6] [8] and gave impetus to further development of the orbital forging presses and their industrial applications. Within several years, research and design works on a larger scale began in Poland (R&D Centre Plasomet in Warsaw, Institute of Metal Forming in Poznań), Japan (GIRI Nagoya), China, USA (VSI Automation Co.), Switzerland (Schmid Ltd.) and United Kingdom (Manchester UMIST, Nottingham University).

3 Orbital Forging Presses with Marciniak Rocking-Die Solutions

The prototype of the Marciniak orbital forging press PXW-100 with capacity 1600 kN and mechanical drive giving variable rocking motion of the upper die was built in 1967 at OBR PLASOMET-CBKPiM, Warsaw. Production of improved Marciniak press for industry, labelled PXW-100 A, began in 1970 at FPA-PONAR PLASOMAT and serial production started in 1973. In 1982, PXW 200 press for cold and warm forming was released with capacity of 2000 kN and bigger working space. The FPA produced more than 250 presses up to 1984 [9]. The PXW presses had also been marketed worldwide and it was the first large scale marketing of the orbital forging presses. Thus, the press was initially used in many of the industrialized countries (e. g. Japan, USA, West Germany, China, and Switzerland) and then was often copied and improved in various countries.

Schematic diagram of the overall design and features of the PXW-100 A press is shown in Fig. 2. It is a straight-side press, which consists of two basic units. The first unit with a hydraulic cylinder (1) in the lower part of press is used to perform up-and-down motion of the lower die and to obtain the required pressure on the surface of worked material. The second unit in the upper part of press gives a rocking motion of the upper die housing (8). Gearbox (15) with hand wheels (16) is responsible for setting the angular speed and direction of rotation of eccentric sleeves (10) and (11). Circular motion of the upper die results from the same direction of eccentric sleeves rotation with the same speed. If the eccentric sleeves rotate with the same speed but in opposite directions than the upper die motion is the straight-line type. Daisy motion is obtained for various angular speeds and opposite directions of eccentric sleeves rotations. The forging angle γ between vertical

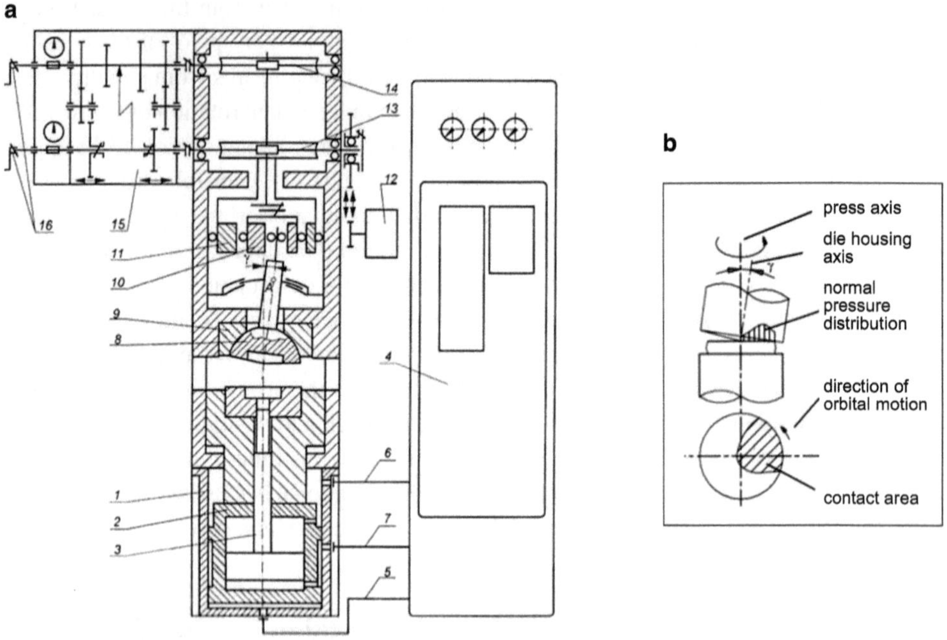

Fig. 2 General presentation of the overall design and features of the PXW-100 A press. **a** *1* main cylinder, *2* piston, *3* ejector, *4* hydraulic drive, *5/6/7* oil pipes, *8* upper die housing, *9* orbital bearing, *10/11* eccentric sleeves, *12* electric motor, *13/14* worm gears, *15* gearbox, *16* gearbox hand-wheels, **b** scheme of orbital forging with circular motion of the upper die

axes depends on setting the angular alignment of the eccentric sleeves by gearbox hand-wheels (16).

Later, other modified machines for orbital forging have been proposed. For example, Japanese NAITO-TEKKO-SHO Co. Ltd. received a licence on the PXW 100 press [4]. They sold Polish presses after some rebuilding. The presses found increasing applications especially in the powder compacting area. In 1973 several PXW 100 presses were installed in China and since then other orbital forging presses were built, based on the Polish design. The Swiss factory "H. SCHMID Ltd." acquired Polish press PXW 100 A in 1976 and started to design and produce a similar press T-200 [4, 10]. They used the same rocking-die mechanism as in the PXW 100 A press but hydraulic and control systems have been gradually improved. Afterward, some other changes were introduced, including the orbital motion mechanism – two eccentric sleeves were driven directly by two independent electric motors. Two larger capacity presses T-300 and T-630 have also been marketed. In 1992, MORI IRON WORKS Co. Ltd. started to build the cold orbital forging presses MCOF-650 with large capacity 6500 kN.

4 Further Applications and Potential Challenges

The main advantages of the orbital forging are associated with the forging force, which is applied only to a limited area or the workpiece surface. Thus, only a part of the workpiece surface is deformed at a given time and friction is significantly lower than in the case of extrusion or upsetting processes. These altogether results in the substantially lower force required for orbital forging than in conventional forging processes. Four types of orbital motion introduced by Z. Marciniak improve processing range and production of parts with big variety of shapes, sometimes impossible to form, like thin discs with large diameter, thin-walled parts, rings or various gears. Investment costs are lower because of lower deformation load, production space and manpower. Die design is usually simple and the dies are easy to change. Orbital forging is also economically competitive for small batch productions because die making is not particularly expensive.

Nevertheless, at the end of 1980's the interest of industry in orbital forging began to decline temporarily. It had been associated with difficult to explain deformation mechanism, relatively low productivity and tool life. However, continuous development of numerical simulation by means of FEM led to renewed interest in the process. It has become possible to analyse a die stress distribution or metal flow, to predict a die failure and to obtain an in-depth understanding of the process, e. g. forging of gears [11, 12]. Some new applications have appeared, e. g. forming of tailored blanks from sheet metal [13] or combining hot forging with cold orbital forging. Some of the key issues still need research: PM compaction, frictional constraints at the two die-workpiece interfaces, explanation of higher workability achieved by rotary forging as comparing with conventional forging, rigidity of dies and press frame and so on.

Currently, mainly suppliers to the automotive industry apply orbital forging. It is still a promising incremental forming process, which could be considered as capable for production of infinitely great range of components.

References

1. Slick, E.E., 1918, The Slick Wheel Mill, The Iron Age, Vol. 102, No. 9, pp. 491–498
2. Slick E.E., 1920, Method of and apparatus for making shafts. U.S. Patent No. 1359625
3. Massey H.F., 1929, Improvements in forging and upsetting machines, British Patent Specification No. 319.065
4. Shivpuri, R., 1988, Past developments and future trends in the rotary or orbital forging process. Journal of Material Shaping Technology, Vol. 6, No. 1, pp. 55–71
5. Marciniak, Z., 1966, Sposób wytwarzania przedmiotów metalowych o zmiennej grubości oraz urządzenie do stosowania tego sposobu, Politechnika Warszawska, Polish Patent 51954, MKP B21J9/02 (in Polish)
6. Marciniak, Z., 1968, Sposób wytwarzania przedmiotów metalowych o zmiennej grubości oraz urządzenie do stosowania tego sposobu, Politechnika Warszawska, Polish Patent 55591, MKP B21J9/02 (in Polish)

7. Marciniak, Z., 1968, Prasowanie obwiedniowe, Mechanik, Vol. XLI, No. 11, pp. 587–589 (in Polish)
8. Marciniak, Z., 1970, Method of producing the metal objects of variable thickness and a device for application of this method, U.S. Patent 3523442 A
9. Nagel, W., 1984, Kształtowanie wahającą matrycą – przegląd stanu – perspektywy, Mechanik, No. 6, pp. 312–314 (in Polish)
10. Schmid Press Release, 2005, Extended application range for orbital forming technology, Heinrich Schmid, Machines, Tools & Dies Ltd., Jona, Switzerland, May 2005
11. Samołyk, G., 2013, Investigation of the cold orbital forging process of an AlMgSi alloy bevel gear, Journal of Materials Processing Technology, Vol. 213, pp. 1692–1702
12. Xinghui Han, Lin Hua, Wuhao Zhuang, Xinchang Zhang, 2014, Process design and control in cold rotary forging of non-rotary gear parts, vol. 214, pp. 2402–2416
13. Merklein, M., Hildenbrand, P., Schneider T., 2014, Application of process–adapted semi-finished parts for the production of thin-walled components via sheet-bulk forming, Proc. 4th AutoMetForm, Freiberg, pp. 58–66

Double Cup Extrusion Test to Evalute Lubricants for Cold Forging

Taylan Altan and Gracious Ngaile

1 Introduction – Double Cup Extrusion Tribo-Test

There are various tribo-tests developed to date used for evaluation of cold forging lubricants [1]. These include: ring compression test, spike test, ball penetration tests, etc. The double cup extrusion test was developed based on experiments, conducted by Geiger [2]. As compared to other commonly used tests such as the ring compression test, the double-cup extrusion test can emulate severe deformation with very high surface enlargement occurring in actual severe cold forming operations.

Figure 1 illustrates the principle of the double cup extrusion test [3]. In this test, the upper punch moves downwards while the bottom punch and the container are kept stationary. The diameters of both punches are the same. The upper cup is formed by a backward extrusion process and the lower cup is formed by a forward extrusion process. The simultaneous action of the two punches inside the cylindrical container generates the two cups. The maximum surface expansion generated in this test may reach as high as 500 %. The ratio of the extruded cup heights ($R = h_1 / h_2$) increases with increasing coefficient of friction (or friction factor m).

Studies on friction conditions in cold forging with this test method have shown the cup height ratio to be extremely sensitive to the friction factor [3]. Thus, by comparing the cup height ratio and punch stroke to the friction factor calibration curves obtained through FEM, the friction factor of various lubricants can be quantified and compared.

Taylan Altan ✉
The Ohio State University, Columbus, Ohio, USA
e-mail: altan.1@osu.edu

Gracious Ngaile
North Carolina State University, Raleigh, NC, USA

A. E. Tekkaya et al. (eds.), *60 Excellent Inventions in Metal Forming*,
DOI 10.1007/978-3-662-46312-3_50

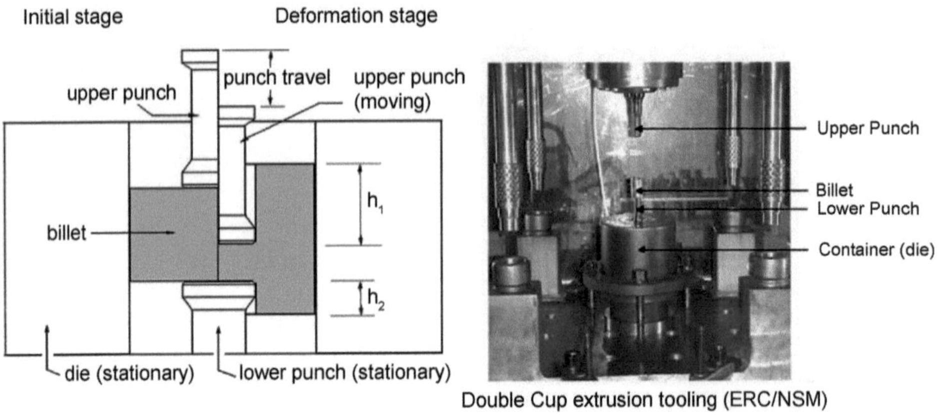

Fig. 1 Principles of the double cup extrusion test tooling [3]

2 Determination of Friction Calibration Curves by FEA

Finite Element (FE) simulations of the double cup extrusion tests were completed in order to understand the parameters (friction, material properties, tool geometry) affecting metal flow in the test and to obtain friction factor calibration curves. The friction factor calibration curves are used later to determine the friction factor of the tested lubricants. The commercial finite element code DEFORM-2D was used for the simulations with the constant shear friction model ($t = mk$). The material properties were determined using the uniform compression test and expressed in the form of a power la, $\bar{\sigma} = K(\bar{\varepsilon})^n$, where K is the strength coefficient and n is the strain-hardening exponent. The FE model is illustrated in Fig. 2. The simulation parameters for a specific material (AISI 8610) are given in Table 1.

Table 1 FE simulation parameters

Condition	Description
Material	AISI 8610
Matl properties	$\bar{\sigma} = K\bar{\varepsilon}^n$: K = 690 Mpa, $n = 0.14$
Billet hight	31.75 mm (1.25 in.)
Billet diameter	31.75 mm (1.25 in.)
Friction model	Shear friction
Friction Factors	$m = 0.00$, $m = 0.02$, $m = 0.03$, $m = 0.04$, $m = 0.05$, $m = 0.06$, $m = 0.065$, $m = 0.07$, $m = 0.08$, $m = 0.09$, $m = 0.10$, $m = 0.15$
No. of elements	1500

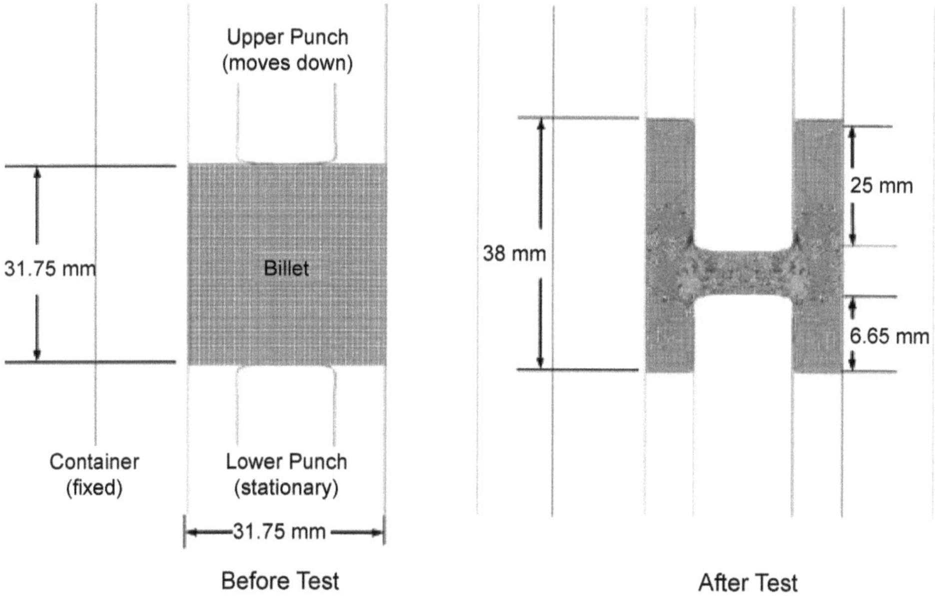

Fig. 2 FE model used in the development of the double cup extrusion test ($m = 0.1$)

The cup height ratio, R, is an indication of lubricity. As the friction factor increases, so does the cup height ratio. Thus, if there was no friction, the cup heights would be the same and the cup height ratio would be equal to one.

Using the nodal coordinates from the FE simulation, the cup height ratio was determined at various stroke lengths for various friction factor (m) values. Thus, friction factor calibration curves, i. e. the cup height ratio versus the punch stroke, are obtained. By matching the cup height ratio and punch stroke from an actual double cup backward extrusion test to the calibration curves, the friction factor for a given lubricant can be determined.

3 Application – Evaluation of Selected Lubricants by the Double Cup Backward Extrusion Test

The experiments were carried out with billets made from AISI 8610 grade alloy steel and 4 different lubricants were tested, using the standard lubrication methods recommended by lubricant suppliers.

The experiments were conducted using a 160 ton CNC hydraulic press with a ram speed of 10 mm/s and a 21 mm punch stroke. By plotting the average cup height ratio (obtained from 20 specimens) and the average punch stroke (obtained from 20 specimens) on the friction factor calibration curves, the average friction factor for each lubricant was

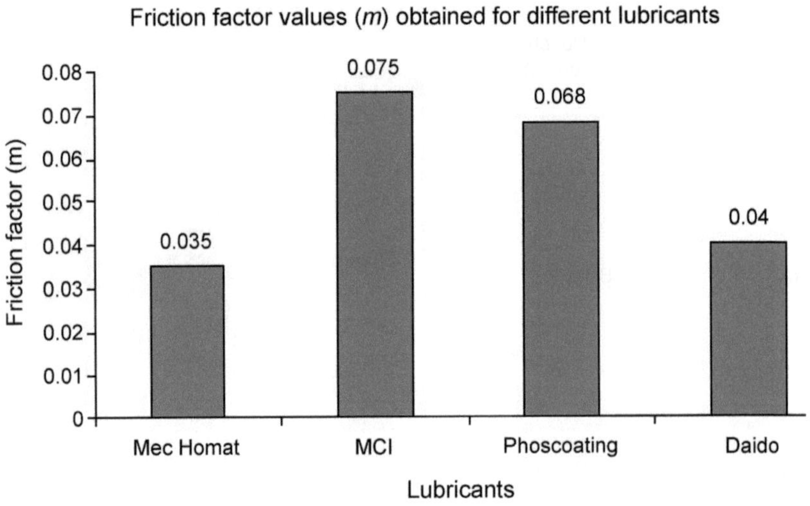

Fig. 3 Friction factor values (m) obtained for different lubricants tested [3]

estimated. The cup height ratios, for some lubricants, varied from one sample to the other and the degree of variation was a function of the lubricant used. Figure 3 presents the friction factors, for the four lubricants tested, in bar graph format. It is clearly seen that the MEC Homat lubricant and the Daido AquaLub lubricant obtained lower friction factors than the zinc phosphate coating + soap under the given conditions (i. e. interface pressure, sliding velocity, surface enlargement, etc.). All tested lubricants were water based and environmentally friendly. However, the exact composition of the lubricants was proprietary and were not divulged by the suppliers.

To study the performance of the lubricants in preventing galling optical micrographs were taken at the upper cup where maximum surface enlargement occurs. The figures showed that with the exception of MCI Z-Coat, the surface morphologies were very similar. The specimens coated with MCI Z-Coat show slightly more scratches in the upper cup than do the other three lubricants. Also both the punches and the container (die) did not show any signs of galling for all the specimens tested.

4 Summary and Conclusions

This study and others [3, 4, 5] clearly illustrated that double cup extrusion process is very useful in the evaluation of cold forging lubricants and coatings. The study also showed that in planning, conducting and evaluating lubrication tests, it is useful to simulate the test, using FEA, in order to determine the effects of various geometrical and process conditions. The application of the test illustrated that for moderate cold forming conditions, as considered in the present study, non-phosphate lubricants that are eco-friendly could replace phosphate coatings.

References

1. Bay, N. "The state of the art in cold forging lubrication", Journal of Materials Processing Technology 46 (1994), 19–40
2. Geiger, R. "Metal flow in combined cup extrusion", reports from the Institute for Forming Technology, Technical University Stuttgart, Nr. 36, 1976
3. Gariety, M., Ngaile. G, and Altan, T. "Evaluation of new cold forging lubricants without zinc-phosphate precoat", Int. J. of Machine Tools and Manufacture, 47(3–4), 2007, 673–681
4. Bay, N., Azushima, A., Groche, P., Ishibashi, I., Merklein, M., Morishita, M., Nakamura, T., Schmid, S., and Yoshida M. "Environmentally bening tribo-systems for metal forming", CIRP Annals, Manufacturing Technology, 59, 2010, p. 760
5. Ghobrial, M.I, Lee, J.Y., Altan T., Bay N., Hansen, B.G."Factors affecting the double cup extrusion test for evaluation of friction in cold and warm forging", CIRP Annals, 42, 1993, p. 347

Extrusion of Scroll Against Counter Pressure

Hidekazu Hayashi and Kozo Osakada

1 Motivation

Air conditioners of scroll compressors have been increasingly mounted on cars from around 1990 because their energy efficiency is high and less noise and vibration are caused. Further, smallness in size of scroll compressor is suitable for reducing the weight of car. A scroll part used in an air conditioner, shown in Fig. 1, is of aluminium alloy with high silicon content and has a spiral fin on a round end plate. Due to this complex shape, scroll parts are usually made by casting with subsequent machining, but due to the low yield ratio of the material and difficulty in machining of the aluminium alloy, manufacturing of scroll by metal forming was desired. NICHIDAI started to develop the method

Fig. 1 Scroll for compressor

The original version of this chapter was revised: The spelling of the first author's name was corrected. The erratum to this chapter is available at: DOI 10.1007/978-3-662-46312-3_68

Hidekazu Hayashi
NICHIDAI Corp, Kyotanabe, Kyoto, Japan

Kozo Osakada ✉
Osaka University, Toyonaka, Osaka, Japan
e-mail: k.osakada@hm.h555.net

A. E. Tekkaya et al. (eds.), *60 Excellent Inventions in Metal Forming*,
DOI 10.1007/978-3-662-46312-3_51

of manufacturing scrolls by using forward extrusion with counter tool around 1993 [1]. It was considered that if the long spiral fin could be extruded to a uniform length and the end surface is flattened, the material and machining costs are saved greatly.

2 Process Idea

When forward extrusion is carried out using a die with multiple exit holes, the lengths of the extruded parts are not equal as shown in Fig. 2a, and machining is necessary to equalize the lengths of the extruded parts.

If a billet is extruded against a floating counter tool supported by pressure as shown in Fig. 2b, it is possible to control the flowing out velocities to attain equal extruded lengths, and if a higher pressure is given to the floating tool, the extruded end surfaces become flat [2].

Figure 3 shows the experimental results of forward extrusion of a boss and a cup simultaneously (a) without and (b) with a floating counter tool supported by pressure. It is found that the counter pressure, which is necessary to equalize the extruded lengths is about 3 % of the flow stress in the case of Fig. 3b, but the extruded ends are not flattened with this low pressure [2].

Figure 4 shows the effect of counter pressure on the ratio of flattened area of the extruded end surfaces. The ratio of the flattened end reach 0.95 when the counter pressure exceeds 40 % of the flow stress of the billet material, and then the ratio increases gradually with the counter pressure [2].

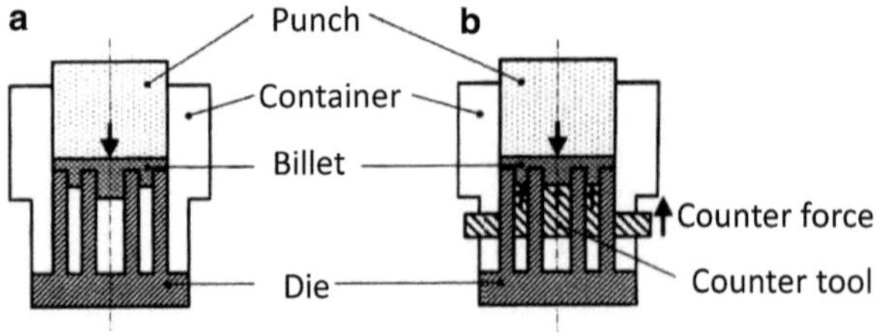

Fig. 2 Explanation of extrusion against counter tool [1]

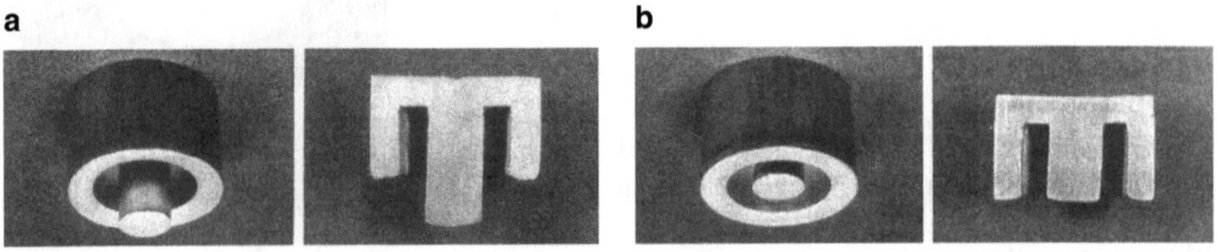

Fig. 3 Forward extrusion of solid boss and cup. **a** Without counter tool, **b** With counter tool supported by pressure

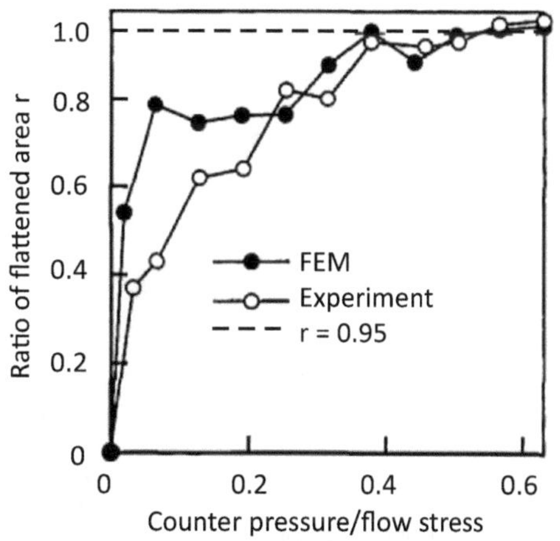

Fig. 4 Minimum counter pressure to equalize lengths

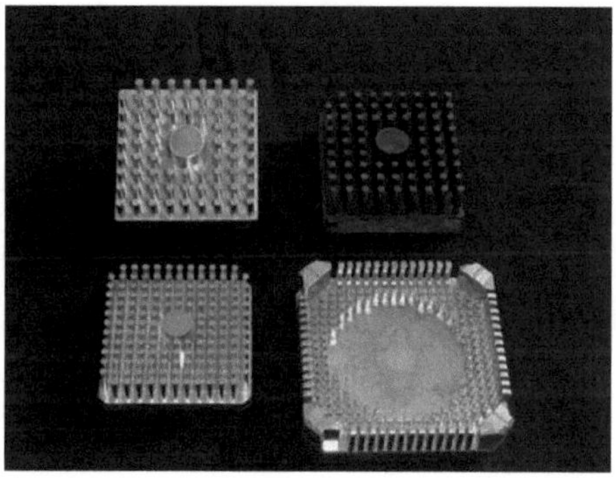

Fig. 5 Heat sinks for computer processor (Home page of Miyamoto-industry [3])

The extrusion method against a floating tool supported by pressure is considered to be effective to manufacture products with many extruded fins. Figure 5 shows the heat sinks for computer processors produced using this method by MIYAMOTO-INDUSTRY [2]. The lengths of the many small fins are equalized and the extruded ends are flattened.

It was considered that a scroll with a spiral fin with a uniform length and flat end surface could be produced by extruding the spiral fin against a floating counter tool supported by pressure.

3 Results and Applications

Figure 6 shows the die structure for extruding a scroll against a knockout tool supported by a counter force. The insert punch, the punch sleeve and the punch are pushed forward simultaneously to extrude the billet through an exit of spiral in shape. During extrusion, the knockout tool is pushed up by the KO pins with a constant force. As extrusion proceeds, the knockout tool retreats downwards by giving a counter pressure to the extruded

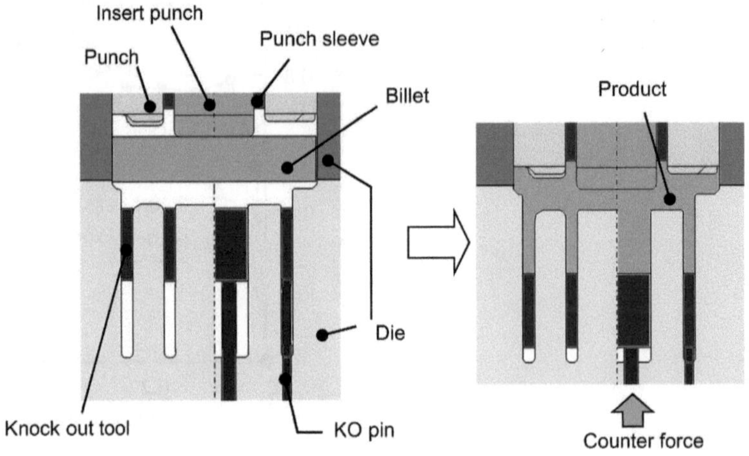

Fig. 6 Arrangement of tools for extrusion against knockout tool supported by counter force

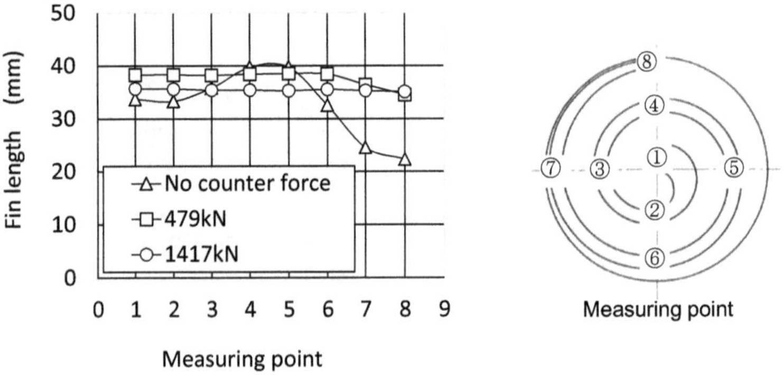

Fig. 7 Effect of counter force on distribution of fin length of scroll

fin. All the tool segments are set in a die set, and the die set is mounted on a 10,000 kN mechanical press, and the counter force is applied by a gas cylinder.

The billet is a thick round plate cut out from a bar of continuously cast Al-(8–12 % Si) alloy and the billet is dipped into a graphite type lubricant and heated to around 400°C before being inserted into the die.

Figure 7 shows the effect of counter force on the distribution of the extruded fin when the counter forces of 0, 479 kN and 1417 kN are applied. The measuring points are indicated in the right side of Fig. 7. The extruded length is uneven when the counterforce is not applied. When a counter force of 479 kN is applied, the fin length becomes almost uniform but non-uniformity is left because the end surface is not flattened well. When a counter force of 1417 kN is applied, the extruded fin length becomes uniform and the end surface is flattened.

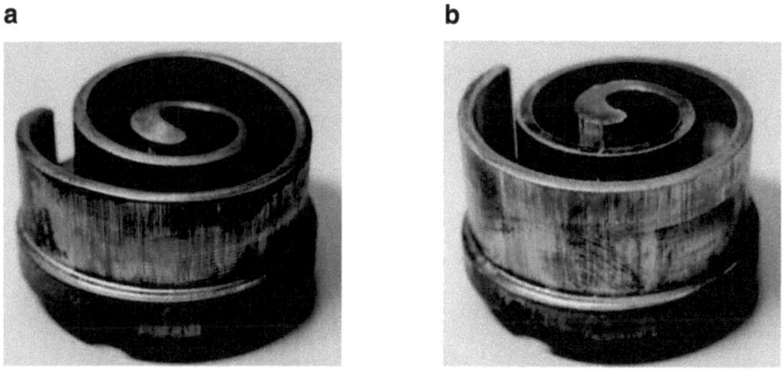

Fig. 8 Extruded scrolls without and with counter force. **a** Without counter force, **b** With counter force of 1417 kN

As shown in Fig. 8a, the extruded spiral fin has uneven height when the counter force is not applied. The length of the spiral fin becomes uniform and the end surface is flattened as Fig. 8b when a counter force of 1417 kN is applied. Although the extruded end surface of the scroll is machined to give a exact length of the fin, the side surface of the spiral fin is left as extruded and thus the volume of material removed by machining is reduced to a great extent compared to the product made by casting.

4 Conclusions

To manufacture a scroll product for an air compressor having a spiral fin, the forward extrusion method with a floating counter tool supported by a constant force is applied. With this method, the length of the long spiral fin is extruded uniformly, and the surface of the extruded end becomes flat. Although the average counter pressure to make a uniform length is as low as about 3 % of the flow stress, the pressure to flatten the extruded end is higher than 40 % of the flow stress.

References

1. Yoshimura, H.: Symposium text of Japan Institute of Light Metals, (1993), 57–64.
2. Osakada, K., Hanami, S. and Arai, N.: Deformation mode in extrusion against counter pressure, J. JSTP, 41–447 (2000), 1026–1030.
3. Home page of Miyamoto-industry, http://www.miyamoto-ind.co.jp/publics/index/11/

High-Performance Permanent Magnets by Cold Forming

Peter Groche and Lennart Wießner

The combination of a further developed Equal Channel Angular Swaging (ECAS) process and subsequent infeed rotary swaging is a promising process chain for the continuous production of permanent magnets. The ECAS process enables a grain refinement as well as an induction of high defect densities in materials. Both effects lead to a higher coercive field strength. In comparison to other SPD-processes lower forming forces are needed and a continuous production is possible. Highly textured magnetically anisotropic samples are realized by a subsequent infeed rotary swaging process.

1 Motivation

In times of rising energy demand, electric mobility and renewable sources of energy are an important step towards a sustainable use of natural resources. For an efficient implementation of these technologies, high-performance permanent magnets for motors and generators are needed. Properties such as high energy density and coercive field strength are favorable. To achieve a particularly high energy density, magnets are used with large amount of rare earths, which have unique electronic, magnetic and optical properties. However, the extraction of rare earths is highly energy-intensive due to geological deposits as well as the chemical similarity and causes intense pollution of the environment.

The characteristic behavior of ferromagnetic materials of permanent magnets in case of magnetization and demagnetization can be described by the hysteresis curve (BH). Here, relationship of the magnetic flux density B in the material and the field strength H is presented. An increase in the field strength H, results in an corresponding increase in the magnetic flux density B. By reducing H to zero, a residual magnetization still remains,

Peter Groche · Lennart Wießner ✉
TU Darmstadt, Darmstadt, Germany
e-mail: wiessner@ptu.tu-darmstadt.de

A. E. Tekkaya et al. (eds.), *60 Excellent Inventions in Metal Forming*,
DOI 10.1007/978-3-662-46312-3_52

which is called as remanence B_r. In order to return the magnetic flux density B to zero, a reverse magnetic field with the coercive field strength Hc has to be applied [1]. In general, a particular high coercive field strength for permanent magnets is highly required.

The production of permanent magnets bases primarily on the fact that the alloy components are mixed in powder form and then melted under high temperatures. Subsequent processing steps include, for example, pressing, sintering and magnetization (sintered magnets), rapid solidification and hot forming (bonded magnets) as well as sintering and precipitation hardening (precipitation-hardened magnets). The powder metallurgy composition, the achieved grain size and grain shape as well as the grain orientation affect the magnetic properties [2, 3]. However, all methods are energy and cost intensive. Furthermore, the geometry of magnets cannot be changed readily.

From literature it is well known that the coercive field strength of a material can be increased by grain refinement. Furthermore, an increase in energy density by the orientation and induction of shape anisotropy of grains is possible. Both phenomena represent a starting point for metal forming processes to adjust and improve the magnetic properties of a material. On the other hand, the proportion of rare earths can be reduced or even substituted without having significant performance losses.

The authors describe in [4] grain refinement of a copper-cobalt alloy by the severe plastic deformation (SPD) process equal channel angular pressing (ECAP). They have found a doubling of coercive field strength at room temperature. The forming of a hard magnetic alloy with ECAP at 600°C also leads to an increase of the coercivity [5]. An increase in coercive field strength up to a certain grain size is observed in soft magnetic materials using the SPD-process High Pressure Torsion [6]. The selective induction of anisotropy is also the subject of several investigations. It is found that, die-upsetting at high temperatures and the backward extrusion of magnetic material lead to a magnetic anisotropy and hence a high energy density [7].

It can be noted that SPD methods appear to be particularly suitable for the production of magnets. Disadvantageous are the facts, that SPD processes are mainly connected to high process forces and discontinuous production. A process chain consisting of continuous severe plastic deformation and a subsequent forming process for realizing a shape anisotropy has not been investigated yet.

In this article the combination of a continuous SPD process and a subsequent induction of anisotropy will be presented.

2 Process Idea

The process idea consists of the further development of a tool concept for a high-speed press for the continuous production of ultra-fine grained material. The basis is a SPD process developed at the Institute of Production Engineering and Forming Machines, called as Equal Channel Angular Swaging (ECAS). This process combines the concept of ECAP with incremental forming. The subsequent forming of nanocrystalline material for in-

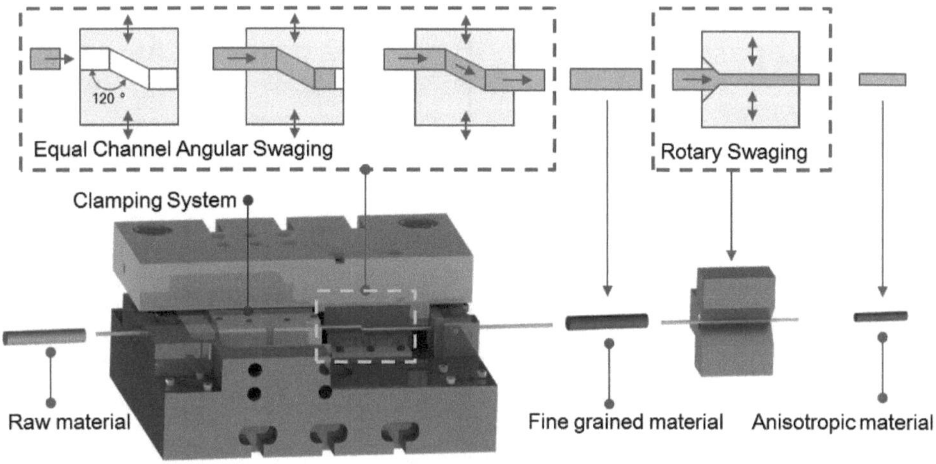

Fig. 1 Process principle of the continuous ECAS-process and subsequent rotary swaging

ducing a shape anisotropy is realized by rotary swaging. The process principle of both forming technologies is shown in Fig. 1. The tool contour is based on the ECAS process presented in [8]. The tools build an eccentric channel with a channel angle of 120°. This enables a two-step deformation within one forming step. A round bar is pushed into the channel by a feeding device and undergoes a shear deformation. Short contact times between the tool and the workpiece occur due to the incremental nature of the forming process, which results in lower friction. As a result of special kinematics of the ECAS process, the friction forces act in the same direction as the feeding. Thus, the supply of material in the process is facilitated in contrary to ECAP. In addition, the axial feeding forces are decoupled from the radial forming forces. The resulting low feeding forces are a basic requirement for transmission of the process in a continuous production. This provides a significant advantage over the established ECAP process [9].

Depending on the desired grain size and deformation potential of the material multiple forming by ECAS of the same sample is possible. In order to texturize the microstructure, the fine grained round bars are further formed by infeed rotary swaging. Elongation of grains in the axial direction is achieved by reduction of diameter. A radial anisotropy with a homogeneous strain distribution is possible by geometry optimizations of the contours of rotary swaging tools.

3 Results and Applications

Görtan uses the ECAS concept on a swaging machine for the incremental tool movement. The swaging frequency of the machine amounts to 30 Hz with a stroke of 0.7 mm. The tools have a channel angle of 120° and a channel diameter of 20 mm. The round specimens

a b

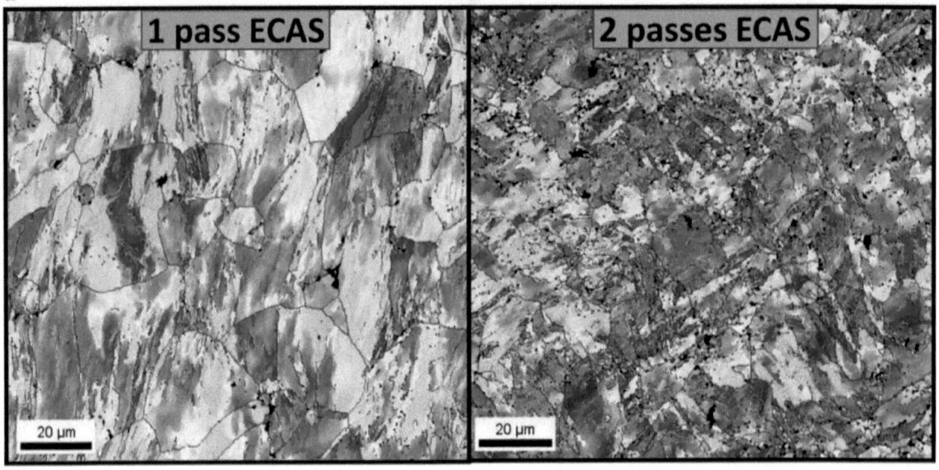

Fig. 2 Microstructure of low-carbon steel (C4 C) after one (**a**) and two passes (**b**) of ECAS [10] (Copyright of Photograph: Shaker Verlag 2014)

are fed with 1 mm/s in the tool system. It is shown that the forming process of low-carbon steel (C4 C) is possible without wrinkles or cracks in the specimens. The measured axial forces in the process are only 60 kN and 104 kN in the second stage, respectively. Radial forces of 108 kN and 278 kN in the second stage are measured. The process forces are quite low in comparison to other SPD methods. Electron back scattered diffraction (EBSD) is used in a high-resolution electron microscope (SEM) to study the microstructure. In Fig. 2, the microstructures are shown after one and after two passes of ECAS. The black lines are representing the high angle grain boundaries (HAGB). The average grain size is 20 µm before forming. After two passes of ECAS the grain size is reduced to 0.55 µm. In addition, an increase of HAGB proportion is observed after the second deformation. These results show that significant grain refinement can be achieved by the ECAS process. Therewith, the method is suitable for increasing the coercive field strength [10].

Further investigations deal with the change of coercive field strength after the ECAS process. The magnetic properties of the samples are determined using a Vibrating Sample Magnetometer (VSM). With the VSM, the magnetic moment is measured as a function of the applied field and temperature. Figure 3 shows the hysteresis curves of low-carbon steel at room temperature as received and after one pass in the ECAS process.

A soft magnetic behavior of low-carbon steel C4 C can be observed. The magnetic moment is increased from 215 to 226 electromagnetic units after one pass in the ECAS process. The crossing of the hysteresis curve of the x-axes indicates the coercive field strength (see Fig. 3). The original material has a coercive field strength of 4 Oersted. It can be seen in Fig. 3, the coercive field strength of C4 C is 7 Oersted after one pass of ECAS. An increase of 75 % can be determined. As Görtan found out in [9] several passes

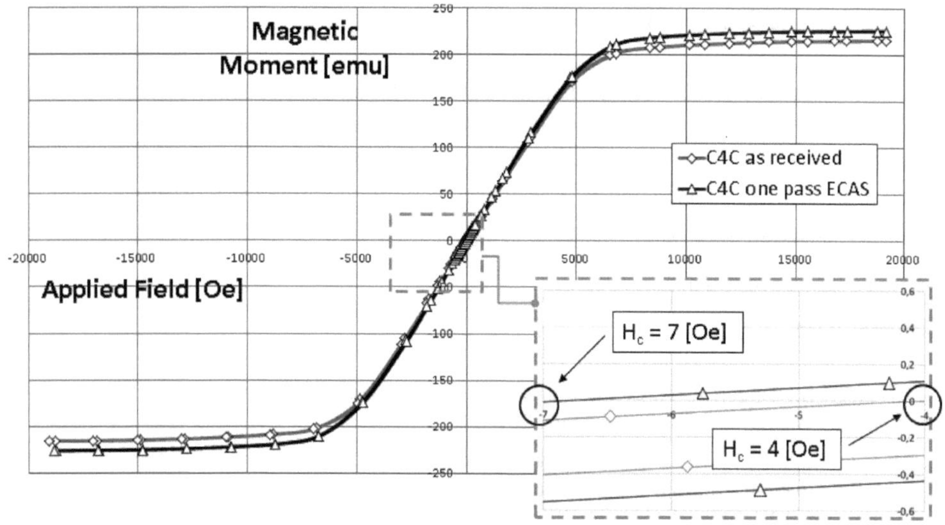

Fig. 3 Hysteresis curves of low-carbon steel (C4 C) and coercive field strength

of ECAS are possible. A further increase in the coercive field strength can be expected. This shows the potential of ECAS for the continuous production of permanent magnets.

4 Conclusion and Outlook

The combination of the newly developed SPD process ECAS and the subsequent rotary swaging process is a promising process chain for the continuous production of permanent magnets with enhanced magnetic properties. This article shows that the grain size of a low-carbon steel can be significantly reduced using ECAS. At the same time, the coercive field strength is increased by 75 %. In comparison to other SPD methods lower forming forces occur which is essential for continuous production. In addition to the magnetic properties a directed adjustment of material parameters such as a high strength and a high ductility is also possible with the newly developed process chain.

In a next step, a new ECAS – tool system is installed on a high-speed press and put into operation. Suitable materials for permanent magnets are formed in one or more passes. Further investigations deal with the induction of a shape anisotropy in axial and radial direction using infeed rotary swaging.

5 Acknowledgement

Financial support by the German federal state of Hesse through its excellence program LOEWE "RESPONSE" is gratefully acknowledged. The authors also thank the DFG for supporting the project "Production of UFG materials by rotary swaging". Many thanks are owed to Prof. O. Gutfleisch and B. Kaeswurm from the Institute for Functional Materials at the TU Darmstadt for VSM measurements.

References

1. Ilschner, B., Singer, R. F., 2010, Werkstoffwissenschaften und Fertigungstechnik. 5. Auflage; Springer-Verlag Berlin Heidelberg; pp. 291–317.
2. Moosa, I., 2013, Powder Metallurgy and its Application in the Production of Permanent Magnets. International Journal of Advanced Research in Engineering and Technology Volume 4, Issue 6, pp. 127–141.
3. Kuhrt, C., 1995, Processing of permanent magnet materials based on rare earth-transition metal intermetallics. Intermetallics 3 pp. 255–263.
4. Suehiro, K., Nishimura, S., Horita, Z., 2008, Change in Magnetic Property of Cu-6.5 mass%Co Alloy through Processing by ECAP. Materials Transactions, Vol. 49, No. 1 pp. 102 to 106.
5. Stolyarov, V. V., Gunderov, D. V., Popov, A. G., Puzanova, T. Z., Raab, G. I., Yavari, A. R., Valiev, R. Z., 2002, High coercive states in Pr–Fe–B–Cu alloy processed by equal channel angular pressing. Journal of Magnetism and Magnetic Materials 242–245, 1399–1401.
6. Scheriau, S., Kriegisch, M., Kleber, S., Mehboob, N., Grössinger, R., Pippan, R., 2008, Magnetic characteristics of HPT deformed soft-magnetic materials. Journal of Magnetism and Magnetic Materials 322, 2984–2988.
7. Gutfleisch, O., Bollero, A., Handstein, A., Hinz, D., Kirchner, A., Yan, A., Müller, K.-H., Schultz, L., 2002, Nanocrystalline high performance permanent magnets. Journal of Magnetism and Magnetic Materials 242–245, 1277–1283.
8. Görtan, M., Bruder, E., Groche, P., Müller, C., 2010, New Severe Plastic Deformation Process to Produce Ultrafine Grained Materials. Steel Research International 81–90, 478–481.
9. Bruder, E., Görtan, M., Groche, P., Müller, C. , 2011, Severe Plastic Deformation by Equal Channel Angular Swaging. Materials Science Forum Vols. 667–669, pp 103–107.
10. Görtan, M., 2014, Severe plastic deformation of metallic materials by equal channel angular swaging: Theory, experiment and numerical simulation, Darmstadt, Shaker.

New Cold Forging Lubricant Replacing Zinc Phosphate Coating

Zhigang Wang and Shinobu Komiyama

1 Motivation

The function of the lubrication coating is to prevent direct contact between the tool surface and the billet material, and to provide a low friction to ensure a smooth relative sliding between the tool and the billet. Zinc phosphate coating was developed in Germany in 1935, and has been contributed for successful cold forging operations for a long time.

As shown in Fig. 1, the treatment process of zinc phosphate coating is complicated and accompanied with waste water by the consumption of enormous amount of water. The sediment of phosphorus compound is hazardous to the environment. These problems were

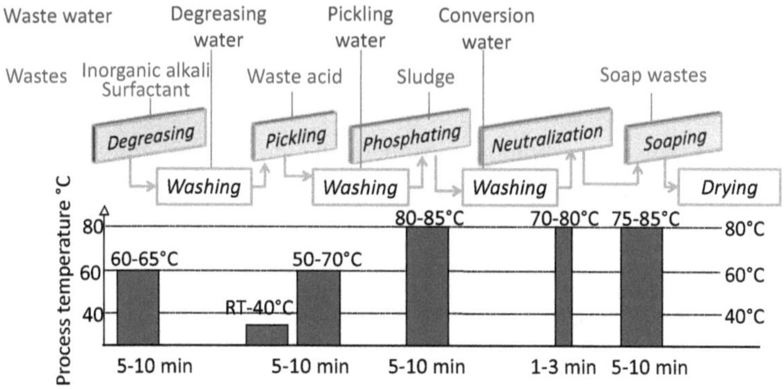

Fig. 1 Problems of phosphate coating

Zhigang Wang ✉ · Shinobu Komiyama
Gifu University, Gifu, Japan
e-mail: zgwang@gifu-u.ac.jp

A. E. Tekkaya et al. (eds.), *60 Excellent Inventions in Metal Forming*,
DOI 10.1007/978-3-662-46312-3_53

Fig. 2 Ideal process of non-reactive coating for cold forging

closed up from the viewpoint of global environment in the 1990's and new non-reactive type coatings are required.

An ideal non-reactive coating may be generated on the surface of steel billet by applying a water-based lubricant and soon dried to leave coating film. It is desirable no control of chemical reaction is necessary and no by-product is produced. The treatment process may be shortened and the coating device is simplified. Figure 2 gives the conceptual treatment process of the ideal non-reactive coating. The process consists of descaling, water washing and lubricant coating with a single step.

However, the tribological situation is very severe in cold forging due to extremely high contact pressure and large expansion of surface area, new lubricant coating needs to keep the ability to protect the surface of the steel material under the large deformation of the steel material and to maintain low friction with the tool.

2 Development of Dry In-Place Coating

Figure 3 shows an illustration of the dry in-place coating, which is formed just before the cold forging operation. This coating film consists of two layers; the upper layer provides low friction with the tool surface and the lower layer protects the tool surface by avoiding direct contact with the billet surface. Before dipping, the billet is heated to a temperature higher than the temperature of the treatment liquid. After dipping, a thin film of the treatment liquid is generated on the surface of the billet. Since the temperature of solid/liquid interface is higher than that of liquid/atmosphere, the temperature gradient in the thin film causes a micro convection and thus makes water evaporation. With the convection, in-

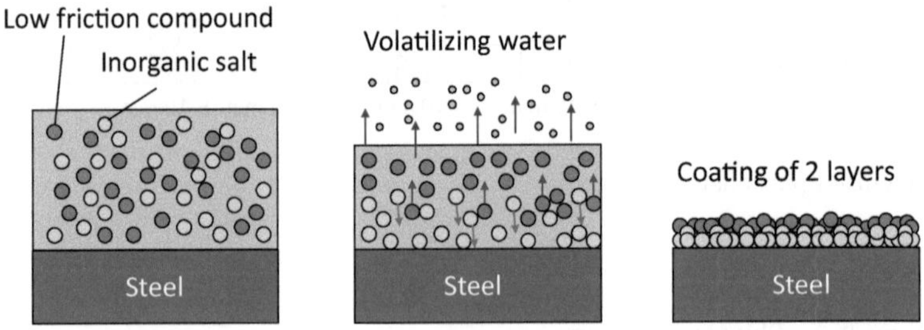

Fig. 3 Illustration of dry in-place coating. **a** Dipping, **b** During drying, **c** After drying

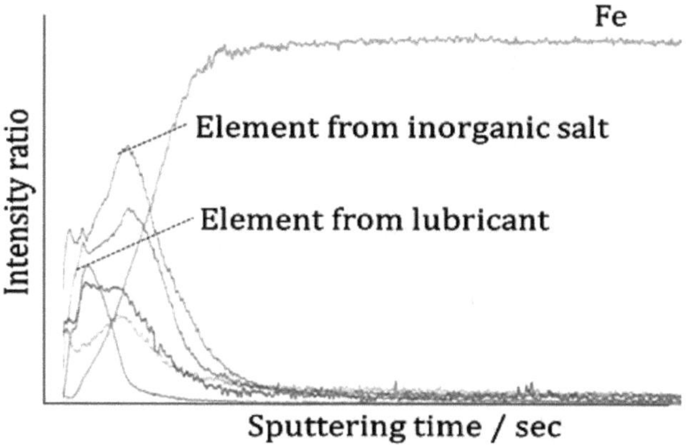

Fig. 4 Depth profile of dry in-place coating film analyzed by GDS

gredients with high affinity to metals and water are separated from the ingredients having high affinity to atmosphere. After drying, the coating film with two layers is obtained as shown in Fig. 3c.

Figure 4 shows the depth profile of chemical elements in the dry in-place coating film analyzed by glow discharge spectrometer. The analysis was carried out while removing the film by sputtering. It can be recognized that the peak of the element from lubricant exists near the surface and the peak of the element from inorganic salt is in the inside of the film.

The main ingredient of the lower layer is a water-soluble alkali metal salt, and the upper layer consists of inorganic or organic lubrication ingredients [1, 2]. Alkali metal salt has a high affinity to the surface of the steel material and has also an excellent resistance against contact pressure and heat, and thus protects the steel surface physically. As inorganic ingredients, graphite, molybdenum disulfide, zinc phosphate, boron nitride and talc can be used. These ingredients have a high resistance to temperature rise and contact pressure and thus can be used under severe forging situations such as closed die forging. Inorganic ingredients have cleavage faces in their crystal structure and a low friction. In contrast, organic ingredients have hydrophobic agents with molecular structure with a few side chains and thus provide a low friction at high temperature because they becomes liquid and the lubrication mode becomes the fluid lubrication. As organic ingredients, mineral oils, fats, soaps, waxes and fluorinated resins can be used.

3 Friction Tests Under Cold Forging Condition

The tribological performance of the developed coating is evaluated mainly by Bowden sliding test, shown in Fig. 5. In Bowden sliding test, the frictional sliding between the spherical tool and the specimen with a lubrication coating is repeated under a constant

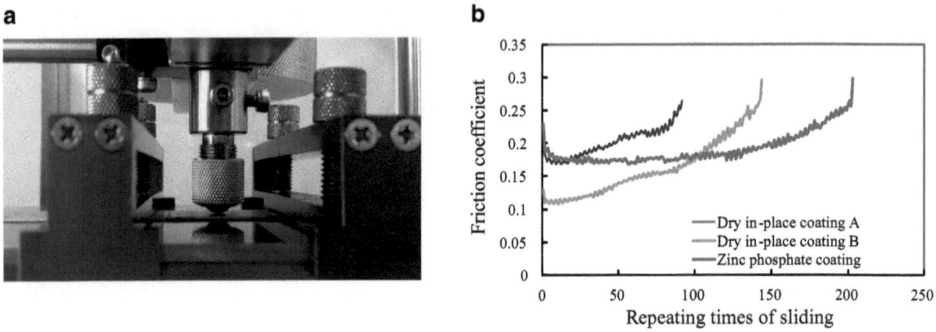

Fig. 5 Bowden sliding test. **a** General view of testing apparatus, **b** An example of testing results

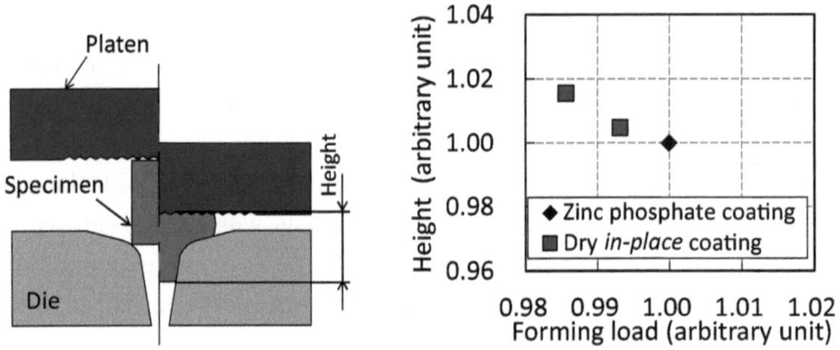

Fig. 6 Spike test [3]

normal load to determine the friction coefficient and wear rate of the coating, and the results are used to optimize alkali metal salts and lubrication ingredients.

Then forging simulation tests are used to investigate the anti-seizure ability of the coating. In order to clarify the performance of the coating under various conditions, ring compression tests, ball penetration tests, backward extrusion tests, spike tests shown in Fig. 6, are used. In spike test, the height of forged specimen and the knockout load are used as evaluating factors. The coating that gives a larger height and lower load, is evaluated to be better.

4 Apparatus for Treatment

The commercial treatment apparatus is necessary to produce the lubrication coating with a cost merit. As shown in Fig. 7, the treatment apparatus consists of the washing area by hot water, the application area of treatment liquid and the drying area to vaporize water. The function of the washing area is not only to remove the iron powder and scale powder generated in shot blasting but also to elevate the temperature of billets to prevent a temperature drop of the treatment liquid. The temperature of the treatment liquid is

Fig. 7 Treatment apparatus. *1–5* Washing zone, *6* Coating zone, *7* Drying zone

Fig. 8 A forging line using the developed coating

generally set to 60 °C to enhance drying efficiency. In the drying area, warm air is normally used to promote drying. The billet coated is transferred and forged by a press immediately.

Figure 8 gives a general view of the forging system using the treatment apparatus.

5 Application to Cold Forging

Figure 9 shows some parts forged with the dry in-place coating. Figure 10 shows the expansion of the range of application of the dry in-place coating [4]. The dry in-place coating was first used for the processes with relatively small expansion of surface area and short sliding distance during forging such as forging of outer racer, and the range of application is expanding for closed die forging of inner racers, tripods and pinion gears. Moreover, the dry in-place coating can be used also to aluminum alloys and stainless steels. One more issue, after forging, the coating can be removed by water washing or alkali degreasing and no toxic waste is produced.

Fig. 9 Examples of forged parts by the developed coating

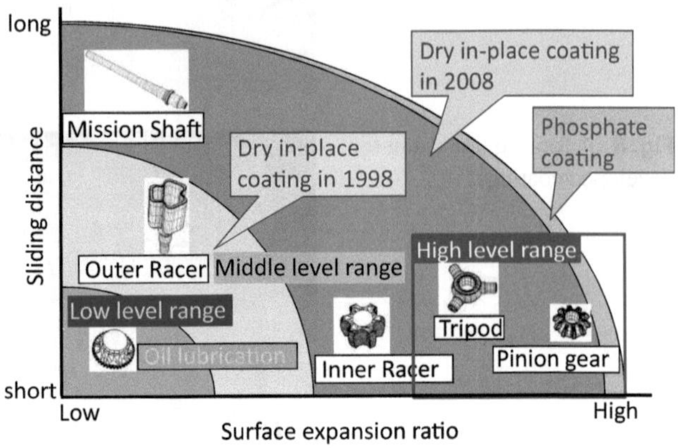

Fig. 10 Application range of the developed coating

6 Current State

The developed coating named PULS started to be sold in the market in 2001, and the number of PULS lines is over 50 in Japan. The PULS line has been introduced in 13 countries, these are China, Korea, Thailand, India, Indonesia, Malaysia, Spain, USA, Italy, France, Germany, Mexico and Poland.

References

1. Japan patent No. 3881129. Water-based lubricant for plastic processing having excellent corrosion resistance and metal material having excellent plastic processability.
2. Japan patent No. 3984159. Water-based lubricant for plastic processing having excellent corrosion resistance and metal material having excellent plastic processability.
3. S. Isogawa, A. Kimura, Y. Tozawa: Proposal of an evaluating method on lubrication. Annals of the CIRP 41(1):263–266.
4. A. Shimizu: Lubrication technology for cold forming in 21st century. Nihon Parkerizing Technical Report, 20(2008), 33–39.

Part IX

Tube and Profile Forming

Incremental Tube Forming

Christoph Becker, Matthias Hermes, and A. Erman Tekkaya

1 Motivation

Lightweight construction is a key factor to success in different industrial sectors due to ecological concerns as well as demands for higher performance [1]. Several approaches exist to realize lightweight construction. In the transportation sector bent tubular structures are often used. To reduce the weight of such structures, it is possible to use load adapted tubular structures. These structures will have less material in section where the corresponding load is low. Based on this approach, the weight of the structures can be reduced. Another possibility is the use of high-strength materials. The main challenge of using high-strength materials is the manufacture, especially the forming, of these materials. Bending of tubes made of high-strength steels leads to high bending moments as well as to a high springback after unloading. To reduce the bending moments in tube bending processes, process combinations are used. One process combination is the combined bend-rolling process [2]. The free-form bending process of a tube is combined with a rolling process which reduces the diameter of the tube. Due to the combination of these processes the needed bending moment can be reduced with an increasing diameter reduction. Another possibility is the usage of an expanding plug [3]. A plug is inserted into the tube. The outer diameter of the plug is bigger than the inner diameter of the tube. This creates an expansion of the tube and the tube bending process is superposed. This combination also leads to a reduction of the bending moment.

The described process combinations are, on the one hand, able to reduce the bending moments and, on the other hand, they can produce bent tubes with a changed diameter. Nevertheless, the described processes did not handle high-strength materials and the di-

Christoph Becker · Matthias Hermes · A. Erman Tekkaya ✉
TU Dortmund, Dortmund, Germany
e-mail: Christoph.Becker@iul.tu-dortmund.de, Hermes.Matthias@fh-swf.de,
Erman.Tekkaya@iul.tu-dortmund.de

A. E. Tekkaya et al. (eds.), *60 Excellent Inventions in Metal Forming*,
DOI 10.1007/978-3-662-46312-3_54

ameter can only be changed to a constant diameter or is restricted concerning the ovality of the tube.

Therefore, an innovative tube bending process with a combined diameter-reducing process will be presented in this paper.

2 Process Idea

The process idea is the combination of a tube spinning and tube bending process [4]. The process principle is shown in Fig. 1. This combination is named incremental tube forming process. A straight tube is fixed on a tube-feeding device which transports the tube along its longitudinal axis through a spinning device. Three spinning rolls are attached to the spinning device and these rotate around the tube. Due to a radial infeed of the spinning rolls the diameter of the tube can be varied. Subsequently to the diameter-reducing spinning process a bending process is superposed. The bending process is initiated by a numerically controlled movement of the bending device. Owing to the free movement of the bending tool a free-formed bending contour of the tube can be achieved. Furthermore, the spinning device can change the infeed of the spinning rolls constantly during the process and, therefore, a tube with variable diameter can be produced.

Besides the already mentioned process capabilities it is also possible to use a mandrel from inside the tube. The mandrel can support the forming step from inside to prevent, for example, folding or cracking. Furthermore, the mandrel can also be used to directly affect the wall thickness of the tube. By changing the gap between mandrel and spinning rolls an exact wall thickness can be applied to the tube and this wall thickness can also be

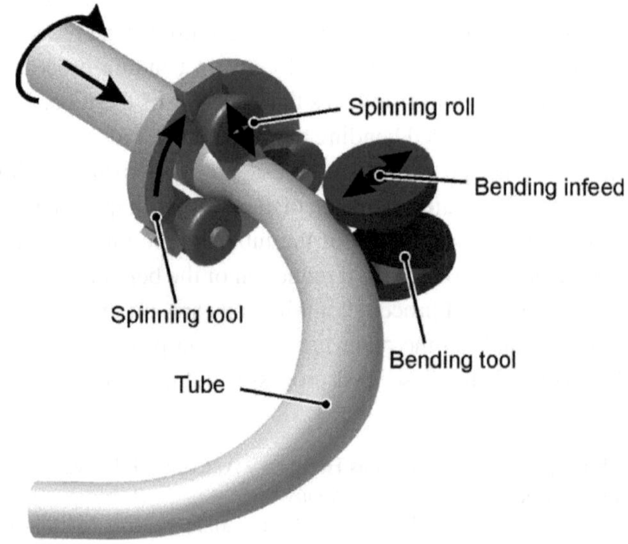

Fig. 1 Process principle of the incremental tube forming process [5]

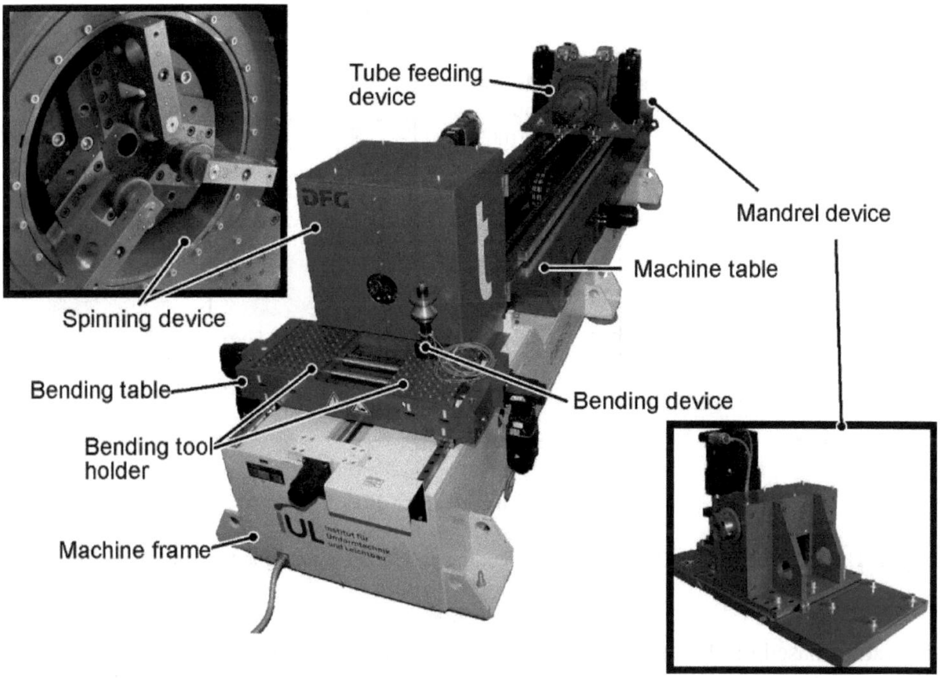

Fig. 2 Machine prototype for incremental tube forming [5]

changed over the longitudinal axis of the tube. The realization of this process idea can be seen in Fig. 2 where the machine prototype is shown.

3 Results and Applications

A process model for the incremental tube forming process has been developed [5, 6]. This process model can determine the needed bending moment as well as the springback after the bending process. The needed bending moment can be reduced down to a twentieth of the bending moment needed for a pure bending operation. This reduction depends, for example, on the rotational speed of the spinning tool, the diameter reduction during spinning, the bending radius, or the feeding speed of the tube. Due to a reduced bending moment the springback after the unloading of the tube is also reduced. This can be seen in Fig. 3. The results are shown for a dual phase steel (DP800) with an initial diameter of 40 mm. For four differently loaded bending radii the corresponding unloaded bending radii are shown. The difference between these radii is the springback.

It is shown that the springback after the incremental tube forming process is significantly reduced compared to a pure bending process due to the stress superposition in the forming zone. Furthermore, it can also be seen that the prediction of the unloaded bend-

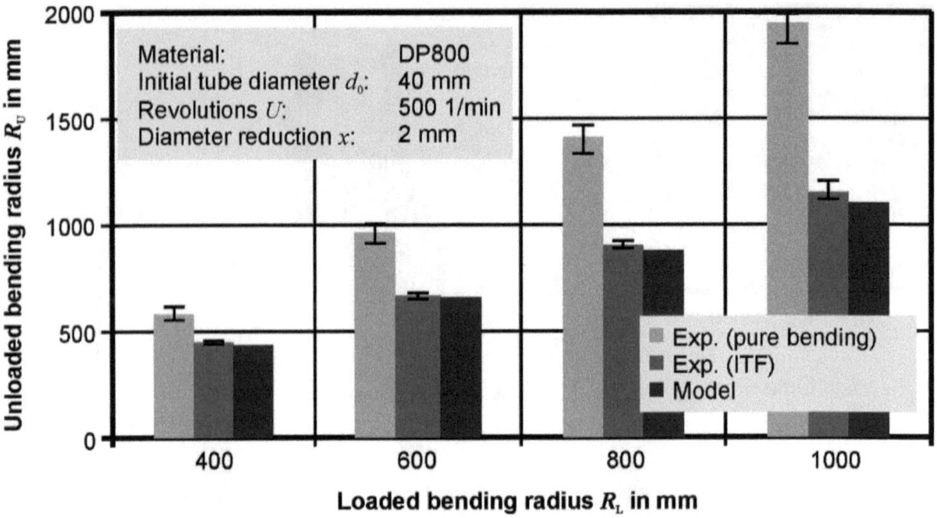

Fig. 3 Unloaded bending radii during pure bending and incremental tube forming (ITF) [5]

ing radius based on the theoretical model is accurate. Therefore, it is possible to predict the reduced springback before starting the experiment which prevents a trial-and-error approach. In addition to the already mentioned process analysis, the influence of the tool setup on the feeding force [7] as well as the tool geometry [8] has been investigated. Furthermore, the influence of process parameters like rotational speed of the spinning rolls or the spinning roll geometry on the surface roughness has been analyzed [9].

4 Potential

The incremental tube forming process offers numerous potentialities, qualifying this process for a large variety of bending and forming tasks:

1. Bending moment reduction: It can lead to a smarter and smaller machine setup. Furthermore, it will reduce cross sectional distortion due to a reduced shear force.
2. Springback reduction: The springback reduction leads to a more precise process result. Furthermore, it extents the process limits since a strong overbending to reach a certain unloaded bending radius is no longer necessary. Another result of the lower overbending is the possibility to reach smaller bending radii.
3. Forming of high-strength steels: Due to the reduced bending moment and springback the bending of high-strength materials is possible.
4. Variation of diameter: The spinning process allows a continuous changing of the diameter along the longitudinal axis of the tube.

5. Variation of wall thickness: By using a mandrel from inside the tube it is also possible to change the wall thickness along the longitudinal axis of the tube.
6. Load-adapted tubes: Due to the variation of the cross section along the longitudinal axis of the tube load adapted tubes can be manufactured. These tubes will have less material where the load is lower and, therefore, weight can be saved in these particular areas.
7. Direct influence on surface roughness: Due to the adaptable rotational speed of the spinning device a defined surface roughness can be applied to the tube [9].
8. Three-dimensional bending: The process allows a three-dimensional free-form bending of the tubes which is caused by a rotation of the feeding device to change the bending plane.

5 Conclusion

The incremental tube forming process is an innovative forming process to manufacture three-dimensionally bent tubes with varying cross sections. This combination leads to a significant reduction of the bending moment and springback. These process phenomena can be explained by a theoretical approach which has been developed for this process. The reduction offers numerous potentialities which have been shown in this paper. This process will enlarge the spectrum of three-dimensionally bent tubes and will also be a possible alternative or addition for already existing processes and process chains. Furthermore, it can be used as a representative for showing the possible effects of a process combination. Therefore, the gained knowledge as well as the theoretical approach can be applied to similar forming operations.

6 Acknowledgement

This process has been investigated in several projects. The work has been kindly supported by the German Research Foundation (DFG) under the grant number TE 508/26-1, the Federal Ministry of Economics and Technology on the basis of a decision by the German Bundestag, and by the European Community's Research Fund for Coal and Steel (RFCS) under the grant agreement number [RFSR-CT-2009-00017].

References

1. Kleiner M., Geiger M., Klaus A., 2003, Manufacturing of lightweight components by metal forming. CIRP Annals – Manufacturing Technology 52 (2), 521–542.
2. Tozawa, Y., Ishikawa, T., 1988, A New Tube Bending Method – Application of 'Bend-Rolling Process', CIRP Annals – Manufacturing Technology, 37 (1), 285–288.

3. Nakamura, M., Maki, S., Nakajima, M., Hayashi, K., 1996, Bending of Circular Pipe Using a Floating Spherical Expanding Plug, Advanced Technology of Plasticity 1996: Proceedings of the 5th International Conference on Technology of Plasticity, Columbus, Ohio, 501–504.
4. Hermes, M., Kurze, S., Tekkaya, A. E., 2008, Verfahren und Vorrichtung zur Umformung eines Stangenmaterials (Method and Device for Forming a Bar Stock), European Patent, EP000002203264B1.
5. Becker, C., 2014, Inkrementelles Rohrumformen von hochfesten Werkstoffen, Aachen, Shaker (Dortmunder Umformtechnik, 79).
6. Becker, C., Tekkaya, A. E., Kleiner, M., 2014, Fundamentals of the incremental tube forming process, CIRP Annals – Manufacturing Technology, 63 (1), 253–256.
7. Becker, C., Staupendahl, D., Hermes, M., Chatti, S., Tekkaya, A. E., 2012, Incremental Tube Forming and Torque Superposed Spatial Bending – A View on Process Parameters, Steel Research International, Proceedings of the 14th International Conference on Metal Forming, 415–418.
8. Becker, C., Isik, K., Bayraktar, A., Chatti, S., Hermes, M., Soyarslan, C., Tekkaya, A. E., 2013, Numerical Investigation of the Incremental Tube Forming Process, Key Engineering Materials 554–557, 664–670.
9. Becker, C., Quintana, G., Hermes, M., Cavallini, B., Tekkaya, A. E., 2013, Prediction of surface roughness due to spinning in the incremental tube forming process, Production Engineering-Research and Development 7 (2-3), 153–166.

Incremental Profile Forming

Goran Grzancic, Christoph Becker, and Matthias Hermes

1 Motivation

In the context of the global shortage of resources, the importance of lightweight design is increasing constantly. Since weight reduction is a major objective of lightweight design, it can be found in various fields of industry [1]. One of the most important industry sectors for lightweight design is transportation [2]. Especially in the automotive or aircraft industry weight reduction is the driving force for innovation and new technologies. In car bodies or aircrafts profiles are mainly used for structural parts, whereas in other industry fields they are also used for non-structural or decorative purposes. Concerning lightweight design different strategies can be followed [3]. One key strategy is component lightweight design. Its main objective is the adaption of material distribution within the component according to the occurring service conditions. Accordingly, profile design changes to load-adapted shapes, mainly characterized by varying cross section geometries along the longitudinal axis. For the manufacturing of such profiles various processes exist. A well-known forming process for profiles with a complex and varying geometry is hydroforming [4, 5]. However, for each profile a corresponding complex forming die is necessary, making the process inflexible with regard to geometry changes [6]. Furthermore tool effort is considerably high in hydroforming compared to other processes. In contrast, kinematical forming processes like e. g. flexible roll forming [7, 8] or incremental tube forming [9, 10] allow the manufacture of several profile geometries with changing cross sections with one tool set. Due to the low tool dependence these kinematical processes feature comparatively high process flexibility. However, the cross section variation is limited, like in e. g. flexible roll forming, where just the profile width and height can

Goran Grzancic ✉ · Christoph Becker · Matthias Hermes
TU Dortmund, Dortmund, Germany
e-mail: Goran.Grzancic@iul.tu-dortmund.de, Christoph.Becker@iul.tu-dortmund.de, Hermes.Matthias@fh-swf.de

A. E. Tekkaya et al. (eds.), *60 Excellent Inventions in Metal Forming*,
DOI 10.1007/978-3-662-46312-3_55

be varied while the basic shape of the cross section stays the same. In order to combine both high workpiece complexity and high process flexibility, a need for new forming procedures as well as advancements of existing processes exists [11]. For this reason, the incremental profile forming process (IPF) will be presented in this article [12].

2 Process Idea

The process idea for the flexible manufacturing of profiles with variable and complex cross section geometries is to have several tools penetrating into the initial profile, leading to the desired shape of the profile. Based on this principle, the incremental profile forming process was designed. The basic principle of the IPF process is given in Fig. 1, where the forming operation is shown exemplarily on the basis of a tube with a circular cross section. Initially, the tube has a feed in axial direction and is formed by several forming tools penetrating radially and independent from each other, leading to the final geometry of the cross section. Additionally, a rotation can be superposed to the forming process in order to vary the tool infeed position in circumferential direction of the tube or to form helical structures.

Basically, IPF is a kinematical forming process where the forming tools have a general and simple shape. In this case the final cross section geometry of the profile is mainly defined by the used NC code. Therefore, a potential change of the profile shape is realizable by the adjustment of the NC code and does not necessarily lead to a tool change. Nevertheless, the process allows the performance of form-closed forming operations where the final cross section geometry is mainly defined by the tool geometry. Here, the design of the forming tools contains a near-net-shaped geometry. Finally, it is also possible to combine the kinematical and the form-closed forming types, leading to a further increase of cross sectional variety.

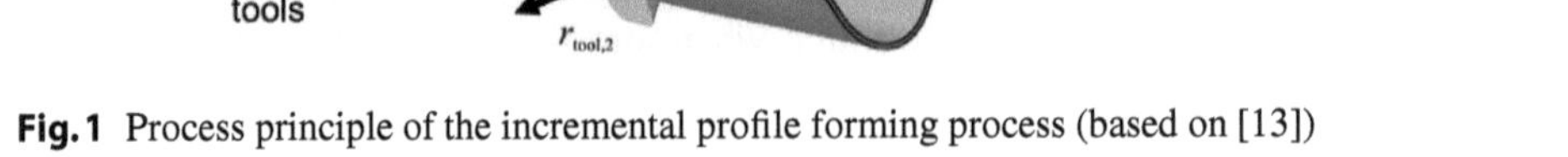

Fig. 1 Process principle of the incremental profile forming process (based on [13])

3 Results and Applications

Based on the previously described process principle, a machine prototype was developed and realized in order to allow further process investigations [13]. An overview of the machine system is shown in Fig. 2 and will be described in the following.

The machine system basically consists of the machine base, a tube feed unit as well as a forming unit. Representing the core element of the prototype, the forming unit is mainly built up of a rotary disc and six tool units. The disc is pivot-mounted in the frame and its rotation is realized by a spur gear at the outer diameter of the disc. The six tool units are radially arranged along the circumference of a centered drilled hole where the initial tube is passed through. A tool unit contains the power unit, providing the linear motion of the forming tools, as well as the forming tool itself.

For the assessment of the producibility of profiles, several experimental and numerical investigations took place. Kinematical, form-closed as well as combined forming operations were performed, giving an overview of the geometric diversity of the process. Additionally, helical structures were realized in both forming types. The manufactured parts are shown in Fig. 3.

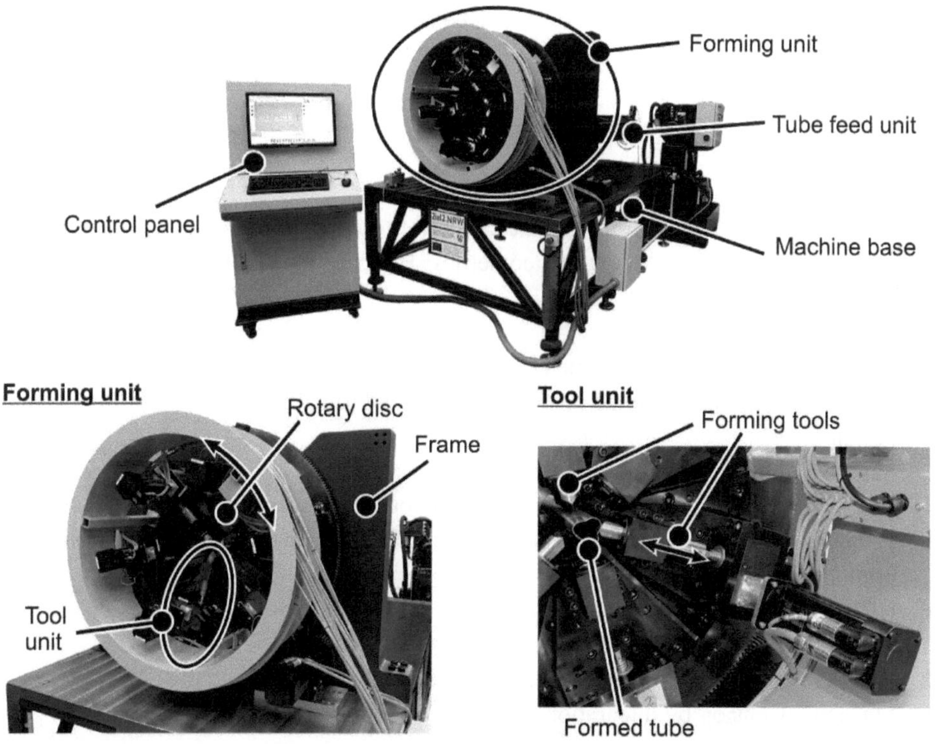

Fig. 2 Machine prototype for incremental profile forming (based on [13])

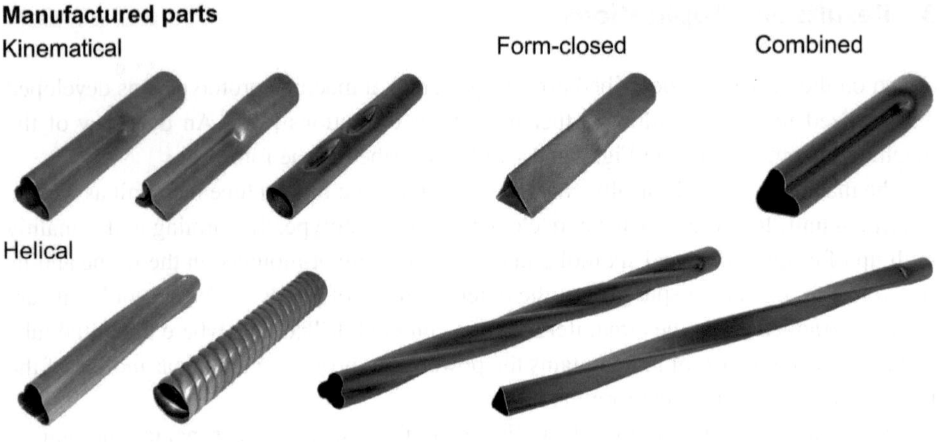

Fig. 3 Manufactured profiles

According to Fig. 3, it is observable that the kinematical forming type is more flexible compared to the form-closed forming type since more possibilities for the variation of cross section geometries exist. The superposition of the rotary movement to the process leads to a further increase of workpiece diversity. Profile examples for realized helical structures are also given in the figure.

4 Potential

Due to the flexible process principle incremental profile forming offers a huge range of manufacturable profile cross section geometries. Hence numerous process potentials can be derived:

1. Variation of cross section geometry along the longitudinal profile axis
2. Manufacture of load adapted profiles, contributing to the global need for lightweight design
3. Simple tool geometries, leading to low tooling costs
4. Efficient tool change, due to the simplicity of tooling concept
5. Parallel use of different tool geometries in one forming operation, leading to a higher profile complexity
6. High process flexibility due to the kinematical forming approach
7. Suitable for small batch sizes due to the incremental approach
8. Form-closed joining of different material combination by creating undercuts
9. Machine extension by integration of bending device for the manufacture of bent profiles with changing cross section shapes [14]

5 Conclusion

Incremental profile forming is an innovative forming method developed for the manufacture of profiles with varying cross section geometries along the center axis. Due to the process principle IPF is characterized by a high degree of flexibility regarding the workpiece complexity, offering a large potential for the manufacture of lightweight design parts, especially for the production of load-adapted profiles. For the further development of the process, a flexible machine prototype with eight numerically controlled axes was designed and realized. The machine system allows the performance of basic process investigations, leading to an increase of knowledge and finally to higher technological maturity of the process. Based on the presented machine prototype, first experimental and numerical investigations with several tool sets took place, showing the diversity of the manufacturable parts.

6 Acknowledgement

The work was funded by the European Union and the federal state North Rhine-Westphalia within the scope of the NRW Ziel2 program 2007–2013 (EFRE: Europäischer Fonds für regionale Entwicklung; Investition in die Zukunft).

References

1. Kleiner M., Geiger M., Klaus A., 2003, Manufacturing of lightweight components by metal forming. CIRP Annals – Manufacturing Technology 52 (2), 521–542.
2. Chatti, S., 2005, Production of profiles for lightweight structures, Norderstedt. Books on Demand GmbH.
3. Tekkaya, A. E., Ben Khalifa, N., Grzancic, G., Hölker, R., 2014, Forming of Lightweight Metal Components: Need for New Technologies, Procedia Engineering, 81, 28–37.
4. Ahmetoglu, M., Altan, T., 2000, Tube hydroforming: state-of-the-art and future trends, Journal of Materials Processing Technology, 98 (1), 25–33.
5. Koc, M., Altan, T., 2001, An overall review of the tube hydroforming (THF) technology, Journal of Materials Processing Technology, 108 (3), 384–393.
6. Dohmann, F., Hartl, C., 1997, Tube hydroforming – research and practical application, Journal of Materials Processing Technology, 71 (1), 174–186.
7. Groche, P., Zettler, A., 2006, Flexibles Profilieren für den Leichtbau, thema Forschung, 1, 26–29.
8. Sedlmaier, A., Henning, R., Abee, A., 2011, Fabrication of load optimized truck members with variable cross sections by flexible roll forming, SCT2011, Düsseldorf, Stahleisen GmbH.
9. Becker, C., 2014, Inkrementelles Rohrumformen von hochfesten Werkstoffen, Aachen, Shaker (Dortmunder Umformtechnik, 79).

10. Becker, C., Tekkaya, A. E., Kleiner, M., 2014, Fundamentals of the incremental tube forming process, CIRP Annals – Manufacturing Technology, 63 (1), 253–256.
11. Staupendahl, D., Becker, C., Weinrich, A., Hermes, M., Tekkaya, A. E., 2012, Innovative Umformverfahren für Rohre, Profile und Bleche aus modernen Stahlwerkstoffen, Stahl und Eisen, 8, 47–54.
12. Hermes, M., Becker, C., Wagner, R., Tekkaya, A. E., 2012, Method and device for incrementally shaping profiled pipes, in particular profiled pipes having cross-sections that vary over the longitudinal axis, Patent application WO2012/000490A2.
13. Grzancic, G., Becker, C., Hermes, M., Tekkaya, A. E., 2014, Innovative Machine Design for Incremental Profile Forming, Key Engineering Materials, 622–623, 413–419.
14. Hermes, M., Becker, C., Tekkaya, A. E., 2013, Process and apparatus for the incremental bending of profile tubes, in particular of profile tubes having cross sections which vary over the longitudinal axis, Patent application WO2013/143512A1

CNC Tube Forming Method for Manufacturing Flexibly and 3-Dimensionally Bent Tubes

Makoto Murata and Takashi Kuboki

1 Motivation

There are many kinds of tube bending methods such as press bending, stretch bending, draw bending and other various bending methods [1–4], and new bending methods have been invented in the 21st century [5, 6]. These bending methods have been developed and are used depending on operation efficiency and other requirements. Bent circular tube products are employed to reduce production cost and weight for manufacturing many kinds of products, such as fluid transportation parts, furniture, transport apparatus, and mechanical parts.

An innovative flexible bending method [7–11] was invented by Murata and developed by Nissin Precision Machines Co., Ltd. The method challenges the conventional image of tube benders and is called MOS bending, named after the researchers, Makoto **M**urata, Shinji **O**hashi and Hideo **S**uzuki. MOS bending is controlled by computer and is applicable for various bending radii with high precision. MOS bending realizes flexible and complex bending with various radii and in arbitrary direction by simple numerically controlled positioning of the bending die in vertical and horizontal directions without any replacement of the dies.

These and other benefits of MOS bending are reviewed here, together with its mechanism. Some examples of the products of MOS bending are presented in this article.

Makoto Murata · Takashi Kuboki ✉
The University of Electro-Communications, Tokyo, Japan
e-mail: kuboki@mce.uec.ac.jp

A. E. Tekkaya et al. (eds.), *60 Excellent Inventions in Metal Forming*,
DOI 10.1007/978-3-662-46312-3_56

2 Process Idea

Figure 1 shows the principle of MOS bending. The composition is very simple with a relatively simple operating sequence. The relative distance between the center of the guide cylinder #3 and the center of the bending die #2 is offset u. A tubular work piece #1 is fed through the die from the guide cylinder by axial force P_L, then is bent as shown in the figure. For the bending unit, the approach distance V between the exit of the guide cylinder and the center of the bending die is not changed. The tube is pushed by the bending die in the vertical direction as indicated by the vertical arrow when the offset u is given in the vertical direction.

The extrusion load P_u depends on the magnitude of the offset u. The bending moment M, which is calculated by

$$M = P_U V + P_L u \tag{1}$$

acts to bend the tube. The position of the die is continuously changed by AC microcomputer-controlled servomotors. The inclination angles in vertical and horizontal directions are controlled by two AC servomotors.

The composition of the machine and a photo of a commercial machine are shown in Figs. 2 and 3, respectively. The AC microcomputer-controlled servomotors change the position of the bending die. A tube #1 is inserted into the guide cylinder #3. When AC servomotor #7 is activated, the tube is pushed toward Z axis by plate #11, which is moved by the motor #7 via the chain #10. The bending die is shifted continuously toward X and Y axes by AC servomotors #5 and #6. The tube is fed to an appropriate length by measuring the feed length by a rotary encoder in order to bend the tube to a target bending angle.

The die offset u determines the bending radius R, and the bending die inclination ϕ should be set to be a suitable value according to the offset u. The relationship between the offset u and the bending radius R is shown in Fig. 4, which was obtained by preliminary experiments. As flexural rigidity, Young's modulus, and other material properties of the tube affect bending radius, the offset u should be selected according to the material and tube dimensions. The other important dimension of the bent tube is bending angle θ of the bending arc, which is dominated by feed length W_t. The relationship between feed

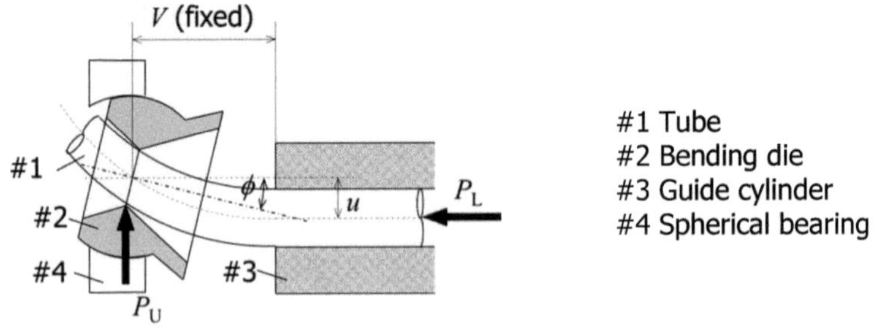

Fig. 1 Principle parts in MOS bending

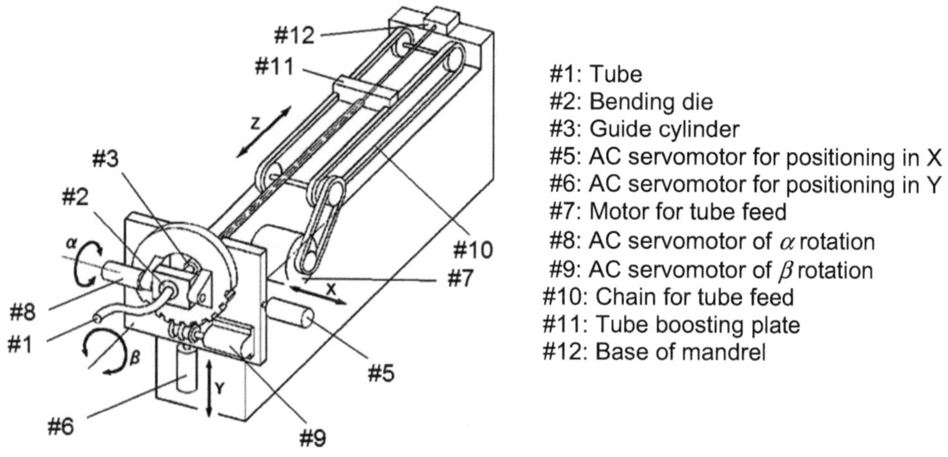

Fig. 2 Main components of CNC bender

Fig. 3 A commercial machine for MOS bending

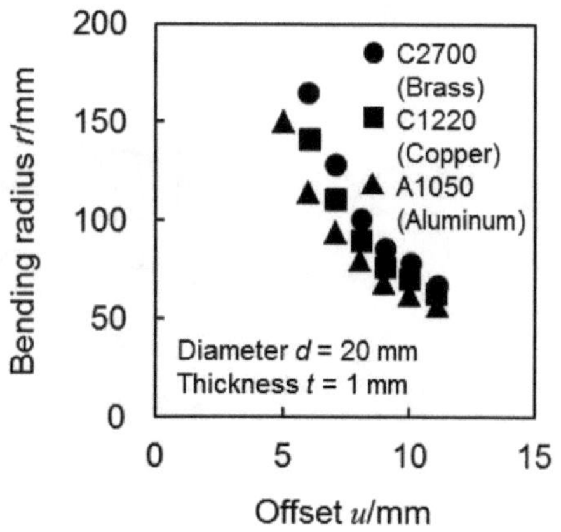

Fig. 4 A relationship between offset and bending radius obtained by experiment

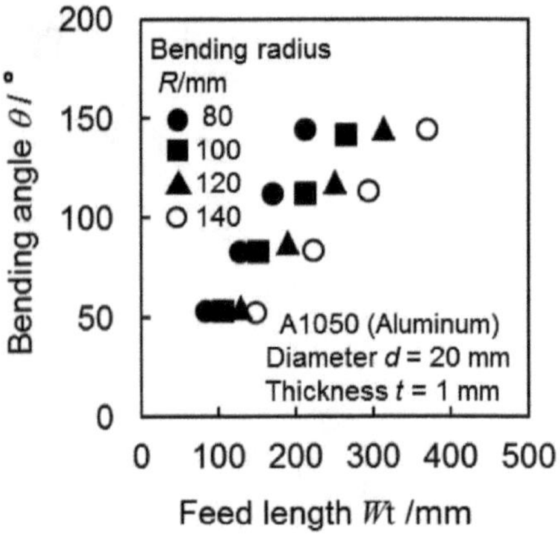

Fig. 5 A relationship between feed and bending angle

length W_t and bending angle θ should be clarified in advance as shown in Fig. 5. Once the suitable offset u and feed length W_t are determined for the target shape, MOS bending machine flexibly bends tubes into arbitrary 3D shapes with arbitrary bending radii, angles and bending directions.

3 Results and Applications

CNC is effectively applied for MOS bending, as the tube-product shape is formed in a segment-by-segment manner in the progress. Each segment is either a straight or an arc with a constant bending radius and bending direction. To form a given arc, offset u, feed length W_t and rotational position in β for die inclination is determined for realization of the target bending radius R, bending angle θ and the bending direction. The relationships between u and R, and W_t and θ, which are introduced in the previous section, are used in the CNC system. Examples of the product formed by the commercial MOS machines are shown in Fig. 6.

Fig. 6 Examples of product formed by MOS bending

4 Benefits

MOS bending has the benefits as follows:

(1) Bending radius can be easily set: Unlike the conventional bending machine, bending dies, around which the tubes are wound, are not used, and thus an arbitrary radius can be obtained flexibly.
(2) Deformation ratio of the tube cross-section is minimized: As the bending section is securely guided by the die hole, accurate roundness of the cross section can be obtained.
(3) Bending radius can be continuously changed: Bending radii, angles and directions can be easily controlled by computer.
(4) Versatile bending forms are available: Bending angle larger than 180 degrees, large-radius bending, and solenoid-form bending are possible.
(5) Sophisticated bending form is obtained: As neither bump nor dent was observed at the boundary between segments, a smooth and continuous surface can be achieved.
(6) Operability of various functions is improved: Inputting the coordinates of bending section is also available. Quiet environment is ensured because of motor-driven operation at high speed.

5 Conclusion

This article reviews an innovative 3D flexible bending method, called as MOS bending, which was invented by Murata in the 1990's. MOS bending machine is composed of relatively simple parts. As it is progressed in a segment-by-segment manner, CNC is effectively applied for MOS bending. MOS bending has many advantages in terms of concise control of bending radius, precision in cross section, flexibility, versatility of bending forms, elegant form of products and operability and so on.

MOS bending was an epoch-making forming process and has an impact even in 21st century. The method has been utilized in Japan, Europe, and the United States. The excellent application includes seat, exterior pieces such as monuments and buffer stops, refrigerator tubes, bicycle parts, automotive parts and other complex components. By means of CNC of utmost simplicity, this new generation of bending method makes it available to the demand from complex tube manufacturing.

References

1. Nakamura, M., Kubota, M., Oki, Y., Maruyama, K., Nakamura, T., 1972, Pipe forming method, 194–195, in Japanese, Nikkan Kogyo Shinbun, Ltd..
2. Hakusui, H., Futahashi I., 1982, J. Jpn. Soc. Technol. Plast., 23-255:290–295, in Japanese.

3. Ochiai, I., 1982, J. Jpn. Soc. Technol. Plast., 23-255:296–302, in Japanese.
4. Takahashi, K., Murata, M., 2006, Trend of Tubular bending, J. Jpn. Inst. Light Metals, 56-5:283–289, in Japanese.
5. Hermes, M., Chatti, S., Weinrich, A., Tekkaya, A., 2008, Three-dimensional bending of profiles with stress superposition, Int. J. Material Form., 1-Issue 1:133–136.
6. Kuboki, T., Takahashi, K., Ono, K., Yano, K., 2013, A new schedule-free mandrel-less bending method for straight/pre-shaped long tubes, CIRP Annals – Manufacturing Technology, 62/1, 303–306.
7. Murata, M., Ohashi, N., Suzuki, H., 1989, New flexible penetration bending of a tube: 1st report, a study of mos bending method, Trans. JSME. ser. C, 55-517:2488–2492, in Japanese.
8. Murata, M., Yamamoto, S., Suzuki, H., 1994, New flexible penetration bending of a tube: 1st report, a study of mos bending method, J. of Japan Society of Technology for Plasticity, 35-398:262–267, in Japanese.
9. Murata, M., 1996, Experimental study of square tube bending by mos bending method, J. of Japan Society of Technology for Plasticity, 37-424:515–520, in Japanese.
10. Murata, M., Aoki, Y., 2014, Analysis of circular tube bending by mos bending method, Proc. 11th INt. Conf. Techn. Plast. (ICTP1996), Nagoya, 893–898.
11. Murata, M., Kuboki, T., Takahashi, K., 2007, Characteristics of tube bending by mos bending machine, 2nd International Conference on New Forming Technology, Bremen, 135–144.

Mechanical Joining of Tubes

Luis M. Alves and Paulo A.F. Martins

1 Motivation

Conventional technologies for connecting tubes are widely utilized in plumbing, air conditioning, refrigeration, process piping and lightweight structures, among other applications. The most well-known technologies are based on commercially available tee fittings, saddle adapters and weld-o-lets for standard geometries and materials, such as carbon steel, stainless steel and copper (Fig. 1a to c). A standard tee fitting (Fig. 1a) has three welds; two in the main tube and one in the branch tube. Saddle adapters or weld-o-lets (Fig. 1b and c) also need to be brazed or welded to the main tube over a pre-cut hole and the attachment to the branch tube is made through a weld or a threated connection.

Custom based solutions are mainly based on nozzle-weld and spin-forming (Fig. 1d and e). Nozzle-weld connections (Fig. 1d) require cutting a hole in the main tube, shaping

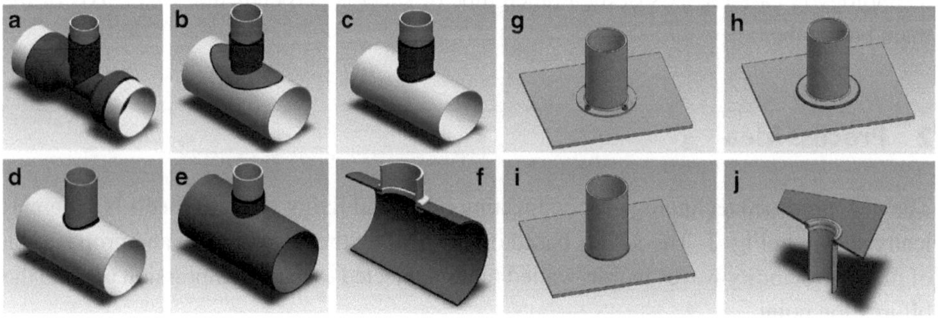

Fig. 1 Conventional and new technologies for connecting tubes and fixing tubes to sheets

Luis M. Alves · Paulo A.F. Martins ✉
Universidade de Lisboa, Lisboa, Portugal
e-mail: pmartins@tecnico.ulisboa.pt

A. E. Tekkaya et al. (eds.), *60 Excellent Inventions in Metal Forming*,
DOI 10.1007/978-3-662-46312-3_57

a contoured end in the branch tube to match the diameter of the main tube and welding along the contour. Spin-forming (Fig. 1e) also requires cutting a hole in the main tube but the difference is that material around that hole is subsequently shaped into a tee fitting where the branch tube will be brazed or welded.

In case of fixing tubes to sheets, the most widespread technologies are based on the utilization of mechanical fixing with fasteners (nuts and bolts or rivets), welding, and structural adhesive bonding (Fig. 1g to i).

No matter the application and the differences between the technologies that are currently available for connecting tubes and fixing tubes to sheets, their universe of applicability is always limited by aesthetic, physical, chemical and mechanical requirements.

The utilization of mechanical fasteners usually suffers from unwanted aesthetic features, corrosion problems or functional difficulties related to the maximum load that nuts, bolts and rivets can support safely. Welding has similar problems to those mentioned for mechanical fasteners plus specific difficulties. These difficulties arise from dimensional inaccuracies and heat-affected zones resulting from the heat-cooling cycles, from weldability of dissimilar materials (e. g. joining steel or aluminium tubes to aluminium or copper sheets), from the production of undesirable fumes and smokes in fabrication, and from expensive and time consuming issues related to quality inspection of the welds. Structural adhesive bonding offers engineers the possibility of joining different types of materials while improving aesthetics by avoiding rivets and bolt heads. However, adhesives require careful preparation of the surfaces where they are to be applied and may experience significant decrease in performance over time and in the presence of hostile environmental conditions (e. g. prolonged expose in moist environments).

The following paragraph presents a cost competitive and environmental friendly joining technology that makes use of axisymmetric or asymmetric plastic instability waves in thin-walled tubes subjected to axial compression for connecting tubes and fixing tubes to sheet panels at room temperature. Figures 1f and j present a glimpse of the new proposed technology in case the axis of the branch tube or sheet is perpendicular to the axis of the main body tube.

2 Process Idea

The new proposed technology for connecting tubes and fixing tubes to sheets is schematically shown in Fig. 2. The idea is based on controlling the development and propagation of plastic instability waves in thin-walled tubes subjected to axial compression beyond the bifurcation point.

In case of tube joining (Fig. 2a) the left and middle schematic drawings show the upper and lower dies that are needed to trigger and propagate inclined, out-of-plane, instability waves between contoured dies at the open and closed positions. The rightmost schematic drawing shows an application of this concept for producing inclined tube joints. The sectional views show the active tool components consisting of upper and lower contoured

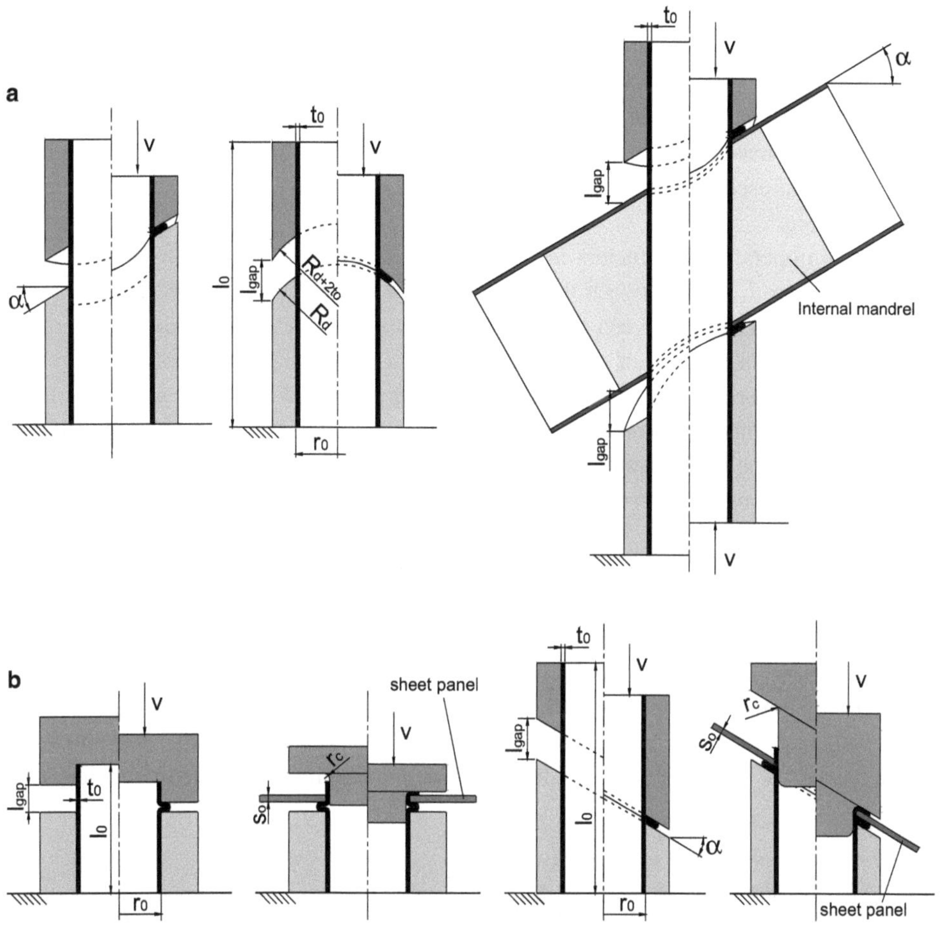

Fig. 2 **a** Sectional schematic views of the tool setup to develop and propagate inclined, out-of-plane, instability waves in tubes and to produce inclined tube joints, **b** Sectional schematic views of the tool setup to develop and propagate inclined, in-plane, instability waves to fix tubes to sheets

dies and internal mandrels (if present). The internal diameter of the dies is dedicated to a specific reference radius r_0 of the main body tube. The radius R_d of the parting out-of-plane surface of the dies together with its inclination α to the axis of the main body tube are dedicated to a specific instability wave or, in case of tube joining, to a specific radius of the branch tube. The difference between the radius R_{d+2t_0} and R_d of the upper and lower parting surfaces is crucial to accommodate the plastic compression bead at the end of stroke. The initial gap opening l_{gap} between the upper and the lower contoured dies controls triggering and propagation of the plastic instability waves namely, the number, width and relative position of the compression beads along the axis of the main tube.

Figure 2b shows two different setups utilized for fixing tubes to sheets that make use of axisymmetric (leftmost setup) or asymmetric (rightmost setup), in-plane, plastic instability waves leading the compression beading and axisymmetric or asymmetric tube end flaring. The plastic instability waves are produced by means of appropriate flat or contoured dies whereas flaring is accomplished by compressing the upper tube end with and appropriate radiused punch in order to expand material outwards and form a single-lap inclined flange.

The main operating parameters influencing the new proposed technology are: (i) the slenderness ratio l_{gap}/r_0 between the initial gap opening and the reference radius of the tube, (ii) the ratio t_0/r_0 between the wall thickness and the reference radius of the tube, (iii) the inclination angle α of the contoured dies, (iv) the mandrel (if present) and (v) the tribological conditions. References [1] and [2] provide detailed information on the proposed technology for connecting tubes with details on the process feasibility window as a function of the main parameters. References [3] and [4] provide similar information for the case of fixing tubes to sheets.

3 Results and Applications

Figure 3a shows applications of the proposed technology for connecting tubes (or half-tubes) in situations where the axis of the branch tube is perpendicular or inclined to the axis of the main body tube. Figure 3b shows applications for fixing tubes to sheets in situations where the axis of the sheet is perpendicular or inclined to the axis of the main body tube.

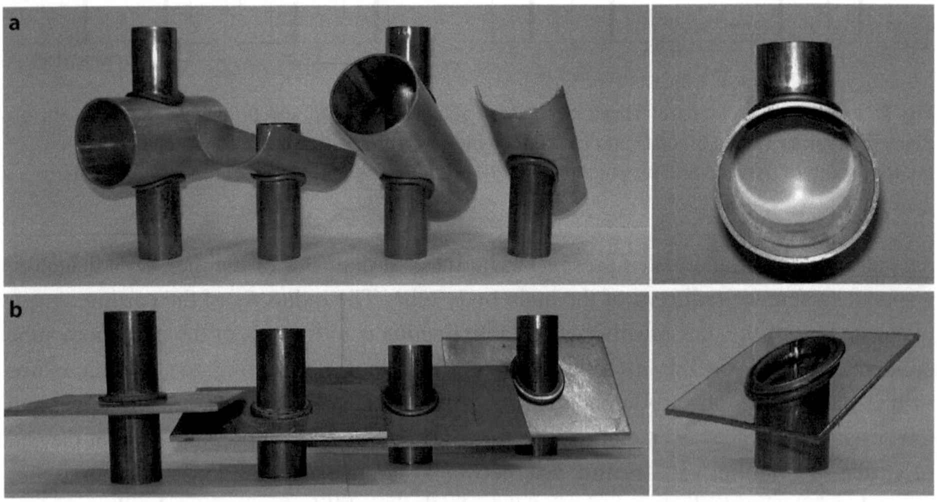

Fig. 3 Application of the new proposed technology for **a** connecting full-size and half-sectioned tubes and **b** fixing tubes to sheets (metallic and polymeric) at different inclination angles

The solution can, for example, be successfully employed in the seat bottom frame of automotive as an alternative to welding in order to reduce costs and eliminate heat distortion problems. All the connections shown in Fig. 3 were performed between welded carbon S460MC steel tubes, seamless aluminium AA6062 tubes, aluminium AA5754 sheets and polycarbonate sheets.

4 Potential

The proposed technology offers significant advantages as compared with conventional solutions based on mechanical fixing with fasteners, welding or structural adhesive bonding:

1. Flexible technology capable of handling small, medium or large batch sizes with different geometries and high levels of repeatability in production line.
2. Environmentally friendly technology that allows savings in raw material and eliminates filler materials and shielding gases.
3. Energy saving technology that eliminates heat-cooling cycles as well as heat affected zones in the regions of the tubes and sheet panels that are joined together.
4. Value added technology that is capable of connecting tubes and fixing tubes to sheet panels made of dissimilar materials.
5. Cost-efficient technology that requires low amount of capital investment because it can be designed to operate with existing machine-tools.

5 Conclusion

Connection of tubes and fixing of tubes to sheets at room temperature by triggering and controlling the propagation of plastic instability waves in thin-walled tubes represents a technological challenge for conventional joining technologies. The proposed technology has potential to increase flexibility of the supply chains in order to meet costumer-defined specifications regarding dimensions, tolerances and materials while addressing growing environmental concerns related to fabrication processes and end-of-life management of products. Because the technology can be successfully employed in fixture conditions that are difficult and costly to achieve by means of conventional processes it can also foster innovative ideas in product development.

6 Acknowledgement

The authors would also like to acknowledge the support provided by MCG – Mind for Metal, Carregado, Portugal.

References

1. Alves L. M., Martins P. A. F., 2012, Tube branching by means of asymmetric compression beading, Journal of Materials Processing Technology 212, 1200–1208.
2. Gonçalves A., Alves L.M., Martins P.A. F., 2014, Tube joining by asymmetric plastic instability, Journal of Materials Processing Technology 214, 132–140.
3. Alves L. M., Martins P. A. F., 2012, Mechanical joining of tubes to sheets along inclined planes, Steel Research International 83, 1135–1140.
4. Gonçalves A., Alves L. M., Martins P. A. F., 2014, Inclined tube-sheet plastically deformed joints, Steel Research International 85, 67–75.

Fabrication of Seamless Metallic Liners for COPV's

Luis M. Alves and Paulo A.F. Martins

1 Motivation

Composite overwrapped pressure vessels (COPV's) are made of a thin metallic liner wrapped with a high strength low density composite. The metallic liner provides shape, toughness, tightness and interface with the gas feeding systems while the overwrapped composite ensures mechanical strength to withstand high pressures and protects the vessel against scratches, indents and other forms of impact damage.

The main advantage of COPV's over similar sized monolithic metallic pressure vessels is a much better strength-to-density ratio due to significant mass reductions. However, there are a number of other ratios related to manufacturing, inspection and certification costs that may offset the strength-to-density advantage of COPV's. The motivation is, therefore, to review the manufacturing processes commonly utilized in the fabrication of COPV's in order to identify which are the parts and fabrication methods that still have much to improve in terms of cost savings. The liners are among the COPV parts that meet this goal.

Nowadays, the state-of-the-art in fabrication of metallic liners for aerospace applications is characterized by two competing fabrication methods that are schematically illustrated in Fig. 1. In the first method (Fig. 1a), the cylindrical central section is rolled and longitudinal welded while the ported and blind end domes are formed by hydroforming or spinning. Both end domes are then circumferential welded to the cylindrical central section to complete the assembly of the liner. The second method (Fig. 1b) consists in machining solid metallic blocks weighting tens or hundreds of kilograms into two halves of the thin-walled liner weighting just a few kilograms. Each half, containing an end dome and part of the central section, are then circumferential welded to complete the assembly.

Luis M. Alves · Paulo A.F. Martins ✉
Universidade de Lisboa, Lisboa, Portugal
e-mail: pmartins@tecnico.ulisboa.pt

A. E. Tekkaya et al. (eds.), *60 Excellent Inventions in Metal Forming*,
DOI 10.1007/978-3-662-46312-3_58

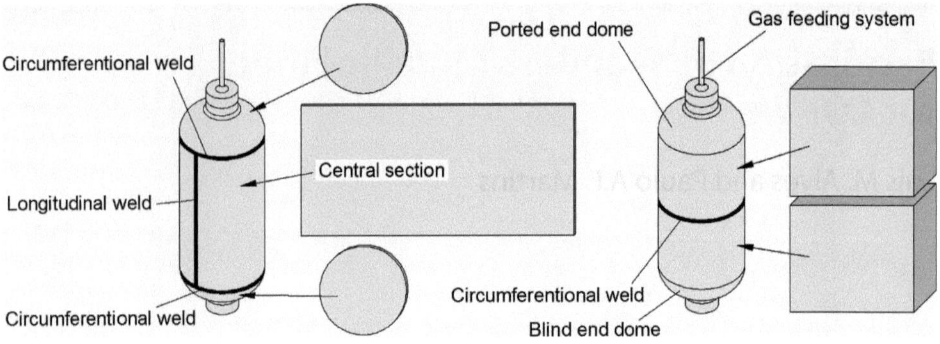

Fig. 1 Schematic representations of the fabrication methods that are currently utilize to produce COPV liners for aerospace applications

The following section presents a new manufacturing process to produce metallic liners for COPV's that are commonly utilized in spacecraft. The process makes use of an innovative tube forming operation that is capable of shaping commercial tubes into seamless metallic liners in a single press stroke.

2 Process Idea

The new manufacturing process to produce metallic liners is carried out by axially pressing the open ends of a tubular preform against two semi-ellipsoidal shaped dies, while providing internal support by means of a recyclable mandrel, until achieving the desired geometry of the liner. Figure 2 presents a schematic of the open, closed and extraction positions of the tooling system, which consists of structural, process and dedicated active parts.

The structural parts comprise a plurality of individual components such as the guiding columns, the die shoe, the ram holder, the support plate and the ram, which are independent of the material and geometry of the liners to be fabricated. The container, the backing ring and the clamping ring are the process parts whereas the inner container, the upper and lower, semi-ellipsoidal, shaped dies and the recyclable mandrels are the dedicated active parts that are dependent on the geometry of the liner to be fabricated (see also Fig. 3a).

The clamping ring fixes the container to the backing ring and die shoe. This part is removed after forming to enable the container to move up with the ram holder in order to extract the metallic liner. The inner container is shrink-fitted into the container (by preheating the container) and its internal diameter is dependent on the geometry of the liners to be fabricated. The shrink-fit interference between the containers is necessary in order to reduce the level of the tensile circumferential stresses generated by the tube forming process.

The upper and lower, semi-ellipsoidal, shaped dies are dedicated to a specific outside radius of the tubular preforms and its profile defines the geometry of the ported and blind

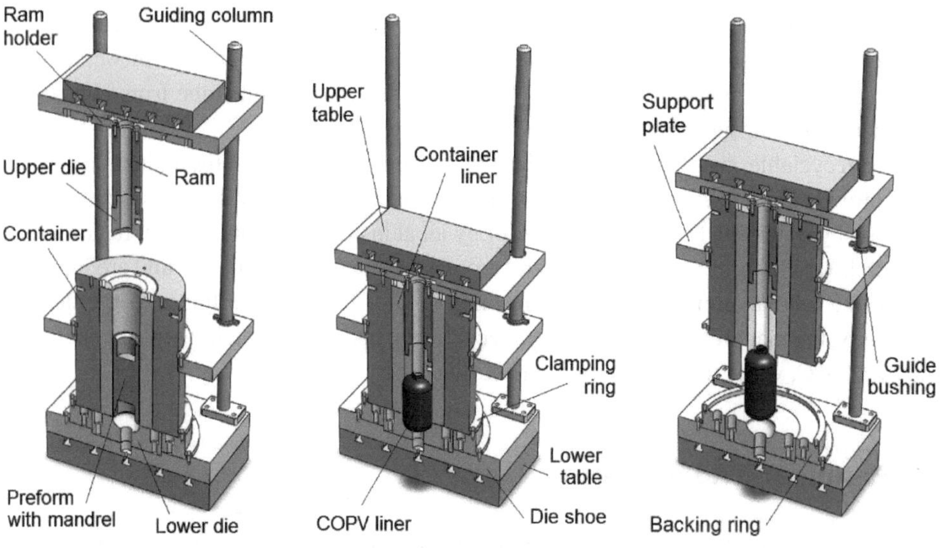

Fig. 2 Schematic representation of the tool system (open, closed and extraction positions) utilized in the new manufacturing process to produce COPV liners

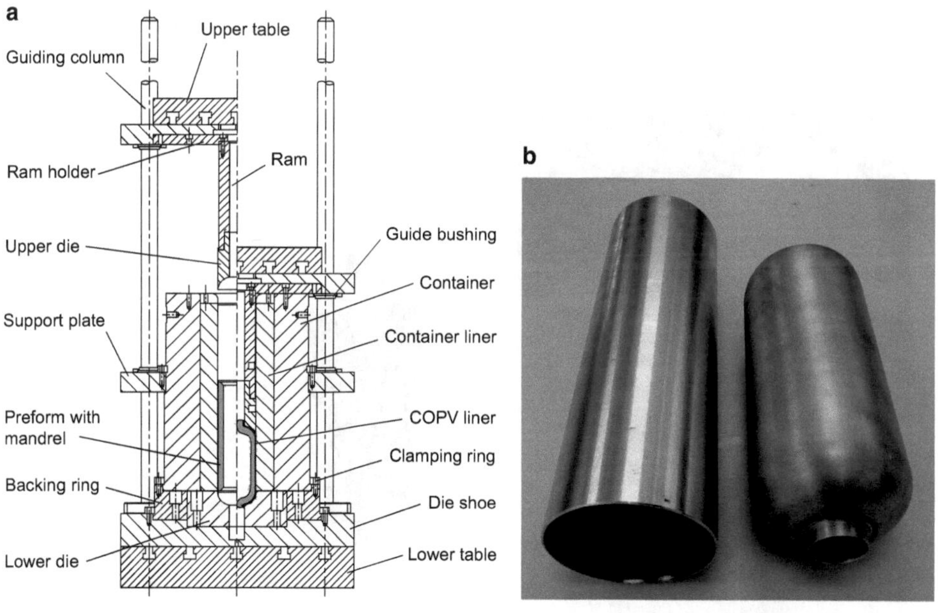

Fig. 3 **a** Two-dimensional schematic representation of the tool system at the beginning and end of the press stroke, **b** Tubular preform and cold formed liner after removing the recyclable mandrel

end domes. The upper die is characterized by its unconventional very sharp edge and is protected against collapse by circumferential tensile stresses through the inner container, which acts as a shrink fit tool part, and by plastic flow that inhibits tube from contacting the sharp edge during forming.

The recyclable mandrels act as dedicated active components of the industrial tool as they provide internal support to the preforms during plastic deformation in order to avoid collapse by wrinkling (at the end domes) or by local instability (at the central region). The mandrels (made from Sn60Pb40) are cast from a low melting point alloy and are installed inside the tubular preforms, prior to forming. After forming the liners to the desired geometry, the mandrels are removed and recycled by heating and melting at a temperature below 200 °C, while leaving the liner intact (Fig. 3b).

3 Results and Applications

Figure 4 shows photographs of metallic liners made from commercial aluminium AA6063-T0 tubes with different diameters (70 mm to 200 mm) and storage capacities to demonstrate the overall feasibility and flexibility of the new manufacturing process. The liners showed in the figure were machined to their final configurations while respecting the specified tolerances, cleaned by chemical passivation and subsequently heat-treated by artificially ageing, after being manufactured by cold forming and mandrels removed. The blind and gas outlet ports are assembled to the open ends of the domes by subsequent

Fig. 4 Photographs showing various liners with different diameters and storage capacity

crimping and welding. Further details on the manufacturing process are available in references [1–3] and the overall concept is protected by an international patent [4].

4 Potential

The new manufacturing process to produce metallic liners for COPV's from commercial tubular preforms has the following advantages against the two competing fabrication methods that are schematically illustrated in Fig. 1:

1. Value added and flexible manufacturing process capable of fabricating seamless liners with different storage capabilities from commercial tubular preforms.
2. Environmentally friendly technology that allows savings in raw material and eliminates welding and expensive inspection requirements prior to the composite overwrapping operation.
3. Energy saving technology that eliminates heat-cooling cycles as well as heat affected zones and residuals stresses in the regions of the liners that are joined together.
4. Precision technology that prevents geometrical misalignment, shrinkage, distortion or cracking originated by longitudinal and circumferential welding of competing fabrication methods.
5. Cost-efficient, non-labour intensive, technology that requires low amount of capital investment and can be design to operate with existing machine-tools.

5 Conclusion

The utilization of COPV's in the aerospace industry goes back to 1970's when the advances in composite materials allowed replacing heavier, thick walled, pressure vessels made of high-strength metals by thin walled metallic liners of aluminium, stainless steel, titanium or inconel wrapped with epoxy or polyamide polymer resins reinforced with high strength carbon, kevlar or glass fibres.

The new manufacturing process allows fabricating COPV liners for aerospace applications with lower costs than existing fabrication methods based on combination of sheet forming and welding or machining and welding. Savings derive from tube forming being a press-working operation that is faster, easier to automate, highly efficient in material consumption and energy, and less demanding in quality and security inspections prior to overwrapping than competing fabrication methods.

6 Acknowledgement

The authors would also like to acknowledge the support provided by OMNIDEA, Aerospace Technology and Energy Systems, Portugal.

References

1. Alves L.M., Silva M.B., Martins P.A. F., 2011, Fabrication of small size seamless reservoirs by tube forming, International Journal of Pressure Vessels and Piping 88, 239–247.
2. Alves L.M., Santana P., Moreira H., Martins P.A. F., 2013, Fabrication of metallic liners for composite overwrapped pressure vessels by tube forming, International Journal of Pressure Vessels and Piping 111–112, 36–43.
3. Alves L.M., Santana P., Fernandes N., Martins P.A. F., 2013, Fabrication of metallic liners for composite overwrapped pressure vessels, International Journal of Advanced Manufacturing Technology 67, 2671–2680.
4. Alves L.M., Martins P.A. F., Pardal T.C., Almeida P.J., Valverde N.M., 2009, Plastic deformation technological process for production of thin-wall revolution shells from tubular billets, European Patent Application PCT/PT2009/000007.

Torque Superposed Spatial bending

Matthias Hermes, Daniel Staupendahl, and Matthias Kleiner

1 Motivation

The demand of three-dimensionally bent tubes and profiles in different industrial sectors increases because of lightweight design, safety issues, space and cost saving. Especially lightweight design becomes more and more important because of economic and ecological reasons. This can be seen in the automobile and aerospace industry as well as in civil engineering, where three-dimensionally bent structures are oftentimes required [1]. The more and more arising challenge is to satisfy all the mentioned demands and to still keep the stiffness of the structure constant or, even better, to increase it. To cope with these challenges profiles with complex cross-sections and special tubes (e. G. Tailored Tubes) are needed, which increases the complexity of bending these semi-finished parts.

In the field of tube and profile bending there are well known processes that offer a high potential for three-dimensional bending of semi-finished products. The problem is that most of these procedures are specialized and optimized for the bending of profiles with circular and simple cross-sections and constant diameters and cross-sections over the longitudinal axis. These processes are applied by standard three-roll-bending machines, the Hexabend [2], the Nissin [3], and the TKS-MEWAG machines and process variants [4]. For 3D bending of more complex profiles, stretch bending is the most suitable procedure at present. This process consists of bending the parts over a curved die with superposed axial tension [5]. Stretch bending is based on the principle of tool-bound shaping and is mainly used in the automotive industry for mass production [6]. The advantages of high shape and measurement accuracy as well as high reproducibility of stretch bending face,

Matthias Hermes ✉
South Westphalia University of Applied Sciences, Meschede, Germany
e-mail: hermes.matthias@fh-swf.de

Daniel Staupendahl · Matthias Kleiner
TU Dortmund, Dortmund, Germany

A. E. Tekkaya et al. (eds.), *60 Excellent Inventions in Metal Forming*,
DOI 10.1007/978-3-662-46312-3_59

however, the disadvantages of low flexibility, expensive tools and machines, and increasing manufacturing costs when bending long profiles and large cross-sections.

2 The TSS Process

The new TSS bending process [7] allows the bending of profiles with arbitrary cross-sections to arbitrary 3D bending contours. The tool set-up of the process and the realized research machine (Fig. 1) consists of three pairs of rolls, which guide and transport the profile through the bending process (axis c), and a roll-based guiding system (bending head) that defines the bending curve in a horizontal plane. This bending axis x is realized by one horizontally located axis. The axis τ ensures the tangential orientation to the c-axis of the bending head. With these three axes it is possible to bend 2D contours, even S-shapes. 3D bending contours are realized by a superposed torque that has two functions. Firstly the torque is necessary to influence the bending plane of the profile's cross section. The resulting necessary twisting axis for the definition of the 3D curve can be exemplary realized by a torsion bearing (α_1), mounted around the three roll pairs, and a compensation axis (α_2), positioned in the bending head. By this mechanism, the bending plane can be changed and a 3D shape is produced by rotating the profile cross section over the longitudinal axis (α_1, α_2). Asymmetrical profiles tend to twist over their longitudinal axis in conventional bending processes due to the difference between the shear center and

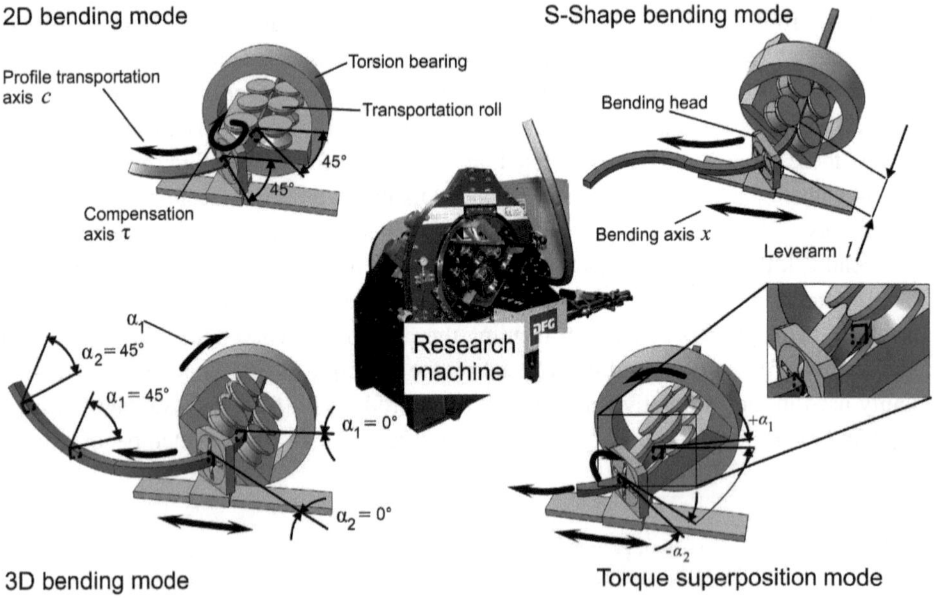

Fig. 1 Process principle and different process modes

the center of gravity of the cross section. With the TSS bending process, it is possible to compensate this twisting by using the second function of the torque when superposing it with the bending moment. This is carried out by creating an offset between the rotational adjustments of the torsion bearing (α_1) and the compensation axis (α_2). Moreover by this superposition of torque twisting of the profiles can be prevented as well as performed with intend.

3 Process Control

For 2D and 3D-Bending of symmetrical hollow section profiles, e. g. square tubes, a process planning method to generate the NC-dada for the machine was developed. The procedure to generate this data is shown in Fig. 2.

Input data are the geometry of the cross section and the bending geometry in a 3D CAD-format and the material data given by a flow curve and important values like the Young's modulus. The first step is to analyze the bending geometry and to subdivide it into different bending radii R_x and for a 3D bend part into different bending planes α_x along the cross section's main axis segment c_x. The next step is to make a semi analytical calculation for each radius R_x to calculate springback and the compensated values R_k, α_k and c_k that have to be bend to get the target contour after unloading. These calculations take into account the machine stiffness and the bending line and bending moment of the profile in the roll tools of the machine. The calculation method is well explained in [8]. Using this data, the bending geometry is redesigned generating a compensated 3D part. The compensated part is then used to generate the NC-data by a kinematic simulation. At the end, it is possible to start a bending test with a compensated NC-data. In Fig. 3 the results of an experiments with the developed prototype bending machine can be seen.

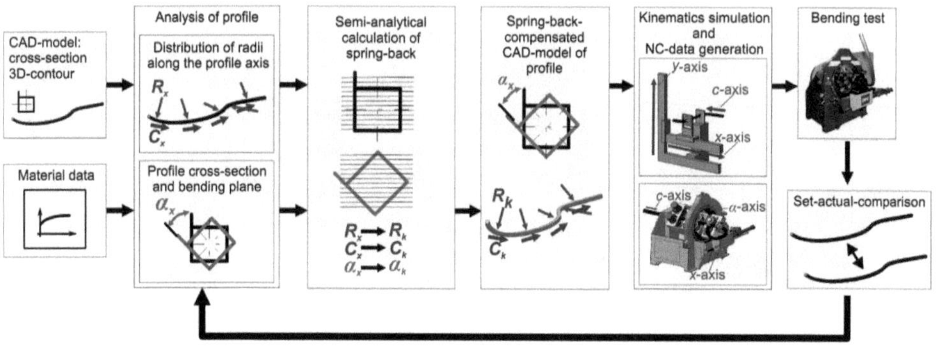

Fig. 2 Process planning tool

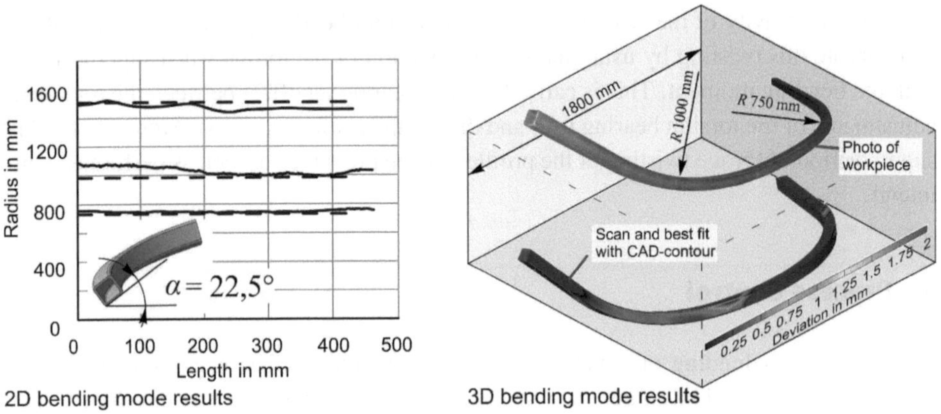

Fig. 3 Result of 2D and 3D bending experiments

4 Industrial Application and Further Development

The recently developed industrial machine can be seen in Fig. 4. This machine has an additional axis that makes it possible to adjust the leverarm distance between the bending head and the transportation rolls. This adjustment makes it possible to bend smaller radii as well as bigger radii without problems of collision of tool elements. A further development can be mounted on the prototype machine as well. The use of an induction bending device is useful for high strength steel tubes for example made of air hardening steel. The process extension with an inductive heating device makes it possible to reduce bending forces and to realize graded structures over the longitudinal axis by an in situ heat treatment of the material. This can be used in vehicle structures for defined crash properties.

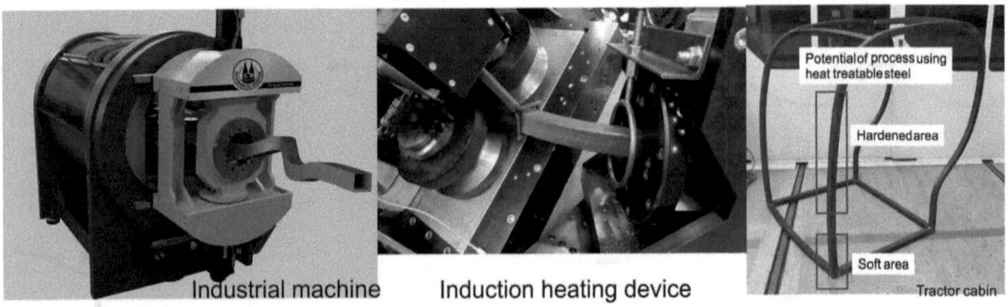

Fig. 4 Process principle and different process modes [9]

5 Conclusion and Outlook

3D bending of profiles with arbitrary cross-sections can be achieved by the TSS bending process. A special machine was developed, which has a high bending flexibility thanks to the kinematic forming principle and a cost efficient tool concept for bending very long lightweight construction profiles to freely definable 3D bending contours. A Process planning tool based on semi analytical calculations and kinematic simulation was developed. With this system a high accuracy can be achieved even at the first bending trial. The machine concept and the process are already introduced in industrial application.

6 Acknowledgement

This process has been investigated in several projects. The work has been kindly supported by the German Research Foundation (DFG) under the grant number TE 508/15-2, the Federal Ministry of Economics and Technology on the basis of a decision by the German Bundestag, and by the European Community's Research Fund for Coal and Steel (RFCS) under the grant agreement number [RFSR-CT-2009-00017].

References

1. Chatti, S., Production of Profiles for Lightweight Structures, 2005, habilitation thesis, Books on Demand GmbH, Norderstedt, Germany.
2. Neugebauer, R., Drossel, W., -G., Lorenz, U., Luetz, N., 2002, Hexabend – A new Concept for 3D-free-form Bending of Tubes and Profiles to preform Hydroforming Parts and Endform Space-frame-components, In: Advanced Technology of Plasticity, Vol. 2, pp. 1465–1470.
3. Murata M., Kuboti T., Takahashi K., 2007, Characteristics of Tube Bending by MOS Bending Machine, In: Proc. of the 2nd Int. Conf. on New Forming Technology, Bremen, Germany, pp. 135–144.
4. Flehmig, T., Kibben, M., Kühni, U., Ziswiler, J., 2006, Device for the free forming and bending of longitudinal profiles, particularly pipes, and a combined device for free forming and bending as well as draw bending longitudinal profiles, particularly pipes, Int. patent with application no. PCT/EP2006/00252, published on 28.09.2006.
5. Corona, E., 2004, A simple analysis for bend-stretch forming of aluminium extrusions, Int. Journal of Mechanical Sciences, 46, pp. 433–448.
6. Geiger, M., Sprenger, A, 1998, Controlled Bending of Aluminum Extrusions, Annals of the CIRP 47/1, pp. 197–202.
7. Hermes, M., Kleiner, M.: Method and device for profile bending. Internationale Patentanmeldung, WO002008113562A1, Veröffentlichungsdatum 25.09.2008.
8. Chatti, S., Hermes, M., Tekkaya, A.E., Kleiner, M., 2010. The new TSS bending process: 3D bending of profiles with arbitrary cross-sections, CIRP Annals – Manufacturing Technology, Volume 59, Issue 1, p. 315–318.
9. N.N.: 3D BENDING – TORQUEFORMER, product information brochure, Schwarze Robitec GmbH, Köln

5 Conclusion and Outlook

[illegible]

Further Development on Tube Hydroforming

Ken-ichi Manabe and Sadakatsu Fuchizawa

1 Motivation

Before fittings for piping or bicycle frames were made by hydraulic bulge forming, they were made by casting and machining or sheet metal stamping and welding. Products fabricated by casting and machining were heavy and their surface was rough. Meanwhile, sheet metal stamping and welding required many manufacturing stages and the products had some problems concerning strength and reliability owing to the use of welding. Research and development activities to solve these problems were started by a group led by Government Industrial Research Institute, Nagoya, Japan (now, National Institute of Advanced Industrial Science and Technology, AIST Chubu) in about 1957. Japan Bicycle Technical Center (JBTC) and several companies participated in the group.

T-forming technology using tube hydroforming was patented in the U.S. [1] in 1940. The above Japanese group undertook basic research on the hydraulic bulge forming of tubes [2–6] and aimed to realize the idea presented in the U.S. patent for practical use and to establish the production technology for manufacturing high-quality products. As a result, they have acquired many patents on forming machines and related technologies [7]. They also took out a U. S. patent (No. 3350905, 1967), British patent (No. 1083354, 1968), and West German patent (No. 1452547, 1971). Patent-licensed companies have improved the technology and succeeded in supplying mass-produced high-quality products, such as butt-welding pipe fittings of various diameters and fittings for bicycle frames, to the market. This T-forming technology is still being used to consolidate small hollow parts.

Ken-ichi Manabe ✉
Tokyo Metropolitan University, Hachioji-shi, Tokyo, Japan
e-mail: manabe@tmu.ac.jp

Sadakatsu Fuchizawa
Utsunomiya University, Utsunomiya-shi, Japan

A. E. Tekkaya et al. (eds.), *60 Excellent Inventions in Metal Forming*,
DOI 10.1007/978-3-662-46312-3_60

As computer technology was developing, it improved tube hydroforming machines and enabled the consolidation of large hollow parts with complicated cross sections. Parts made by tube hydroforming have a light weight and high rigidity. Meanwhile, global warming due to greenhouse gases such as CO_2 has recently become a serious problem. To reduce the amount of CO_2 in the exhaust gas from a car, the reduction of the car weight is strongly required. Tube hydroforming can meet this requirement. Further challenges of tube hydroforming have been done for developing advanced forming technology of low-formability-material tubes such as high strength steel and magnesium alloy, and for novel machine development, machine downsizing, and production efficiency improvement.

2 Tube Hydroforming Machine

In the early stage of development of the tube hydroforming machine, from a number of experimental data, successful process window and process design guideline were established and then appropriate mechanically synchronized axial feeding system was developed.

Figure 1 shows an example of the developed machine published in Industrie-Anzeiger. In this machine, internal pressure and axial feeding were controlled by hydraulic system, which had higher flexibility compared with the original mechanical control system.

After the tube is set in a split die, the die is closed by a clamping-pressure cylinder, and the sealing of both ends of the tube by a rapid-closure cylinder follows. Then, high-pressure water is supplied into the tube through a pressure intensifier. The tube deformed by the internal high pressure and axial feeding from both ends of the tube. A pressure equalizer is used equalize the amount of the axial feeding from both end of the tube. To fabricate sound products by tube hydroforming, control of the internal pressure and axial feeding is the most important factor. Too high internal pressure causes bursting and too

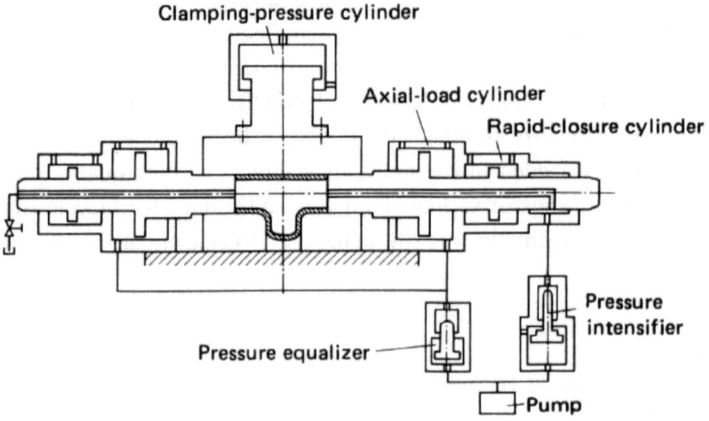

Fig. 1 Liquid bulge forming machine [8]

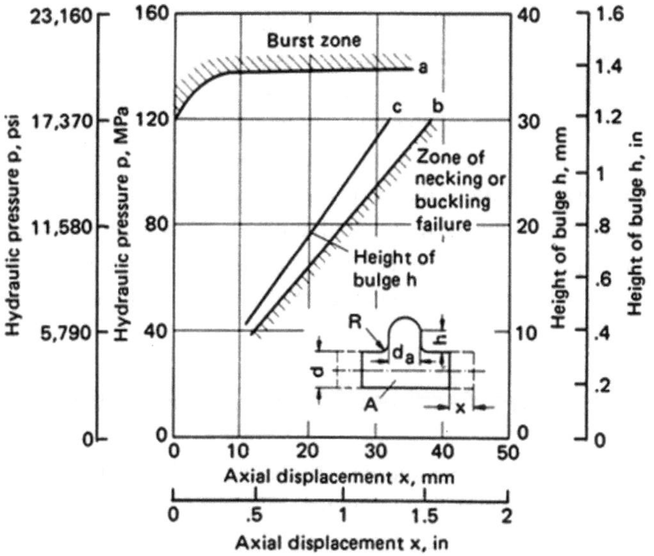

Fig. 2 Forming diagram of asymmetrical bulging [8]

Fig. 3 Comparison of forming procedure of hanger lug [9]. **a** Sheet metal forming (stamping + welding), **b** Hydraulic bulge forming of tube

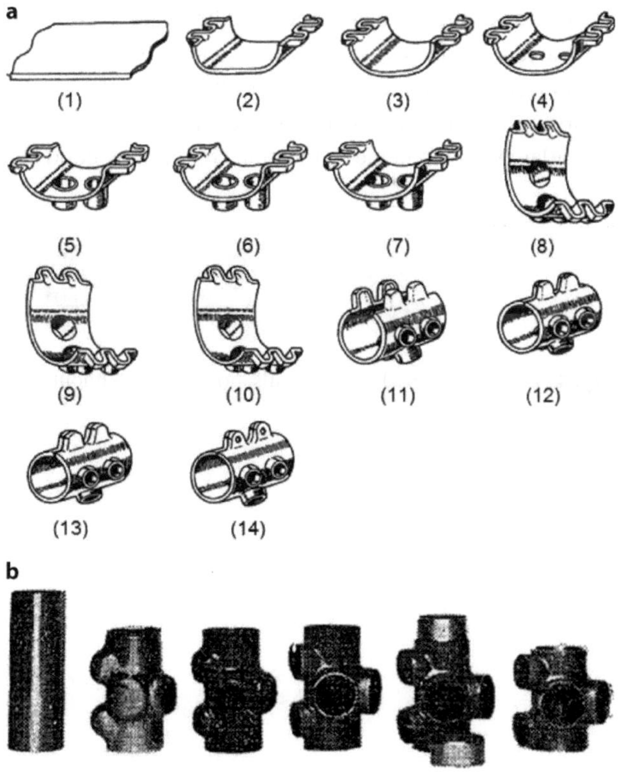

much axial feeding results in buckling or wrinkling. An example of the forming diagram of asymmetrical bulging is shown in Fig. 2. The loading path c should not exceed the bursting limit a or the buckling or wrinkling limit b.

Figure 3 shows comparison of the forming procedure of a hanger lug (fitting for a bicycle frame). Sheet metal forming involves many procedures of stamping and welding, whereas hydraulic bulge forming makes it possible to fabricate a lug from a thin tube without stamping and welding. Hydraulic bulge forming has greatly reduced the number of forming steps, improved productivity, and reduced the cost. High-quality lightweight products made by tube hydroforming were revolutionary at the time.

3 Tube Hydroforming Machine with Pressure Pulsation

In the tube hydroforming process, the workpiece in the die cavity is strongly pressed against the die wall by the internal high pressure. When the lubrication between the workpiece and die is poor, the workpiece essentially adheres to the die surface and becomes difficult to move. In this case, the shape along the die cannot be completed. To improve this situation, a new method in which the internal pressure is made to pulsate at several Hz using a DVD servo pump was proposed [10]. The pulsation of the internal pressure causes fluctuation of the force pressing the tube to the die wall and makes the tube material easily flow along the die surface. Figure 4 shows the effect of internal pressure pulsation. The workpiece bursts in the conventional hydroforming machine but bursting is avoided in the hydroforming machine with pressure pulsation. The pulsation of the internal pressure has a similar effect when the lubrication condition is good [11, 12].

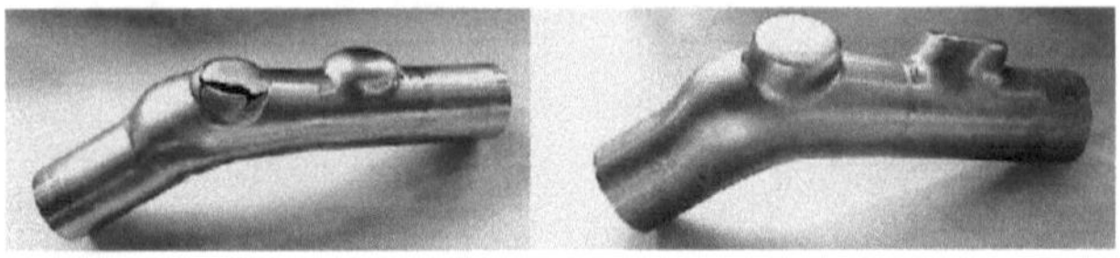

Fig. 4 Effect of internal pressure pulsation [13]. **a** Without pulsation, **b** With pulsation

4 Hot Hydroforming of Aluminum Alloy Tube

Aluminum tubes are often used to reduce the weight of automobiles, but there is a problem that the formability of aluminum at room temperature is low compared with that of steel. In general, its formability improves at elevated temperature, but there are many problems that must be solved, such as the long heating time, temperature control, oxidation, and lubrication.

Recently, these problems have been solved by applying various techniques and hot hydroforming technology for aluminum alloy tubes has been developed [14, 15, 16]. Subframes of automobiles made using this technology are shown in Fig. 5. Right- and left-side members of the front subframe and front and rear cross members of the rear subframe are made by the hot hydroforming of aluminum alloy tubes without welding. These members have very complicated cross sections that cannot be achieved at room temperature. Owing to such complicated cross sections, it is possible to avoid interference among the many parts in a small engine room. The tube material is a newly developed Al-Mg alloy. To shorten the heating time, direct electric conduction heating is used. After preheating, a tube is set in a preheated die and deformed by internal pressure and axial feeding. High-precision products with high uniform wall thickness are obtained by considering the design of the shape in preforming, controlling axial feeding, and dividing the forming process into stages.

The maximum expansion ratio of the peripheral length is over 70 % and the amount of springback is small. The pressure medium is air. A member can be deformed at a low internal pressure (about 3 MPa) of one percent of the forming pressure at room temperature because of the low flow stress at elevated temperatures. The clamping pressure becomes

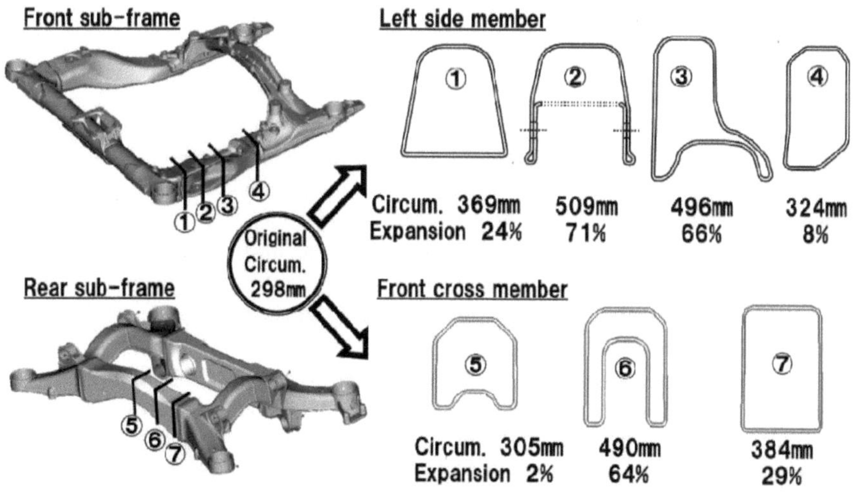

Fig. 5 Hot hydroforming of aluminum alloy tube [15]

about 1/10 of that at room temperature, and the hydroforming machine can thus be downsized. The problem of grain growth and the generation of cavities in the tube material at elevated temperatures have been overcome by the development of materials.

By the above-described development of hot hydroforming technology, the rigidity of the subframe is improved and the numbers of components and welding processes are reduced; the reduction of weight and the lowering of cost are also achieved. This aluminum subframe is 38 % lighter than the steel subframe. Compared with the aluminum subframe made by the conventional process, the number of parts is reduced by 67 %, the welding length by 55 %, the weight by 10 %, and the cost by 20 %. This subframe was adopted in the Honda Legend launched in 2004.

5 Conclusion

Tube hydroforming is a comprehensive forming technology that utilizes a number of related technologies such as machine and die design, machine control, process design including the development of tubes for hydroforming, and computer simulation. It enables the reduction of automobile weight by decreasing the number of components and removing the need for welding. The reduction in the number of forming steps helps lower the manufacturing cost. Recently, a tube expansion ratio of 200 % has been achieved at room temperature by using a movable die without annealing [17]. There have been attempts to apply this technology to the axle housings of automobiles. Tube hydroforming has been used as a weight-reducing technology mainly in the automotive industry. Its application is expected in various fields because of its great potential.

References

1. U. S. Pat. No. 2203868, 1940.
2. Ogura, T., Ueda, T., 1958, Production of Stepped Hollow Shafts by Liquid Bulging (I) – Relationship between the Hydraulic Pressure and the Bulge (in Japanese), Reports of the Government Industrial Research Institute, Nagoya, (Nagoya-kogyo-gijutu-shikenjo-hokoku), The Government Industrial Research Institute, Nagoya, Japan, 7-2, 89–94.
3. Ogura, T., Ueda, T., 1958, Production of Stepped Hollow Shafts by Liquid Bulging (II) – Relationship between Hydraulic Pressure and Length of Stepped Part (in Japanese), Reports of the Government Industrial Research Institute, Nagoya, (Nagoya-kogyo-gijutu-shikenjo-hokoku), The Government Industrial Research Institute, Nagoya, Japan, 7-10, 719–723.
4. Ogura, T., Ueda, T., Ishikawa, M., 1959, Production of Stepped Hollow Shafts by Liquid Bulging (III) – Lubricants and Frictional Force in Liquid Bulging (in Japanese), Reports of the Government Industrial Research Institute, Nagoya, (Nagoya-kogyo-gijutu-shikenjo-hokoku), The Government Industrial Research Institute, Nagoya, Japan, 8-9, 603–608.
5. Ogura, T., Ueda, T., Ishikawa, M., 1962, Production of Stepped Hollow Shafts by Liquid Bulging (IV) – Anti-Friction Performance of Lubricants (in Japanese), Reports of the Gov-

ernment Industrial Research Institute, Nagoya, (Nagoya-kogyo-gijutu-shikenjo-hokoku), The Government Industrial Research Institute, Nagoya, Japan, 11-3, 131–136.

6. Ogura, T., Ueda, T., Ishikawa, M., 1962, Production of Stepped Hollow Shafts by Liquid Bulging (V) – Adaptability of Electric-Welded Pipe for Liquid Bulge Forming (in Japanese), Reports of the Government Industrial Research Institute, Nagoya, (Nagoya-kogyo-gijutu-shikenjo-hokoku), The Government Industrial Research Institute, Nagoya, Japan, 11-9, 524–528.
7. Japanese Pat., No. 411728, 1963/No. 417730, 1963/No. 418055, 1963/No. 480529, 1966/No. 482147, 1966/No. 511493, 1967/No. 730010, 1973, etc.
8. Ogura, T., Ueda, T., Takagi, R., 1966, On the Application of a Hydraulic Bulging Process (in German), Industrie-Anzeiger, 88, 769–772, 1001–1004. [Lange, K., 1985, Handbook of Metal Forming, McGraw-Hill, 23.24–23.25.]
9. Okamoto, N., 1965, Role of Deformation Processing in Bicycle Industry in Japan (in Japanese), J. Japan Society for Technology of Plasticity, 6–50, 135–140.
10. Japanese Pat., No. 2999757, 2000.
11. Hama, T., Asakawa, M., Fukiharu, H., Makinouchi, A., 2002, Finite Element Analysis of Hammering Hydroforming of Automotive Part (in Japanese), Proc. 53rd Japanese Joint Conference for Technology of Plasticity, 223–224.
12. Hama, T., Asakawa, M., Fukiharu, H., Makinouchi, A., 2003, Investigation of Effect of Pulsation of Hydraulic Pressure by Static-Explicit FEM (in Japanese), Proc. 54th Japanese Joint Conference for Technology of Plasticity, 339–340.
13. OPTON, Co., Ltd., Catalog.
14. Fukuchi, F., Hayashi, N., Ogawa, T., Yokoyama, O., Hori, I., 2005, Development of aluminum automotive parts technologies for mass-production (in Japanese), J. Japan Institute of Light Metals, 55-3, 147–152.
15. Kiyama, K., Kitano, Y., Nakao, K., 2006, Forming technology of aluminum extruded material for automobile (in Japanese), J. Japan Institute of Light Metals, 56-1, 63–67.
16. Ishihara, Y., Yamamoto, D., 2008, Development of production engineering of hot tube bulge forming and its application to sub-frame (in Japanese), Report of the Iron and Steel Institute of Japan Meeting, CAMP-ISIJ, Vol. 21, 302–305.
17. Wada, M., Mizumura, M., Iguchi, K. Kaneda, H., 2014, Large-expansion hydroforming technology achieving three-times expanding, Procedia Engineering 81 (Proc. 11th ICTP), 2217–2222.

Part X
Further Developments

In-Situ Measurement of Loading Stresses by Means of X-ray Diffraction with Multi-State Sheet Specimen

Alper Güner and A. Erman Tekkaya

1 Motivation

For accurate finite element simulations, description of the material behavior plays an important role. The anisotropic plastic flow of sheet metals is modeled with anisotropic yield conditions which need different number of material parameters. In order to increase the flexibility of these yield conditions the needed number of material parameters should also be increased. The quality of the identified material parameters directly influences the quality of the numerical predictions. In order to obtain these parameters there exist different characterization tests like uniaxial tension test, plane strain tension test, Miyauchi test and biaxial tension test [1]. Inhomogeneity of the deformation due to the edge effects is the main problem in most of the test specimens. The forces needed for analytical stress calculation act on the whole inhomogeneous deformation zone and therefore the stresses calculated with these force values differ from the real stresses [2]. There are mainly three solutions to this problem in the literature: The analytically calculated stresses can be corrected by empirically obtained factors [3], the specimen geometry can be optimized [4] or inverse analysis can be utilized to exploit inhomogeneous deformation fields [5].

These strategies do not attack the main problem that the stresses cannot be measured directly. An alternative solution is to measure the stresses by X-ray diffraction (XRD) locally on the specimen surface [6]. In order to increase the efficiency and accuracy of the proposed method a new specimen geometry was developed at the Institute of Forming Technology and Lightweight Construction (IUL) and this strategy has been applied for yield locus determination.

Alper Güner · A. Erman Tekkaya ✉
TU Dortmund, Dortmund, Germany
e-mail: alper.guener@iul.tu-dortmund.de, Erman.Tekkaya@iul.tu-dortmund.de

A. E. Tekkaya et al. (eds.), *60 Excellent Inventions in Metal Forming*,
DOI 10.1007/978-3-662-46312-3_61

2 Main Principle and Experimental Set-Up

X-ray diffraction is a standardized method used to obtain residual stresses in metallic materials since 1950's [7]. In this method, the distances between the atoms of metal crystals are measured. Measured distances are compared to the undeformed lattice spacings and a strain value is calculated which is the elastic strain. With the assumed elastic constitutive relations residual stresses are calculated directly by the help of the measured elastic strains.

In order to measure the loading stresses, exactly the same method is utilized (Fig. 1). Elasto-plastic material behavior dictates that the elastic strains continue to grow also during plastic flow. For the case of an applied stress, σ_{app}, the total strain is composed of elastic and plastic parts. If the elastic strains are obtained by XRD, the applied stresses can be found with the known elastic modulus. This strategy can be combined with an optical strain measurement system and the whole stress-strain history can be captured without measuring globally acting forces and overall tool displacements.

In order to apply this methodology an X-ray diffractometer is mounted on a universal testing machine (Fig. 2). By the help of a cross-table located between the diffractometer and the carrying structure the whole diffractometer can be moved in horizontal and vertical directions. By this way multiple measurements can be performed at different spots on a

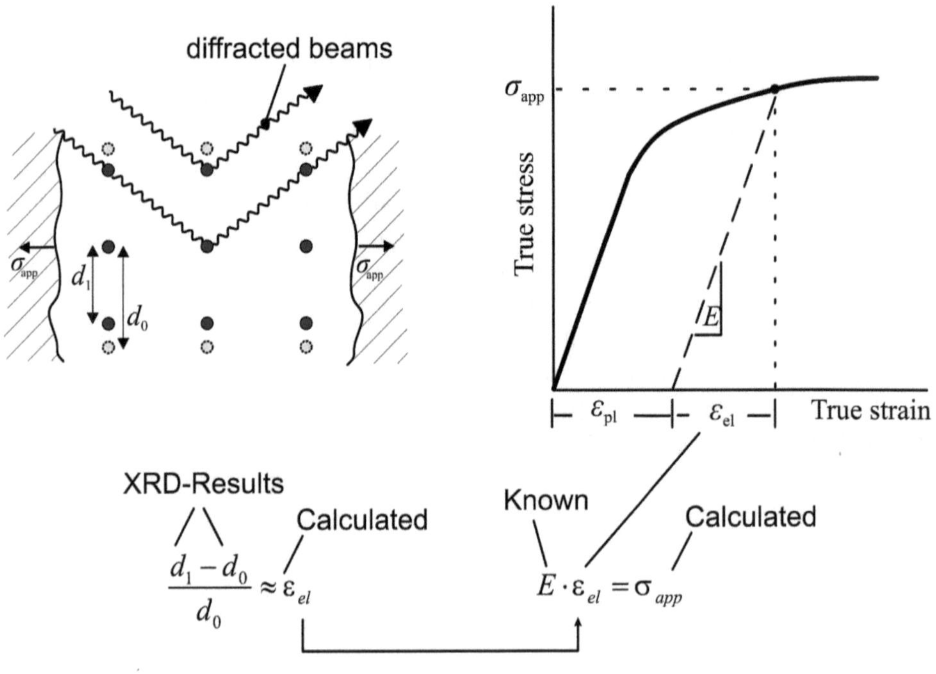

Fig. 1 Principle of measurement of loading stresses with X-ray diffraction

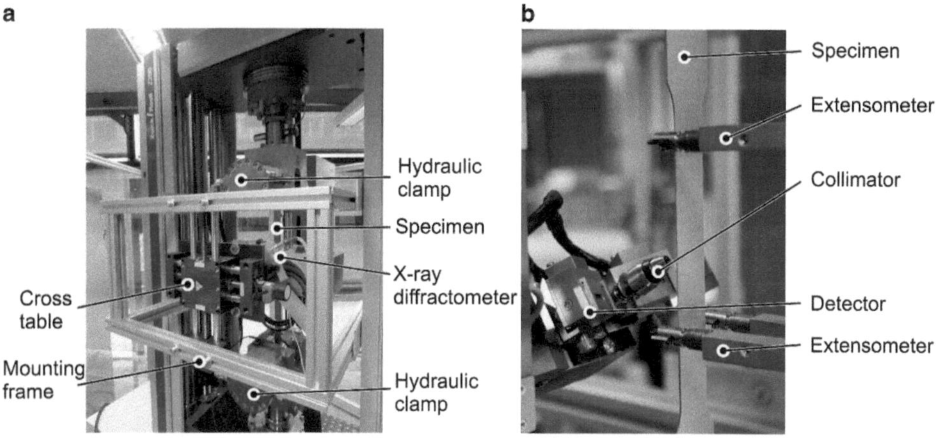

Fig. 2 **a** X-ray diffractometer on the universal testing machine, **b** uniaxial tension test specimen in front of the goniometer [8]

specimen. An optical measurement system is mounted on the other side of the specimen. With this system the strain distribution on the whole specimen surface is captured. The tilt angle Ψ can be varied between +45° and −45°. Diffracted beams are identified by two linear detectors which are located on both sides of the collimator. Hereby, Cr-tubes are used with a maximum power of 300 W and wavelength of 0.2289 nm.

In order to perform the XRD measurements the test should be paused for period of 4 to 6 minutes depending on the material. In this time the specimens are diffracted from different angles. When the measurement is completed the specimen is loaded to the next desired level of strain and the test is paused again. This sequence goes on until the maximum deformation level is reached. For that reason the tests take considerably more time as compared to conventional testing methods. For safety purposes transparent Plexiglas panels with a wall thickness of 8 mm are located around the universal testing machine which is enough because of the low energy level of the used XRD unit.

3 Multi-State Specimen and Results

Stopping the experiments due to XRD measurements cause a stress relaxation in the materials. This shows itself as sudden force drops due to viscoplastic behavior of metallic materials. Therefore the measured stresses are the relaxed stresses and this stress drop should be compensated. As a first step the stress relaxation should be characterized. For that purpose interrupted uniaxial tension tests were performed with 10 minutes of relaxation times. It is found that the stress relaxation is independent of the strain hardening level and orientation of the specimen. For the analyzed sheet materials AA5182 and DC06 a stress relaxation value of 10 and 35 MPa were found respectively. The relaxation curves were then used to calibrate a viscoplastic material model, namely the Norton-Hoff model. By this way, the stress relaxation under other deformation states like plane strain tension and shear could be obtained. The second step is to design a new specimen geometry which suits itself for the XRD measurements. The aims in the design of the geometry are as follows:

- The new specimen should generate uniaxial tension, plane strain tension and shear deformation states.
- Mentioned deformations should be homogeneous in a region of 5 mm diameter. By this way the XRD measurements can be performed within this homogeneous region.
- The specimen should be large enough to allow the goniometer to move freely in front of the specimen.
- The specimen must be cut out with existing methods like laser cutting.
- Conventional universal testing machines should be used to test the material.

To fulfill these requirements the specimen shown in Fig. 3a is designed. The new specimen geometry has mainly 3 kinds of gauge zones. The long arms found at the upper right and lower left corners realize uniaxial tension. The zones with 25 mm width next to the uniaxial zones generate plane strain tension in the core of these zones. Finally, in

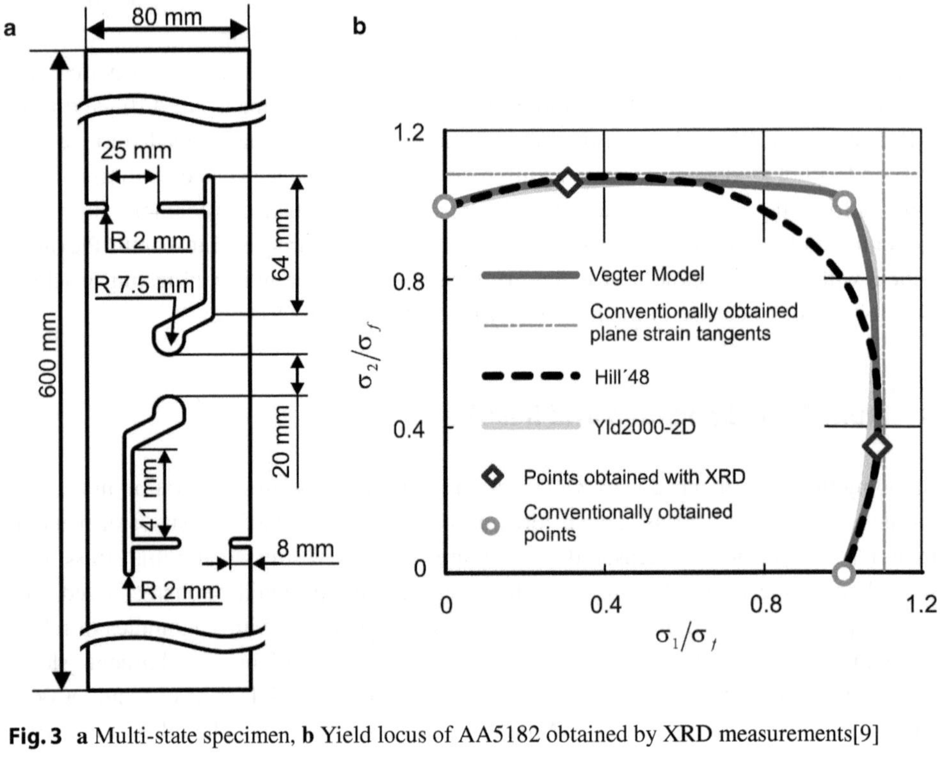

Fig. 3 **a** Multi-state specimen, **b** Yield locus of AA5182 obtained by XRD measurements[9]

the middle of the specimen shear deformation is achieved due to the point symmetrical design. Deformation states in those regions are verified with finite element simulations and optical strain measurements.

This specimen is used to locate the stress components on the yield loci of materials (Fig. 3b). As mentioned before the strains were measured with an optical strain measurement system. The tests were performed until fracture of the material and at each 1 mm of tool displacement XRD measurements were performed on the mentioned deformation zones. Obtained stress values were also compared with the conventionally obtained stress states. It was seen that the conventional values are slightly higher than the XRD measurements. It is worth to note that with this method two stress components on the surface can be measured. Therefore the yield locus can be identified directly. This is not the case for conventional plane strain tension tests, where the second principal stress cannot be measured since there is no force sensor to measure the horizontally acting forces.

4 Conclusion

Measurement of loading stresses with XRD for yield locus determination solves the problem of inhomogeneous deformation fields since the measurements can be performed in the middle of gauge zones where the preferred stress and strain states act ideally free from edge effects. This new method of local and in-situ stress measurement can lead to new opportunities in the field of sheet metal characterization and forming. Ongoing research focuses on generation of alternative specimen geometries with multiple stress and strain states, understanding of stress relaxation in sheet metals and application of XRD on textured materials. There is also a potential of the proposed strategy to be used in metal forming applications. By this way the stresses on the surface of workpieces can be measured during a forming process and a better understanding of the processes can be achieved. This information can also be used as an additional verification parameter for the numerical analyses.

5 Acknowledgement

The authors would like to thank the German Research Foundation (DFG) for their financial support of the work reported here, which is acquired within the scope of the research project PAK250.

References

1. Kuwabara T 2007 Advances in experiments on metal sheets and tubes in support of constitutive modeling and forming simulations Int. J. Plast. 23 385–419
2. An Y G, Vegter H and Elliott L 2004 A novel and simple method for the measurement of plane strain work hardening J. Mater. Process. Technol. 155–156 1616–22

3. Flores P, Tuninetti V, Gilles G, Gonry P, Duchêne L and Habraken A M 2010 Accurate stress computation in plane strain tensile tests for sheet metal using experimental data J. Mater. Process. Technol. 210 1772–9
4. Hanabusa Y, Takizawa H and Kuwabara T 2013 Numerical verification of a biaxial tensile test method using a cruciform specimen J. Mater. Process. Technol. 213 961–70
5. Güner A, Soyarslan C, Brosius A and Tekkaya A E 2012 Characterization of Anisotropy of Sheet Metals Employing Inhomogeneous Strain Fields for Yld2000-2D Yield Function Int. J. Solids Struct.
6. Iadicola M A, Foecke T and Banovic S W 2008 Experimental observations of evolving yield loci in biaxially strained AA5754-O Int. J. Plast. 24 2084–101
7. Noyan I and Cohen J B 1987 Residual Stress: Measurement by Diffraction and Interpretation, New York: Springer Verlag
8. Güner A, Zillmann B, Lampke T and Tekkaya A E 2014 In-situ measurement of loading stresses with X-ray diffraction for yield locus determination Int. J. Automot. Technol. 15 303–16
9. Güner A, 2015 In-situ stress analysis with X-ray diffraction for yield locus determination, PhD Dissertation, TU Dortmund, Shaker Verlag, ISBN 978-3-8440-3414-1

Smart Hot Stamping for Ultra-high Strength Steel Parts

Ken-ichiro Mori

1 Motivation

An effective approach for improving the fuel consumption of automobiles is to reduce the weight of automobiles [1]. Because high strength steel sheets are considerably cheaper than aluminium alloy sheets, the use of high strength steel sheets for automotive body-in-white parts remarkably increases. Although the high strength steel parts have superior mechanical properties, stamping operations become difficult due to the high strength. As the strength of the sheets increases, the forming load increases, and thus the springback becomes large, particularly for ultra-high strength steel sheets. In addition, small formability, short tool life and delayed fracture are problematic. Specially, cold stamping operations of ultra-high strength steel sheets having a tensile strength above 1.2 GPa have not been established yet [2].

By heating sheet metals, the stamping operation becomes easy due to the decrease in flow stress and the increase in formability. Hot stamping processes of quenchable steel sheets are attractive for forming of ultra-high strength steel parts [3]. The sheets are heated in the furnace, then the heated sheets are stamped, and finally the formed sheets are held at the bottom dead centre of a press for die quenching. The stamping load is considerably small, the springback is prevented, and the formability is improved. In addition, parts having about 1.5 GPa in tensile strength can be produced by die quenching.

The present hot stamping processes of ultrahigh strength steel parts have the following disadvantages (see Fig. 1):

Ken-ichiro Mori ✉
Toyohashi University of Technology, Toyohashi, Japan
e-mail: mori@plast.me.tut.ac.jp

A. E. Tekkaya et al. (eds.), *60 Excellent Inventions in Metal Forming*,
DOI 10.1007/978-3-662-46312-3_62

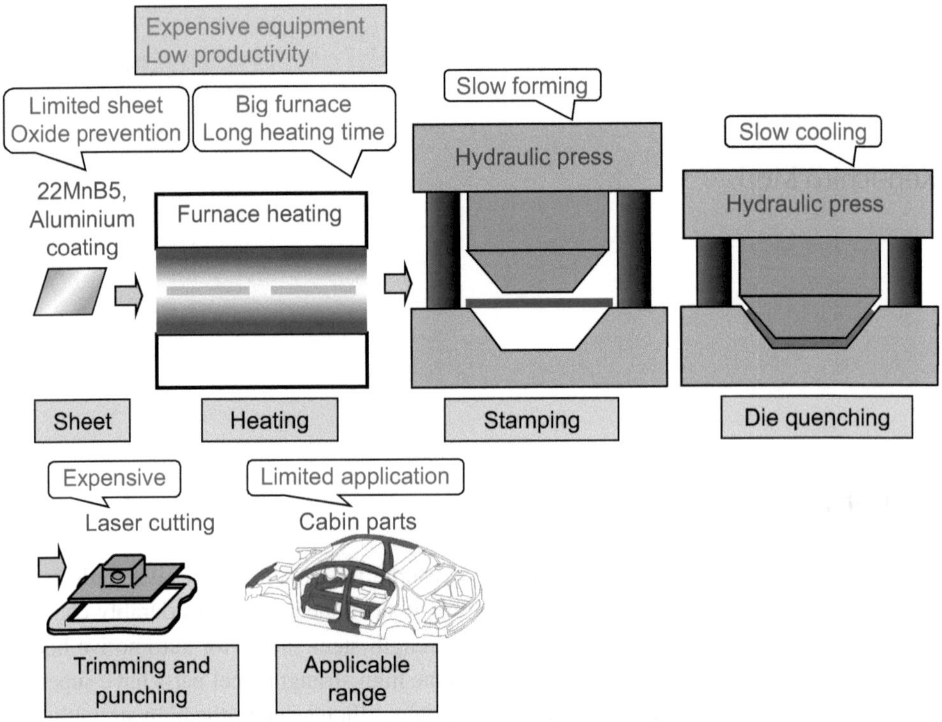

Fig. 1 Disadvantages in present hot stamping processes

1. The equipment consisting of a furnace, press, laser cutting machine, etc. is large and expensive.
2. The productivity is low due to die-quenching and slow hydraulic press, i. e. 2 or 3 shots per one minute.
3. The applicable range is limited to reinforced parts of automobile cabins and bumpers.
4. For non-coated sheets, shot peening is required, and aluminium-coated sheets for preventing the oxidation at high temperatures are comparatively expensive.
5. Laser cutting of formed parts for trimming and punching becomes costly.

2 Hot Stamping Using Resistance Heating

The heating approach of the steel sheets in hot stamping is crucial. Although the drop in temperature in bulk workpieces used for the forging is comparatively small, that for the sheet metals after the ejection from the furnace is very rapid due to large surface area. In addition, oxide scale on the surface of the sheet is remarkable. The size of the furnace becomes large due to slow formation of the intermetallic compound on the surface of the

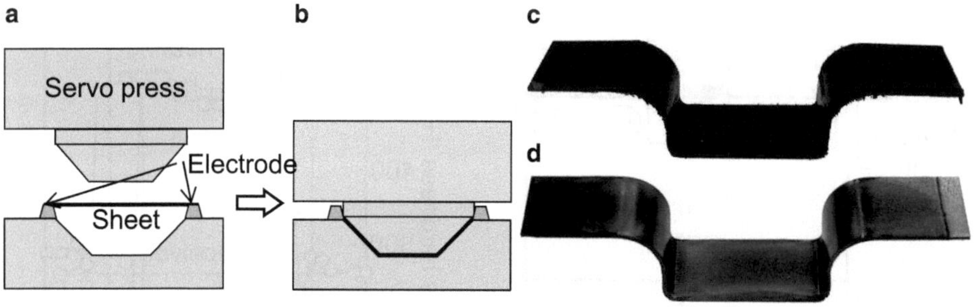

Fig. 2 Hot stamping using resistance heating and hat-shaped sheets. **a** Resistance heating, **b** Forming and die quenching, **c** Furnace-heated sheet, **d** Resistance-heated sheet

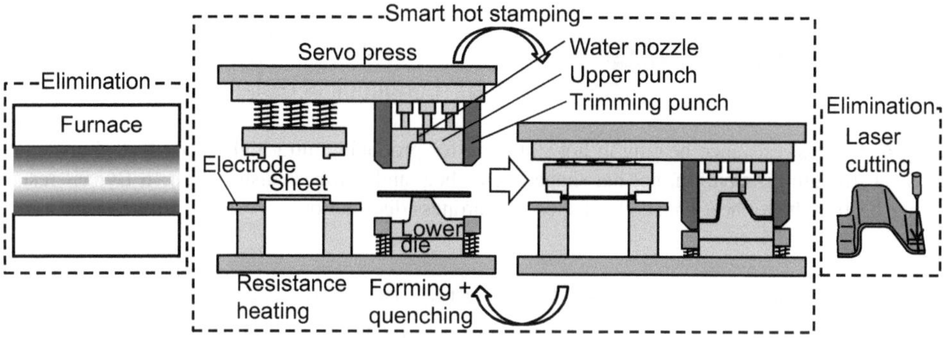

Fig. 3 Smart hot stamping having resistance heating, water quenching, servo press and hot trimming

aluminium-coated sheet during heating. For rapid laser heating, it is difficult to obtain a uniform temperature in the sheet due to local heating.

Mori et al. [4] proposed a hot stamping process using rapid resistance heating to improve the productivity. The sheets are heated in only 2 s to 900 °C required for quenching. Hot stamping using resistance heating and the hat-shaped sheets are shown in Fig. 2. The oxide scale of the stamped parts is hardly generated due to rapid heating [5]. Big furnaces are replaced with a power supply. The efficiency of resistance heating is higher than that of induction because of direct passage of current through sheets. A smart hot stamping system using resistance hearting is illustrated in Fig. 3.

3 Tailor Die Quenching for Producing Part Having Strength Distribution

Parts keeping a balance of strength and toughness can be produced by controlling the die quenching operation in hot stamping. In addition, cold shearing can be used for hot-stamped parts having non-quenched trimmed and punched zones instead of laser cutting

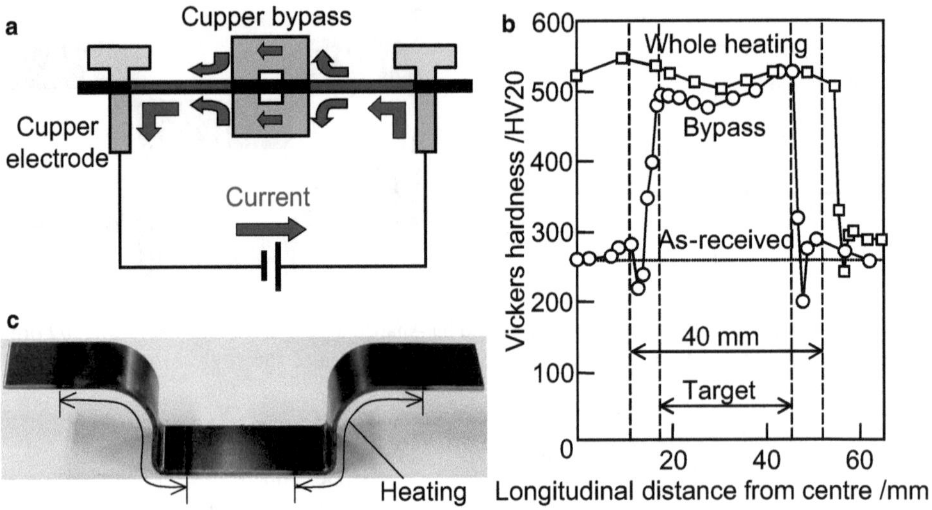

Fig. 4 Bypass resistance heating in hot stamping for producing ultra-high strength steel products having strength distribution, hot hat-shaped bent sheet and hardness distribution. **a** Bypass resistance heating, **b** Hot heat-shaped bent sheet, **c** Hardness distribution

having low productivity. Hein and Wilsius [6] have partially prevented quenching in hot stamping by reducing the cooling rate with partially heated tools, and Mori and Okuda [7] have used grooved tools for local reduction in cooling rate. Since quenchable steel sheets 22MnB5 generally used in hot stamping are not hardened for slow cooling such as air cooling, cooling during die quenching is partially delayed to prevent the martensite transformation. These treatments are called tailored tempering.

A tailored die quenching process using bypass resistance heating in hot stamping for producing ultra-high strength steel parts having a strength distribution was developed [8]. In the resistance heating of the sheet shown in Fig. 4, non-electrified portions by contact with copper bypasses with a low resistance were not heated, whereas portions not in contact with the bypasses were heated by the passage of current and were quenched by holding tools at the bottom dead centre. Bypass resistance heating was stable as partial heating, and the electrical power loss was small.

4 Hot Spline Forming of Gear Drum

Gear drums used in automobile transmission are mainly produced by bulk forming from plates instead of conventional forging from billets. A plate is drawn into a cup, and then the side wall of the cup is formed into a gear shape by ironing. The forging from billets is inappropriate to the production of parts having small thickness such as gear drums due

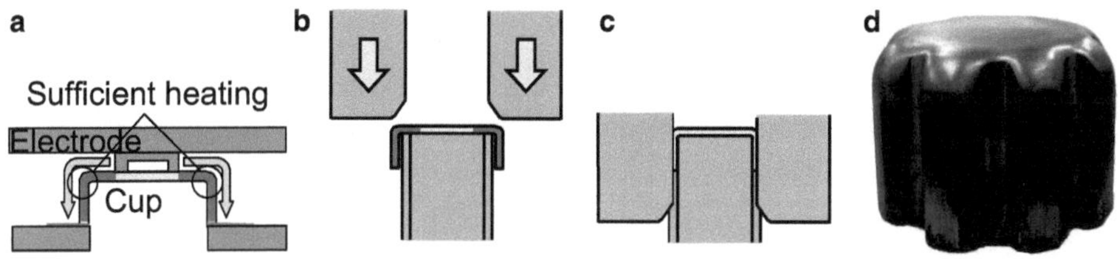

Fig. 5 Hot spline forming of gear drum having resistance heating of side wall. **a** Resistance heating, **b** Heating at 900°C, **c** Die quenching, **d** Produced drum

to large change in shape, and thus the application of forging from plates having thicker thickness than sheets gradually increases.

To improve the formability in the spline forming of ultra-high strength steel gear drums, the side wall of a cup formed into a gear shape is heated by resistance heating (see Fig. 5) [9, 10]. The bottom and edge of the side wall are in contact with the upper and lower electrodes, respectively. When the thickness of the side wall is kept uniform by applying ironing in the deep drawing of the cup, the side wall is uniformly heated by the electrification, i. e. the cross-sectional area of the side wall is uniform in the current direction. In addition, no heating of the bottom of the cup has the function of preventing the rapture in the bottom during the spline forming. The applicable range of the resistance heating is extended to spline forming. The side wall of the resistance-heated drawn cup is ironed and then die-quenched. Since resistance heating is very rapid, the cup is hardly oxidised.

5 Hot Stamping of V-Shaped Hollow Axle Beam Using Air Filled into Sealed Tube and Resistance Heating

Hot stamping is employed for forming not only steel sheets but also steel tubes. Since oil and water are conventionally used as pressure media in hydroforming of tubes, the heating temperature is limited. To increase the forming temperature, air is employed as pressure media. Although hot tube forming using air is similar to the hydroforming, the limitation of the heating temperature is removed.

V-shaped hollow axle beams used for suspensions of automobiles are conventionally formed by the cold hydroforming of tubes, and then the formed beams are quenched to increase the strength. The axle beam is produced by a hot stamping process using air filled into a sealed tube shown in Fig. 6 [11]. The formability is improved by heating, and the additional heat treatment is eliminated by die quenching.

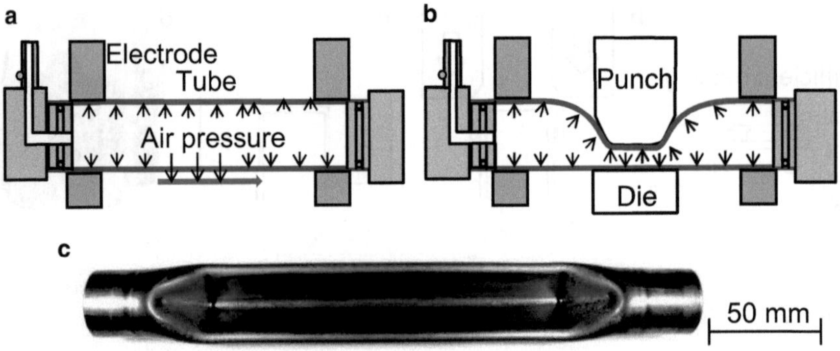

Fig. 6 Hot stamping of V-shaped hollow axle beam using air filled into sealed tube and resistance heating. **a** Resistance heating, **b** Forming and die quenching, **c** V-shaped hollow axle beam

References

1. Kleiner, M., Geiger, M., Klaus, A., 2003, Manufacturing of lightweight components by metal forming. CIRP Annals – Manufacturing Technology 52 (2), 521–542.
2. Abe, Y., Ohmi, T., Mori, K., Masuda, T., 2014, Improvement of formability in deep drawing of ultra-high strength steel sheets by coating of die. Journal of Materials Processing Technology 214 (9), 1838–1843.
3. Karbasian, H., Tekkaya, A.E., 2010. A review on hot stamping. Journal of Materials Processing Technology 210 (15), 2103–2118.
4. Mori, K., Maki, S., Tanaka, Y., 2005, Warm and hot stamping of ultra high tensile strength steel sheets using resistance heating. CIRP Annals – Manufacturing Technology 54 (1), 209–212.
5. Mori, K., 2012, Smart hot stamping of ultra-high strength steel parts. Transactions of Nonferrous Metals Society of China 22 (2), s496–s503.
6. Hein, P, Wilsius, J.B., 2008, Status and innovation trends in hot stamping of USIBOR 1500 P. Steel Research International 79 (2), 85–91.
7. Mori, K., Okuda, Y., 2010, Tailor die quenching in hot stamping for producing ultra-high strength steel formed parts having strength distribution. CIRP Annals – Manufacturing Technology 59 (2), 291–294.
8. Mori, K., Maeno, T., Mongkolkaji, K., 2013, Tailored die quenching of steel parts having strength distribution using bypass resistance heating in hot stamping. Journal of Materials Processing Technology 213 (3), 508–514.
9. Mori, K., T. Maeno, T., Fukui, Y., 2011, Spline forming of ultra-high strength gear drum using resistance heating of side wall of cup. CIRP Annals – Manufacturing Technology 60 (1), 299–302.
10. Mori, K., Maeno, T., Nakamoto, S., 2014, Hot spline forming of ultra-high strength steel gear drum using resistance heating. Key Engineering Materials 622–623, 201–206.
11. Maeno, T., Mori, K., Adachi, K., 2014, Gas forming of ultra-high strength steel hollow part using air filled into sealed tube and resistance heating. Journal of Materials Processing Technology 214 (1), 97–105.

Technologies for Forming and Foaming of Aluminium Foam Sandwich

Bernd Viehweger and Alexander Sviridov

1 Motivation

Recently, new materials are increasingly used, which have not only the lighter weight and better machining properties, but also excellent special properties e. g. crash safety. It is an objective of today's researchers at many institutions to develop these materials. For a successful application in products, economical manufacturing processes are required. New production technologies have to be developed, as conventional processes cannot process these new materials. Furthermore, the processing is hindered by high capital investments for the purchase of necessary equipment.

This article reports on the research project "Crash-safe front modules for rail vehicles of aluminium foam". The team of the project consists of user, manufacturer and developer. The partners are Bombardier Transportation, Wilhelm Schmidt Co., AMIC Angewandte Micromesstechnik Co. and Chair of Design and Manufacturing of the Technical University of Cottbus. It is the aim of the project to produce the front module for ITINO – Trainset of aluminium foam sandwich.

A production technology of the front modules of aluminium foam sandwich (AFS) was developed at the Chair of Design and Manufacturing together with the project partners. Due to the annual requirements of front modules, small quantities must be producible with the manufacturing process. The technology proof testing has been carried out step-by-step first with the test geometry, then with the front module segment and finally with a demonstrator front module in real size. The technical and economic feasibility of front modules of this innovative material should be demonstrated. The focus of this study was an investigation of forming and foaming strategies, the development of joining procedures as well as its testing.

Bernd Viehweger ✉ · Alexander Sviridov
Brandenburg University of Technology, Cottbus, Germany
e-mail: Bernd.Viehweger@t-online.de

A. E. Tekkaya et al. (eds.), *60 Excellent Inventions in Metal Forming*,
DOI 10.1007/978-3-662-46312-3_63

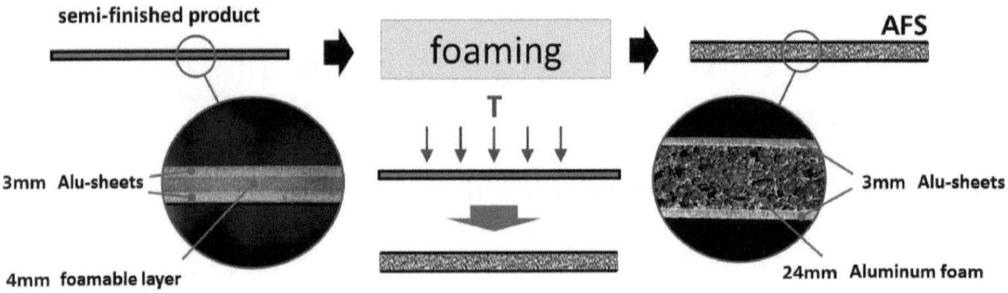

Fig. 1 Aluminium foam sandwich before and after foaming

2 Aluminium Foam Sandwich

Aluminium foam sandwiches consist of a mix of three layers with the core layer made of pressed aluminium powder with blowing agent (TiH_2). Conventional aluminium sheets (Al-alloy 6060) are applied as cover sheets. The three layers are roll-bonded into a sandwich and with the help of cutting, forming and foaming operations, it can be processed in a construction element or part of assembly. The foaming of the sandwich sheets is carried out in special furnaces in which they are heated to foaming temperature. As a result, the core layer expands to its final thickness. After the foaming operation, an aluminium foam sandwich arises, consisting of two aluminium cover sheets and a core layer of aluminium foam (Fig. 1).

3 Technology Development

The production of AFS-parts requires primarily a material-adapted design of the component shape based on the properties of the new material. Through simulation, the component behaviour and the crash behaviour of the front module of aluminium foam sandwich were determined and the material thickness was defined. The simulation showed that cover plates with the thickness of 3 mm and an aluminium foam core layer of 24 mm satisfy the requirements for crash safety as well as the required loads.

For the first tests, a three-dimensional-geometry has been developed with the characteristic shape of some shaped elements of the front module (test geometry). For the forming of this test geometry, a forming tool was developed. It consists of punch and die which were milled from a solid (Fig. 2). The result of the tests showed that the forming of the complex geometries of unfoamed sandwich plates is possible.

In a further process step, the formed test geometry was foamed in a furnace. The foaming process takes place close to the melting temperature of the cover plates. The part has a low strength and can change its shape during foaming. Thus, a foaming tool is required. The massive forming tool is unsuitable for this purpose. For the foaming, a

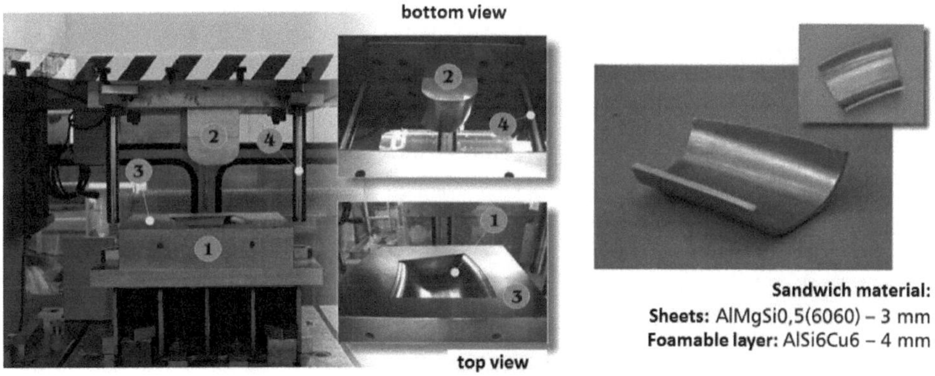

Fig. 2 Forming tool and formed part. *1* die, *2* punch, *3* master plate, *4* guide pillar

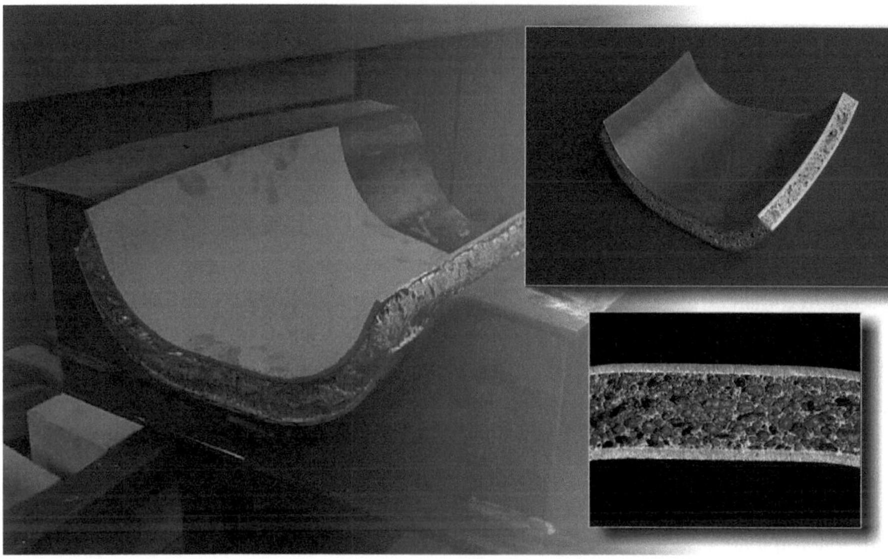

Fig. 3 Foaming tool and foamed part

separate foaming tool made of steel sheet was produced. In the next step, the test geometry was foamed in the furnace within this tool (Fig. 3).

The economic efficiency is one of the most important criteria for the series production of components of aluminium foam sandwiches. Therefore, another objective was to optimize the process and the tools. The optimized technology and tools were tested for the production of the front module segment. The front module segment is the a-pillar of the original front module on a scale of 1 to 5.

A tool in shell construction has been used to avoid the expensive milling of the punch and die. The forming tool consists of 10 mm steel sheets which are formed by 3D bending.

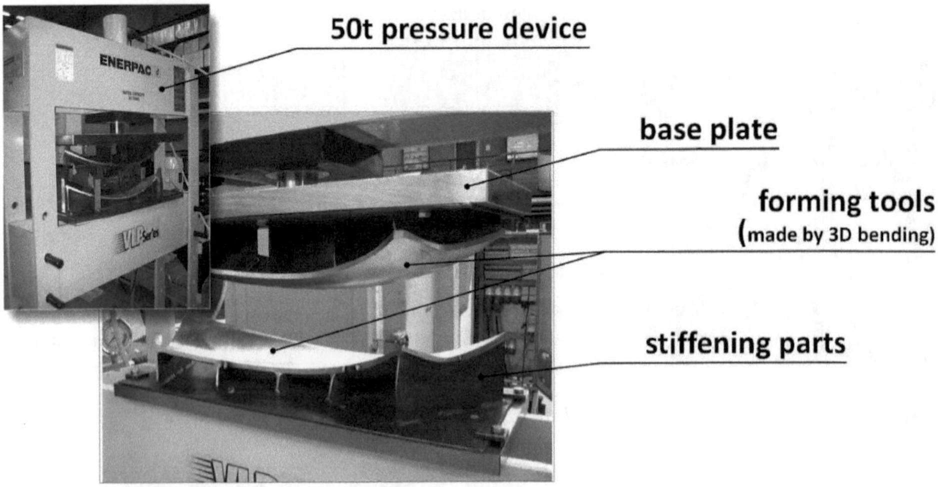

Fig. 4 Forming tool for production of front module segment

This tool production leads to lower production costs compared to milling from the solid and achieves better economic efficiency (Fig. 4).

The sandwiches have been preheated before forming up to 400 degrees to reduce the necessary forming force. Thus, a simple 50 t-pressure device with very low hourly rates can be applied instead of an expensive press. This pressure device can be supplied with very low investment; it is easily transportable and requires no complex preparation for installation.

The optimized forming tool can also be used for the foaming process. Due to the shell construction, it can be heated quite quickly and is therefore used as foaming tool. Therefore, the production of a separate foaming tool is not necessary. The front module segment was formed with this technology, it was foamed in the forming tool and it shows high dimensional accuracy.

4 Production of the Demonstrator

The results obtained from the prototypes were applied to manufacture a demonstrator – a half-front module in original size. Due to the size, the front module was divided into several components. The components were formed and foamed by the developed technology. After foaming, they were cropped at the edges and joined to the front module. For the joining of the aluminium foam sandwich components, the standard welding technology (MIG, WIG) is applicable, because only the cover plates are welded. Advantageously, the welding did not damage the aluminium foam layer.

The demonstrator produced as part of the project is shown in Fig. 5. It has been demonstrated that the front module of aluminium foam sandwich can be economically manufactured. The advantages of the technology are:

Fig. 5 Demonstrator "Front Module"

- forming with a simple forming device – no forming press is necessary
- application of cost-effective tools
- elimination of a separate foaming tool

Depending on quantity, the production costs in comparison to the conventional construction of glass fibre reinforced components are the same or even 10 % less. Moreover, a front module of aluminium foam is almost 20 % lighter in comparison to the conventional construction. Further advantages are significant higher crash safety, vibration damping and insulation, better recycling and cheaper repair options.

5 Conclusion

The economic use of new innovative materials requires the development of the economic production process. As an example, an economic process route for the production of the front module of the rail vehicles of aluminium foam sandwiches was developed and tested. The results are better component properties like increased crash safety, better strength, vibration damping, insulation, recycling and reduced production costs in comparison to the conventional glass fibre reinforced plastic solution.

6 Acknowledgement

The investigations were carried out in the BMBF–InnoRegio project and were coordinated by RIO e. V. – Regional innovation Alliance. The authors highly appreciate the support and coordination of BMBF and RIO e. V.

References

1. Seeliger, H.-W., 2004, Aluminium Foam Sandwich (AFS) Ready for Market Introduction. Advanced Engineering Materials, 6/2004
2. R. Neugebauer, R., Hipke, T., Hohlfeld, J., Thümmler, R., 2004, Ein Weg zu leichteren Bauteilen. Konstrukteur, 12/2004
3. Baumeister, J., Lehmhus, D., Commercially available products made of PM aluminium foams – status and prospects, Banhart, J., Fleck, N.A., Mortensen, A. (Ed..): Cellular Metals: Manufacture, Properties, Applications, Verlag MIT Publishing, Berlin 2003
4. Aluminiumschaum, Aluminium Zentrale e. V., Merkblatt W 17 1. Auflage, 1999

Plastic Consolidation of Metal Matrix Composites by Pressure Cycling

Glenn S. Daehn

1 Motivation

The properties of materials are developed based on their constituents and microstructure. One of the most direct ways to control the structure and properties of materials is by the direct mixing of two or more phases to control the key properties such as stiffness, strength, density, thermal conductivity and so on. This is the key idea behind the creation of composite materials. There are numerous methods for creating composites and numerous morphologies. Two limiting morphological cases are composites reinforced with continuous fibers or strands. This gives very directional properties. At the other end of the spectrum are random particulate composites. These can have very isotropic properties. Often the most interesting properties are derived when there are high volume fractions of a reinforcing phase, but this can be difficult to produce in practice.

To fabricate composites from pressurizing metal with ceramic particles, very high pressures can be needed to make the deformable particles conform to one another, and allow reinforcement rearrangement, which are essential to produce a nearly dense composite. The problem is similar to that of developing a hardness indent in a metal that was studied in detail by Tabor [1]. These and similar studies show that pressures must exceed about 3 times the flow stress of the metal in order for the metal particles to conform to the non-deforming shapes. This makes the consolidation of composites based on relatively hard metals difficult as high pressures are required and very hard dies and large presses are therefore required to plastically densify even simple shapes. This contribution shows how high pressure cycling can develop high composite density from highly reinforced composites. This can reduce press-forces to accessible levels.

Glenn S. Daehn ✉
Ohio State University, Columbus, USA
e-mail: daehn.1@osu.edu

A. E. Tekkaya et al. (eds.), *60 Excellent Inventions in Metal Forming*,
DOI 10.1007/978-3-662-46312-3_64

A number of papers starting with classic observations of Roberts and Cottrell [2] have shown that materials can behave in a nearly-viscous manner if plastic deformation is created by some form of internal strain mismatch. Several groups have shown that local internal plastic deformation can be developed by phase transformations [3] or thermal expansion mismatch [4] and this can give superplastic characteristics to the composite materials on repeated cycling of the process which induces plasticity [5]. This all can be shown to be a result of the application of the Levy-Mises equations.

2 Process Idea

The Levy-Mises equation teaches us that during plastic deformation, the principal plastic strain components are proportional to the deviatoric stress components. This can be useful in many operations that rely on plastic deformation in processing. If a plastic deformation is already ongoing a remote hydrostatic pressure can produce plastic flow, this can allow the matrix to flow easily, if multiple pressure cycles are produced. Figure 1 shows how local shear strains can be created in an elastically heterogeneous material under the

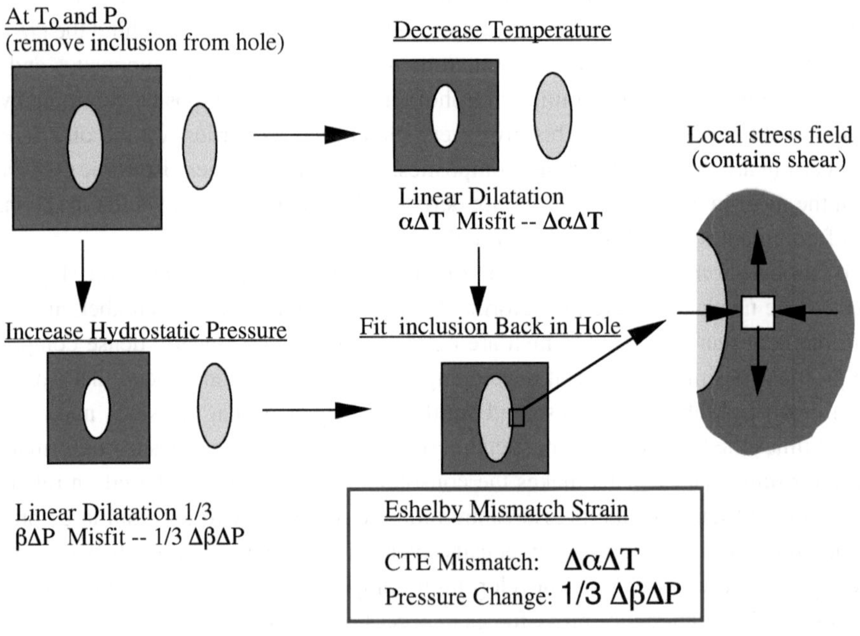

Fig. 1 Schematic diagram of how shear stresses, large enough to drive plastic deformation can be created from simple isotactic pressure in an elastically heterogeneous solid. This takes place because compatibility requires that the hoop strain will not be equal to the radial strain in the matrix at the particle-matrix interface. Differences in compressibility (β) or coefficient of thermal expansion (α) can drive local shear strain

application of remote pressure. The strain metric of interest is $\Delta\beta\Delta P$ where $\Delta\beta$ is the difference in compressibility between the two phases and ΔP is the pressure change on a cycle.

The key idea is that $\Delta\beta\Delta P$ is large enough, pressure cycling may allow composite consolidation to high density in a manner that is much more effective than pressure alone.

3 Results and Applications

This idea has been proven to be effective and a summary is presented here and details can be found elsewhere [7–12]. In the pressure cycling consolidation of green metal matrix composite compacts, the dominant mechanism involves the pressure changes inducing plastic deformation in the metal phase at particle interfaces and the applied pressure biases the resolution of these strains to provide net densification with each pressure cycle. Figure 2 shows a typical illustrative result [8] in the model lead-alumina system. This clearly shows that even highly reinforced composites can develop much higher densities if subjected to pressure cycling versus constant static pressure at the largest value.

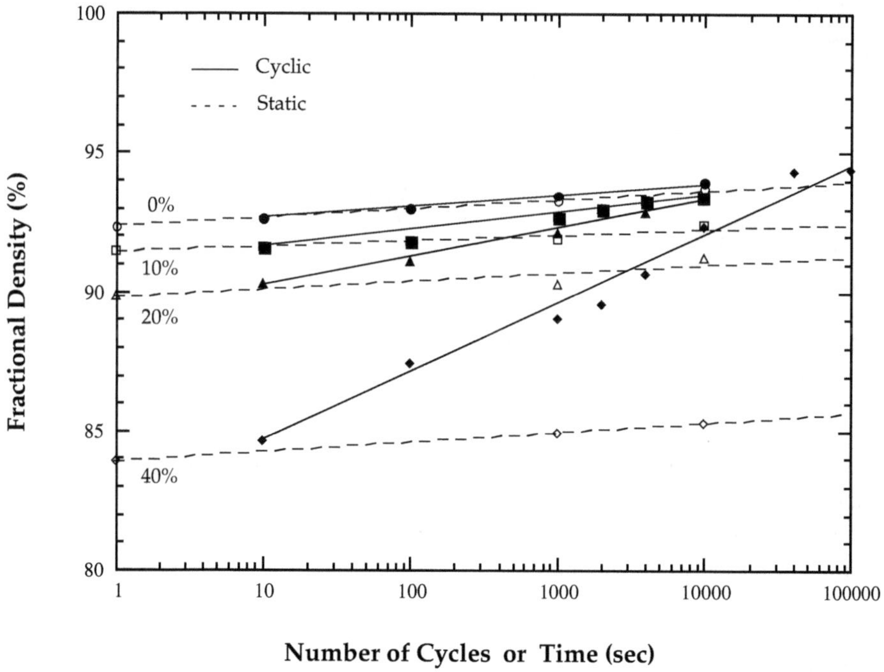

Fig. 2 Consolidation of lead-alumina composites at 1 hz and under static conditions with a maximum or constant stress of 414 MPa at varied volume fractions between zero and 40 % alumina. [8]. Symbols refer to alumina content as shown on the dotted line

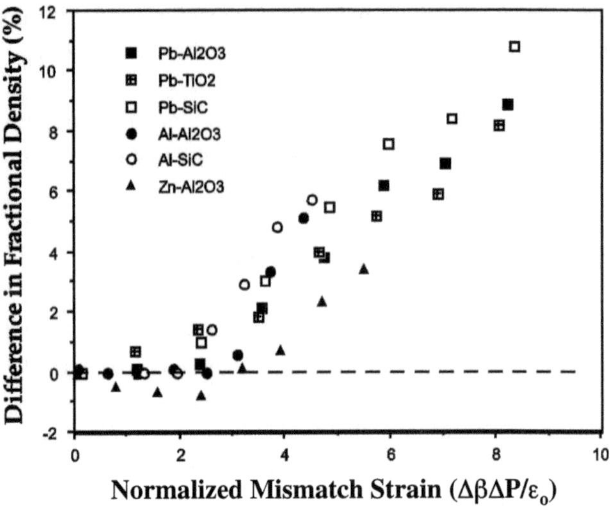

Fig. 3 The difference in fractional density between the static and cyclic compaction as a function of the normalized mismatch strain in six composite systems. All cyclic data were obtained after 100,000 pressure cycles. ε_0 represents yield strain

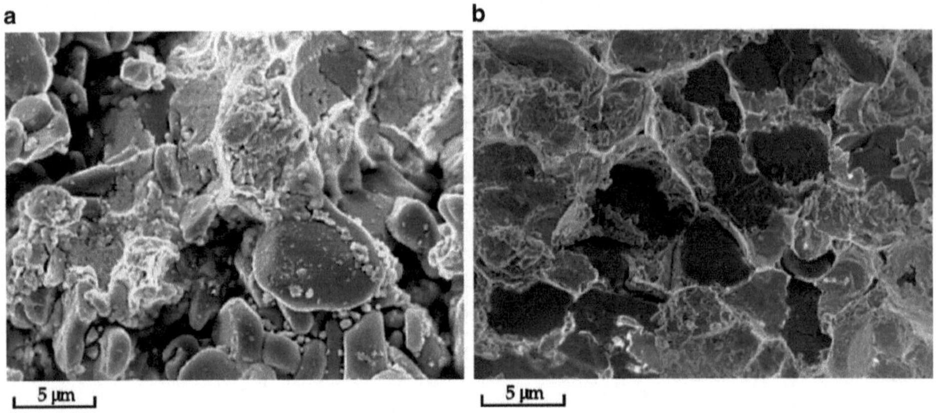

Fig. 4 Fracture surfaces of Pb $\pm 40\,\%$ Al_2O_3. **a** Green compacts formed by static pressure of 414 MPa, **b** or 105 cycles of 0-414 MPa [8]

Figure 3 shows that the mismatch strain is indeed the important figure of merit for consolidation and that this effect is robust over many systems. The materials subjected to pressure consolidation develop much higher densities and strengths. This is also shown in their fracture surfaces, shown in Fig. 4.

4 Potential

This approach has been quite powerful for the consolidation of highly reinforced metal matrix composites, as may be important for wear – resistant components. There are many ways to enhance this further. For example heating and cooling with differential thermal expansion, or phase transitions, can drive plastic deformation. However this requires time for heat transfer. Repeated pressure cycling is fast and effective but can be challenging with respect to developing presses and tools that are resistant to fatigue. New ways such as repeated shocks to develop pressures may be developed; also these methods may be useful for creating net-shape composites.

5 Conclusion

Simple cyclic pressures acting on elastically heterogeneous materials can produce internal plastic strains. These strains in turn can enhance consolidation of metal matrix composites. This is a potentially simple way to improve composite properties.

6 Acknowledgement

The key ideas that this paper has largely been the results of collaboration between the author and R. H Wagoner with a number of students and post-doc's involved including Ching-Yao Huang, G. Jiang, Y. Fu, and others. The US-NSF funded much of this research.

References

1. D. Tabor. *The Hardness of Metals*, Oxford, England: Oxford University press, 1951.
2. A. C. Roberts and A. H. Cottrell, Phil. Mag., 1956.
3. O. A. Ruano, J. Wadsworth and O. D. Sherby, Metall. Trans. A, 1982.
4. S. H. Hong, O. D. Sherby, A. P. Divecha, S. D. Karmarkar and B. A. MacDonald, J. Compos. Mater. 1988, 22, 102–123.
5. G.S. Daehn and G. González Doncel, *Metallurgical Transactions* **20 A,** pp. 2289–355 (1989).
6. G.S. Daehn, *Chapter 15 Comprehensive Composite Materials*, edited by A. Kelly and C. Zweben, New York, NY: Elsevier, pp. 419–445 (2000).
7. G. Jiang, W. Wu, G.S. Daehn and R. H. Wagoner, *Acta Mater*. **48 (17),** pp. 4331–4335, (2000).
8. C. Y. Huang and G.S. Daehn, Acta Mater, 44, pp. 1035–45 (1996).
9. C-Y Huang and G.S. Daehn, Acta Mater, 45, pp. 4283–96, (1997).
10. G. Jiang, G.S. Daehn and R. H. Wagoner, Scripta Mater. 44, pp. 287–292, (2001).
11. Y. Fu, J. J. Lannutti, R. H. Wagoner and G.S. Daehn, Materials and Metallurgical Transactions, 33 A, 183–191, (2002).
12. G. Jiang, G.S. Daehn and R. H. Wagoner, Powder Metallurgy, 46, 78–82, (2003).

Process-Integrated Heat Treatment of Hot Forged Components

Adis Huskic, Mohammad Kazhai, and Bernd-Arno Behrens

1 Motivation

For the production of hot forgings with outstanding mechanical properties a heat treatment from the hot forging heat can be used. Hereby, economic and ecological advantages can be achieved in contrast to conventional hot forging process chains, which consist of various heating and cooling cycles. Depending on the type and the initial state of the material used, different forging properties and microstructures can be set up by an adjusted temperature control [1, 2].

To avoid cost-intensive trial-and-error experiments in process design, finite element software systems are increasingly used. Commercial finite element systems were some time ago limited in simulating the phase transformation due to thermo-mechanical treatments of forgings. In recent years significant developments were made in this field. So far, numerous studies for simulating the phase transformation at the Institute of Forming Technology and Machines of the Leibniz University of Hannover were performed [3, 4].

For the numeric calculation of the phase transformation during cooling there are various approaches. The basis of the numerical calculation of phase transformations is a coupling between the thermal, mechanical and metallurgical relations. For the calculation of heat treatment processes the spatial and temporal tension, temperature and microstructure developments as well as their interactions are necessary. The description of the phase transformation kinetics can be realized on different approaches. Regarding the target microstructure, approaches such as the approach of Johnson-Mehl-Avrami were used for the calculation of a diffusion controlled phase transformation, which hereby describes the phase transformation of the austenite into the ferritic, perlitic and bainitic microstructure [7, 8]. For a diffusionless phase transformation the model of Koistinen-Marburger is

Adis Huskic · Mohammad Kazhai · Bernd-Arno Behrens ✉
Leibniz University of Hannover, Garbsen, Germany
e-mail: behrens@ifum.uni-hannover.de

A. E. Tekkaya et al. (eds.), *60 Excellent Inventions in Metal Forming*,
DOI 10.1007/978-3-662-46312-3_65

often used. This model describes the transformation from the austenitic to the martensitic phase [9].

In the context of this contribution results of experimental and numerical investigations on the integrated heat treatment of high stressed forgings are presented. The investigations were carried out on the demonstrators "removed shaft" and "common-rail" produced by full-forward cup-backwards hot extrusion process and hot die-forging with a subsequent hot deburring.

2 Experimental Test Setup and Numerical Simulation

The experimental studies on the integrated heat treatment were performed on a fully automated hot forging cell in order to achieve reproducible results from the experimental studies. The hot forging cell consists of a screw press SPR500 (Co. Lasco), an industrial robot KR 16-2 F-S (Co. Kuka) and a chamber furnace N41/H (Co. Nabertherm) as well as a cooling furniture developed by the Foundation Institute of Materials Science (IWT) Bremen (Fig. 1).

The demonstrator "removed shaft" was produced by full-forward cup-backwards hot extrusion. The cylindrical raw part made of the **h**ighly strength **d**uctile **b**ainitic (HDB) steel 22MnSiCr6-6-5 was first heated to a temperature of 1200°C then upset in the preliminary step for descaling and finally extruded in the main forming step. The demonstrator "common-rail" was hot die-forged by using a cylindrical raw part and subsequent hot deburred. Both demonstrators were finally specific heat treated in the cooling furniture from the hot forging heat.

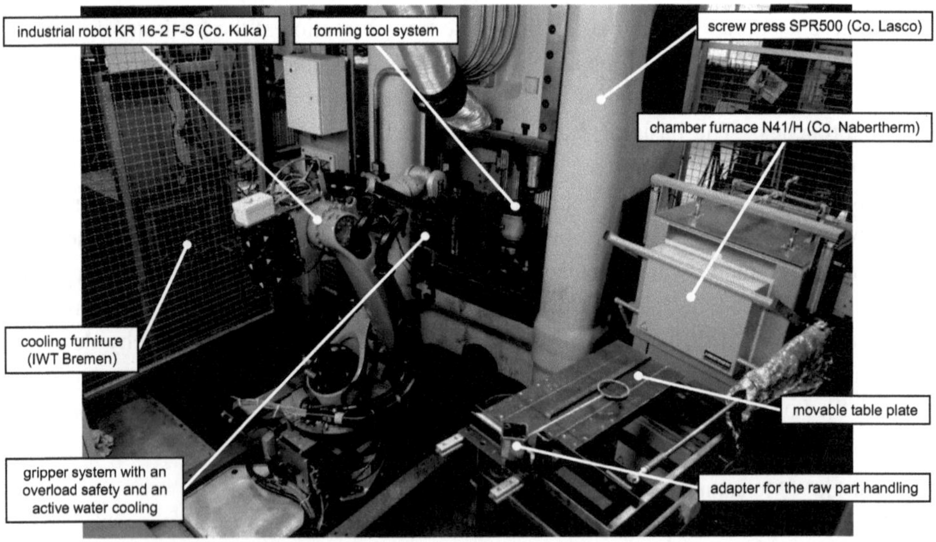

Fig. 1 Experimental setup – a full automated hot forging cell [10]

The cooling strategies are based on deformation-continuous-cooling-transformation (D-CCT) diagrams, which are determined at the Department of Ferrous Metallurgy of the RWTH Aachen University [11]. With the experimental determined D-CCT-diagrams only a rough cooling strategy can be defined because of the high and locally different plastic strains occurring in hot forging processes.

The finite element software Simufact.forming was used to execute the numerical simulations. In order to describe the kinetics of the phase transformations the models of Koistinen-Marburger and Johnson-Mehl-Avrami were used. While the Koistinen-Marburger's model defines the diffusionless phase transformation from austenite to martensite, the Johnson-Mehl-Avrami's model characterizes the diffusion controlled phase transformation from austenite to ferrite, perlite and bainite. The numerical simulations of the phase transformations are based on the cct- (continuous-cooling-transformation) and the ttt- (time-temperature-transformation) diagrams, which were determined at RWTH Aachen. Not only these diagrams were implemented into the material data of the simulation software, but also the thermo-mechanical like flow curves at different temperatures. To reduce the calculation time, the symmetrical properties of the model were used.

3 Results

The microstructure of the approximately four kilograms demonstrator "removed shaft" was set up by a compressed air cooling of one bar for a period of 100 seconds and a subsequent air cooling under free convection to ambient temperature (Fig. 2). The results of the microstructure analyses show a homogeneous distribution of bainitic structure. This was confirmed by measurements of the macro hardness, which is about 400 HV30.

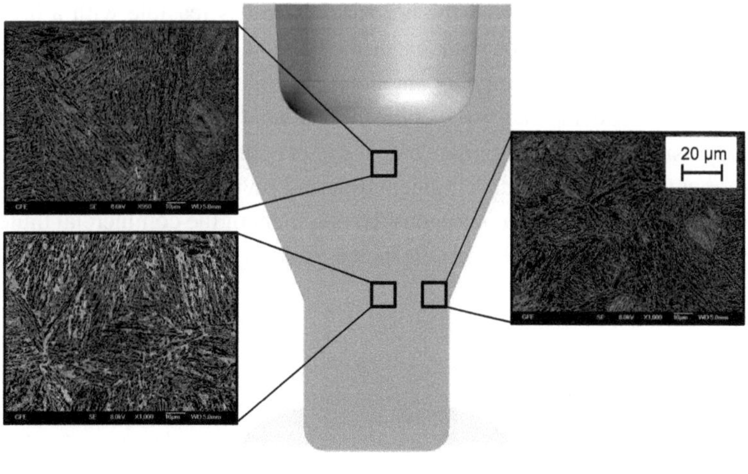

Fig. 2 Bainitic microstructure of a "removed shaft" from the HDB-steel 22MnSiCr6-6-5

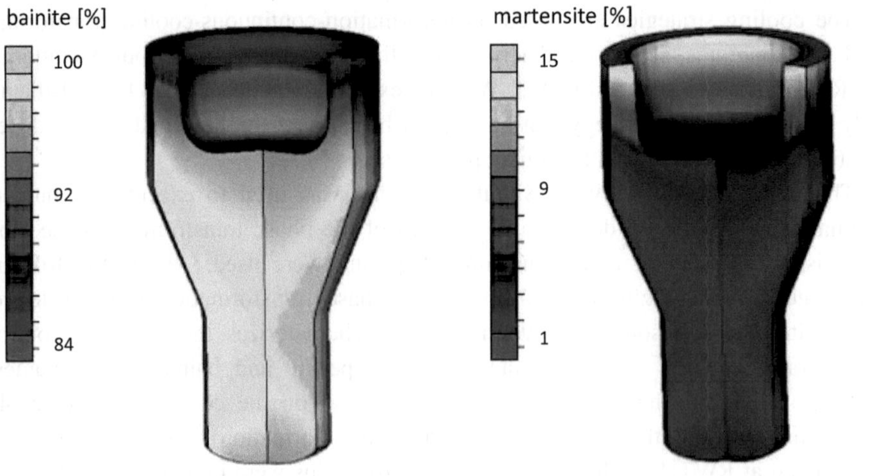

Fig. 3 Numerically calculated phase distribution of a "removed shaft" from the HDB-steel 22MnSiCr6-6-5

Figure 3 shows the numerically determined distribution of the microstructure of the demonstrator "removed shaft" from the HDB-steel 22MnSiCr6-6-5 after the heat treatment described above. The results show that a predominantly bainitic microstructure is present in the shaft and in the shoulder area. In the upper cup area a small proportion of martensite could be determined. These small proportions of martensite could also be identified in experimental studies.

4 Conclusion

The integrated heat treatment from the hot forging heat is feasible with a high process reliability and reproducibility for both, experimentally and numerically. This process development enables benefits mainly for the small and medium-sized forging industry in Germany, such as reducing the production time, reduction of energy and manufacturing costs. By using commercial finite element systems further competitive advantages in terms of shortening the process development time and saving expensive trial-and-error experiments can be achieved. The continuous development of the commercial finite element systems due to realistic and reliable results increases the competitiveness of manufacturing companies.

5 Acknowledgement

The results were achieved in the research project "EcoForge", which was executed under the leadership of the "Forschungsvereinigung Arbeitsgemeinschaft Wärmebehandlung und Werkstofftechnik e. V. (AWT)". It was funded by the AiF from the fund "Leittechnologien für kleine und mittlere Unternehmen" that is part of the "Programm zur Förderung der industriellen Gemeinschaftsforschung und -entwicklung (IGF)". This program is issued by the "Bundesministerium für Wirtschaft und Energie" due to a resolution of the German Bundestag. The final report can be requested at the office of the AWT (Paul-Feller-Str. 1, 28199 Bremen).

References

1. Bach, Fr.-W., Doege, E., Schmidt-Jürgensen, R., Dähndel, H., van Well, M., 2003, Auswirkungen einer Integration der Wärmebehandlung auf die Prozesskette zur Zahnradherstellung durch Präzisionsschmieden, International Conference on Accuracy in Forming Technology ICAFT 2003, Berichte aus dem IWU Band 22, Chemnitz
2. Bach, Fr.-W., Behrens B.-A., Gretzki T., Hassel T., Odening, D., 2009, Integrierte Wärmebehandlung komplexer Präzisionsschmiedebauteile mittels einer prozess- und geometrieangepassten Zwei-Phasen-Spraykühlung, International Conference on Accuracy in Forming Technology ICAFT 2009, Berichte aus dem IWU Band 52, Chemnitz
3. Behrens, B.-A., Olle, P., Götze, T., 2008, Simulation der Mikrostruktur und des Verzugs geschmiedeter Bauteile, UTF Science; Ausgabe 4/2008, Meisenbach Verlag, Bamberg
4. J. Ditmann. 2002, Mikrostruktur und mechanische Eigenschaften von geschmiedeten Bauteilen, Leibniz Universität Hannover, Dissertation, Hannover
5. Denis, S., Gautier, E., Simon, A., Beck, G., 1985, Stress-phase transformations interactions – principles, modelling, and calculation of internal stresses, Materials Science and Technology
6. Reti, T., 2002, Residual Stresses in Carburized, Carbonitrided, and Case-Hardened Components, In: Totten, George E., Maurice A. H. Howes und Tatsuo Inoue (Herausgeber): Handbook of Residual Stress and Deformation of Steel, S. 190–208, ASM International, Materials Park, Ohio
7. Avrami, M., 1940, Kinetics of phase change, Journal of Chemical Physics, S. 1103–1112
8. Johnson, W. A., Mehl, R. F., 1939, Reaction kinetics in process of nucleation and growth, Transaction AIME 135, S. 416–458
9. Koistinen, D.P., Marburger, R.E., 1959, A general equation prescribing the extent of the austenite-martensite transformation in pure iron-carbon alloys and plain carbon steels, Acta Metallurgica, S. 59–60
10. Behrens, B.-A., Bonk, C., Frischkorn, C., Grbic, N., Huskic, A., Kazhai, M., Moritz, J., Neumann, A., Schrödter, J., Bouguecha, A., 2014 , Aktuelle Forschungsergebnisse am Institut für Umformtechnik und Umformmaschinen, 21. Umformtechnisches Kolloquium Hannover – Industrie und Wissenschaft – Gemeinsam die Zukunft gestalten, S. 9–32
11. Fischer, M., Dickert, H. H., Bleck, W., Huskic, A., Kazhai, M., Hadifi, T., Bouguecha, A., Behrens, B.-A., Labanova, N., Felde, A., Liewald, M., Egorov, F., Garbrecht, M., Brinksmeier, E., Reimche, W., Bruchwald, O., Frackowiak, W., Maier, H. J., Bucquet, T., Hinrichs, B., Fritsching, U., Hoja, T., Hoffmann, F., Zoch, H.-W., 2014, EcoForge: Energieeffiziente Prozesskette zur Herstellung von Hochleistungs-Schmiedebauteilen, Journal of Heat Treatment and Materials HTM, S. 209–219

[illegible]

References

[illegible]

Micro-Tube Hydroforming System Based on Floating Die Assembly

Gracious Ngaile and James Lowrie

1 Motivation

The demand for miniaturized products has dramatically increased in the last ten years. To cope with this demand, different micro manufacturing techniques that are capable of producing miniature parts had to be developed. These processes include, micro-casting, micro-EDM, micro-ECM, micro-forming, micro-mechanical- assembly etc., [1–6]. In developing these processes, a lot of effort has been put in scaling down conventional macro manufacturing machines/processes. There are, however, numerous challenges in scaling down conventional macro manufacturing processes to a micro level. Proportional scaling is difficult to achieve because most of the manufacturing processes are composed of multitude of variables (process, geometric, material, environment, etc.) that vary in time and space, in a complex manner. In scaling down a conventional forming machine/tooling, the system must have very tight tolerances and precise positioning. This can be exacerbated by tool deflections caused by forming forces as well as vibrations of the tools during the forming process. Furthermore, it can be difficult to predict the forming forces and workpiece material behavior due to an inability to quantify the size effect present in the process. Needless to say, there are process specific problems which are unique to the metal forming process that is being miniaturized [7].

Take an example of a tube hydroforming process where a tubular part is produced by injecting fluid into the tube. The fluid pressure is then raised as the tube ends are being fed until the tube conforms to the die cavity. The fluid pressure which is responsible for the deformation of the tube cannot be scaled down as the diameter of the tube is reduced to a micro size. Since the pressure to deform the tube is a function of tube diameter to tube wall thickness ratio (D/t ratio), the fluid pressure needed for large or micro tubing of the

Gracious Ngaile ✉ · James Lowrie
North Carolina State University, Raleigh, NC, USA
e-mail: gngaile@ncsu.edu

A. E. Tekkaya et al. (eds.), *60 Excellent Inventions in Metal Forming*,
DOI 10.1007/978-3-662-46312-3_66

same D/t ratio will be in the same level. Furthermore, while the force exerted on a hollow punch could be scaled down proportionally as the tube size is decreased to a micro level, developing a punch that is capable of supplying pressurized fluid, feed the material, as well as ensuring a reliable seal at the punch-tube-die interface is a major challenge [7–11]. Due to the small nature of the tubular blank, the positioning of the punch must be extremely accurate to achieve an even distribution of pressure on the tube ends. Depending on the size of the micro punch, drilling a tiny hole in the punch for fluid supply has proved to be difficult if not impossible.

These challenges call for new micro-manufacturing processes that are based on different principles from the conventional macro manufacturing processes. An innovative micro tube hydroforming (THF) process that is based on floating a die assembly in a pressurized fluid is the subject of this paper.

2 Process Idea

Micro- THF system based on floating die assembly is a novel technique for hydroforming metallic micro tubular components. It is based on decoupling the actions occurring at the punch-tube-die junction of a conventional THF system. Figure 1 shows a typical macro-THF system where the punch-tube-die junction has to fulfill several requirements namely; material feed, sealing, and high pressure fluid supply. All these requirements are mutually depended and thus presenting a huge challenge in scaling down the conventional THF to a micro level. Figure 2 shows a schematic diagram of the floating based THF system.

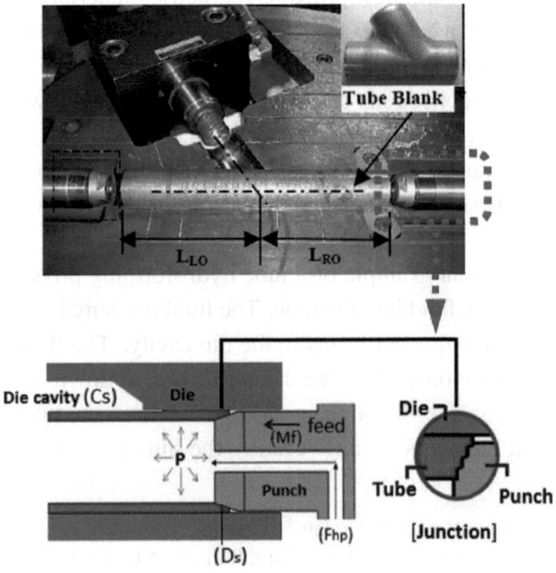

Fig. 1 Punch-tube-die junction in conventional macro – THF [9]

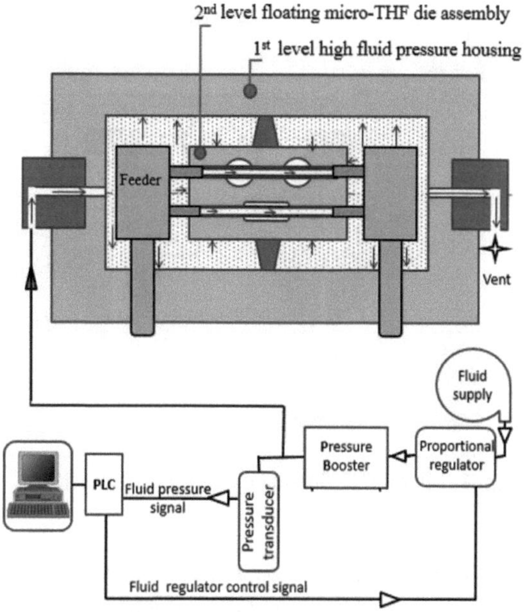

Fig. 2 Floating based micro-THF system [8]

The 1st level contains a high pressure chamber and the 2nd level contains a floating die assembly. The system is called "floating" because the die assembly is submerged in a pressurized fluid chamber. The unique feature of this THF system is that the functions of the tool-tube interfaces (punch-tube and die-tube interfaces) are decoupled to better suit the needs of a micro-scale process. Let we examine the functions of the punch-tube junction in the conventional macro THF system: The punch is responsible for supplying and maintaining the high pressure in the working fluid as well as feeding the tube into the die cavity. In order for the punch-tube interface to function at all, a reliable seal must be created that moves with the material of the tube so that both expansion and feeding can take place simultaneously. In the floating dies assembly, the function of sealing is decoupled from the punch-tube interface and added to the die-tube interface (Figs. 2 and

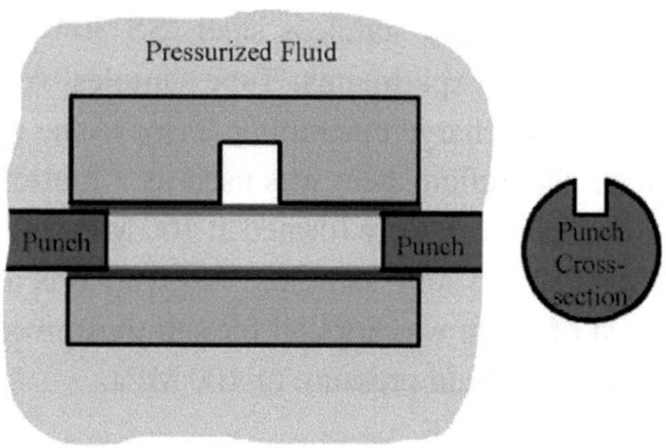

Fig. 3 Floating die assembly surrounded by pressurized fluid [7]

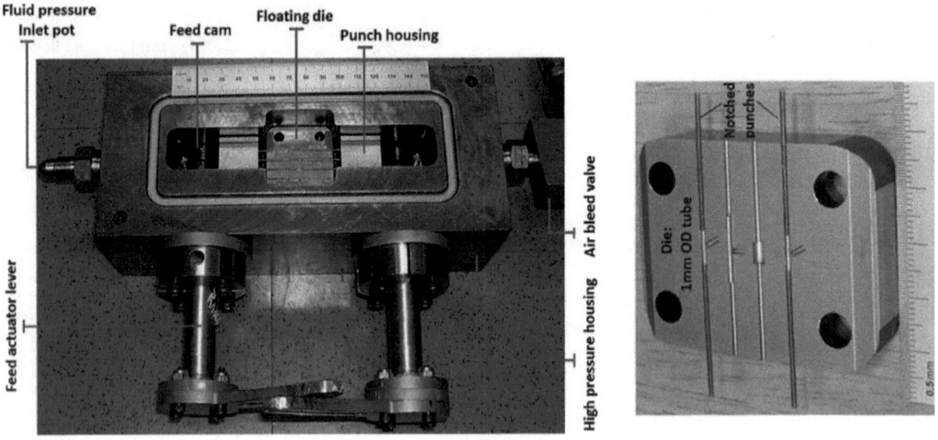

Fig. 4 Micro-THF floating die assembly setup [9]

3). The sealing could be achieved by wrapping the tube with a Teflon sheet or the tube could be coated by a solid film which would act both as a seal and a lubricant. The design should be such that fluid is prevented from entering the die cavities. This decoupling gives more freedom in the punch design and avoids many of the issues involved with creating a reliable seal with the punch.

The floating based micro-THF system was fabricated and tested. Figure 4 shows the major components of the system which consists of high fluid pressure housing, floating micro-THF die assembly, cams for material feed, feed actuator, fluid pressure inlet pot, and notched punches. The fluid is supplied to the die assembly through a computer controlled pressure intensifier. To clamp the bottom and top high pressure housing chamber, a 150 ton press is used. Feed cams were designed with a maximum feed of 3 mm [9].

3 Results and Applications

Figure 5 shows parts that were hydroformed using the floating based micro-THF system. Annealed stainless steel (SS 304) tubular blanks 1 mm OD and 2 mm OD were used in the experiments. Tube samples were cut to 12 mm and 20 mm lengths using wire electro-discharge machining. Prior to the experiments, the samples were cleaned by acetone and Teflon sheet was used as a lubricant. Figure 5a shows 1 mm OD × 20 mm long samples which were formed at 100 MPa and a feed of 1 mm. Figures 5b and c shows 2 mm OD × 12 mm long samples. No feed was applied for samples shown in Fig. 5b, while a feed of 1 mm was applied for samples given in Fig. 5c. All the tests were carried out at a maximum fluid pressure of 100 MPa.

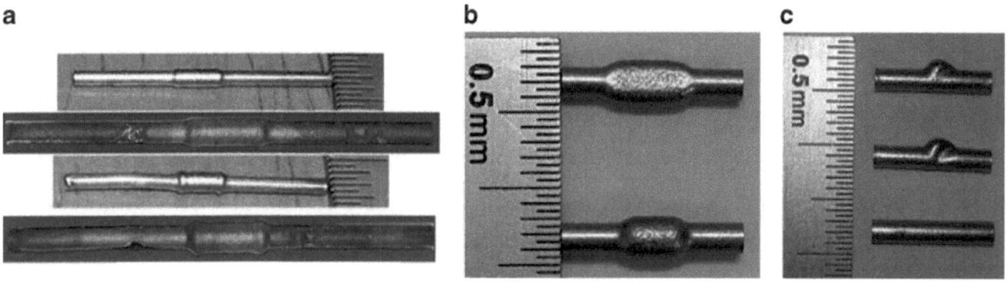

Fig. 5 Micro hydroformed parts [9]

4 Potential

The advancement of micro-THF technology has been hindered by, among others, the lack of robust micro-die systems that could facilitate hydroforming of complex parts that require both expansion and feeding. The development of the floating based micro-THF system capable of tube expansion and feeding has the potential to open avenues for producing metallic micro-tubular components of various geometric complexities that were infeasible. There are numerous fields where micro-tubular components could be applied. For example, micro tube hydroformed parts could be implanted into the human body as part of a circulatory implant, or small shunt. They could also be useful in micro fluidic devices or micro heat exchangers, due to their thin walls and potential for very complex shapes. Additionally, due to the work hardening involved in tube hydroforming there is also the potential that they could be used as micro-shafts in micro-electro-mechanical systems (MEMS).

Decoupling the sealing function from the micro punch provides the potential for designing micro punches of different geometric configurations. For example, the initial tubular cross-section no longer needs to be circular. This increases the range of micro-tubular configurations that could be hydroformed. Furthermore, this methodology lends itself to the utilization very small punch diameters, on the order of microns, as long as they are able to withstand the feeding load.

5 Conclusion

The micro-THF system based on the floating die assembly is capable of hydroforming complex micro tubular components irrespective of the size of the tube. The tooling design concept is based on decoupling the requirements at the die-tube-punch junction as applied in conventional THF systems. The system was successfully tested on SS 304 tubes of outer diameter 1 mm or 2 mm.

6 Acknowledgement

The author would like to acknowledge the National Science Foundation, through which this work was funded under Project No. CMMI # 0900148. Any opinions, findings, and conclusions or recommendations expressed in this material are those of the author and do not necessarily reflect the views of the National Science Foundation. The author would also like to thank Mr. Steve Cameron for fabricating the dies.

References

1. Razali, A.R., Qin,Y., A Review on Micro-manufacturing, Micro-forming and their Key Issues, Procedia Engineering 53 (2013) 665–672
2. Geiger,M., Messner, A., Engel, U., Production of micro-parts – size effects in bulk metal forming, similarity theory, *Production Engineering*, Vol. IV/1 (1997), p. 55–58
3. Krishnan, N., Cao, J., Kinsey, B., Parasiz, S., Li, M., Investigation of deformation characteristics of micro-pins fabricated using micro-extrusions, P*roceedings of IMECE, 2005.*
4. Qin, Y., Micromanufacturing Engineering and Technology*, Elsevier science (2010)*
5. Hartl, C., Micro-Hydroforming; Chapter 9, Micro-manufacturing Engineering and Technology, *Elsevier science (2010)*
6. Wagner, S.W., Analysis of a Non-Traditional Micro Tube Hydroforming Process, Doctorial Dissertation, Michigan Technological University, US., 2013
7. Lowrie, J., Development of a Micro-Tube Hydroforming System, Master thesis, Department of Mechanical and Aerospace Engineering, North Carolina State University, US., 2014
8. Ngaile, G., Development of MicroTube Hydroforming System, *Proceedings of the 8th international conference on MicroManufacturing, Victoria, Canada, March 2013*
9. Ngaile, G. and Lowrie, J., New Micro Tube Hydroforming System Based on Floating Die Assembly Concept, *ASME Journal Micro and Nano-Manufacturing,* December 2014, Vol. 2, doi:10.1115/1.4028320
10. Ngaile, G. and Lowrie, J., Punch Design for Floating Based Micro-Tube Hydroforming Die Assembly, *Proceedings of the 9th international conference on Micro Manufacturing, Singapore, March 2014*
11. Lowrie, J., Ngaile, G., Scalability of Conventional Tube Hydroforming Processes from Macro to Micro/Meso, accepted, *Proceedings of the 9th international conference on Micro Manufacturing, Milano Italy, March 2015*

Tube Drawing with Tilted and Shifted Die

Adele Carradò, Farzad Foadian, and Heinz Palkowski

1 Motivation

Cold drawing is one of eldest metal forming operations where a cross-sectional area and/or the shape of a rod, bar, tube or wire is reduced by pulling through a die. Drawing can be performed to reduce the diameter and/or wall thickness of tubes as well as its tolerances, to improve the surface finish, or to increase the strength of the product [1]. For industrial production of seamless tubes, the first forming operation often is extrusion or piercing of the billet, for example by rolling. Due to vibrations of the mandrel, tolerances in positioning of the die and billet, as well as potential temperature differences within the billet, this step inherently results in variations of thickness over length and circumference, eccentricity, and ovality. These geometrical irregularities can be found downstream through the complete process route [2]. Eccentricity, the deviation of the wall thickness in a tube from its average value in circumferential direction, in the pre-tube leads to an inhomogeneous material flow over the circumference in the following cold drawing steps and generates a complex RS pattern which can affect the mechanical and fatigue behavior of the final tube [3]. To quantify the effects, eccentricity was measured before and after drawing. Tube manufacturers and moreover customers using high priced materials have a great interest in reducing the mass of the product as far as possible, so e. g. by improving the preciseness in thickness. As a result of pressure from increasing competition, copper tube manufacturers want to minimize cost through reduced product wastage caused by defects and specifically the defect of tube eccentricity [4]. In this short paper, an innovative

Adele Carradò
Institut de Physique et Chimie des Matériaux de Strasbourg, Strasbourg, France

Farzad Foadian · Heinz Palkowski ✉
TU Clausthal, Clausthal-Zellerfeld, Germany
e-mail: heinz.palkowski@tu-clausthal.de

A. E. Tekkaya et al. (eds.), *60 Excellent Inventions in Metal Forming*,
DOI 10.1007/978-3-662-46312-3_67

method to control the eccentricity will be presented, introducing results achieved with SF copper tubes [5].

2 Process Idea

The main idea of this process is to control the eccentricity during the tube drawing. For this reason, the tube was drawn with tilted and/or shifted dies. The choice of drawing parameters has a direct effect on the eccentricity. A guiding die was combined with the main die (see Fig. 1a) to facilitate the tube drawing with shifting. To tilt the tubes, for laboratory investigations specially prepared die holders with defined skewing were used (Fig. 1b). Depending on the position of the minimum/maximum thickness to the tilted or shifted die, one can obtain a decrease or even increase in wall thickness. Under laboratory conditions, shifting was said to be positive (+) when the tube's minimum wall thickness was in the direction of shifting (Fig. 1c). Positive tilting was defined with the maximum wall thickness of the tube in the direction of tilting (Fig. 1d).

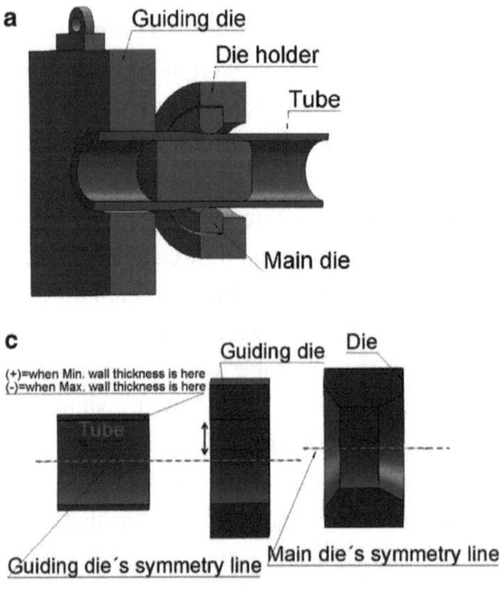

Fig. 1 **a** Schematic of the guiding and the main die, **b** tilted die holder, definition of positive and negative position by shifting (**c**) and by tilting **d**

3 Results and Discussion

The relative changes in eccentricity for different tubes with different tilting angles for two following drawing steps are shown in Fig. 2. A positive value of the relative change in eccentricity means that the absolute eccentricity is reduced; a negative value indicates an increase of the absolute value of eccentricity. For both drawing steps there are similar trend lines [5].

The important point which can be seen in this figure is that, when the maximum of the tube is placed on the upper side and in the direction of the tilting (so called positive setup), there is an increase in eccentricity. However, when the minimum side of the tube was placed in upper position and in the direction of the tilting (using negative setup) there is a reduction in eccentricity. Moreover, increasing the tilting angle increases the intensity of the effect. In order to decrease the eccentricity, the best results appear with $-5°$ tilting. Under the selected conditions this was a limiting value. Higher angles led to an instable process and failure by rupture during drawing. If an increase in eccentricity is preferred, the positive angles should be used. The equations for the trend-line for the first and second drawing steps are given below:

$$\text{First drawing} \quad E = 0.106 - 0.093 \times \theta$$

$$\text{Second drawing} \quad E = 0.045 - 0.0915 \times \theta$$

where E is the eccentricity (%) and θ is the tilting angle (°). It can be seen that for both drawing steps the characteristics of the tubes is the same. It can be concluded that the effect of tilting within a certain range can be used to change eccentricity strongly. Another (side-) effect occurring at drawing with a tilted die is that the position of the maximum and minimum wall thickness can rotate into a new position. In Fig. 3 the wall thickness over the circumference was measured in eight different positions for drawn tubes without tilting (0°) and with $-5°$ tilting, respectively. As can be seen, in the case without tilting the position of the maximum and minimum wall thickness is stable. Drawing with a

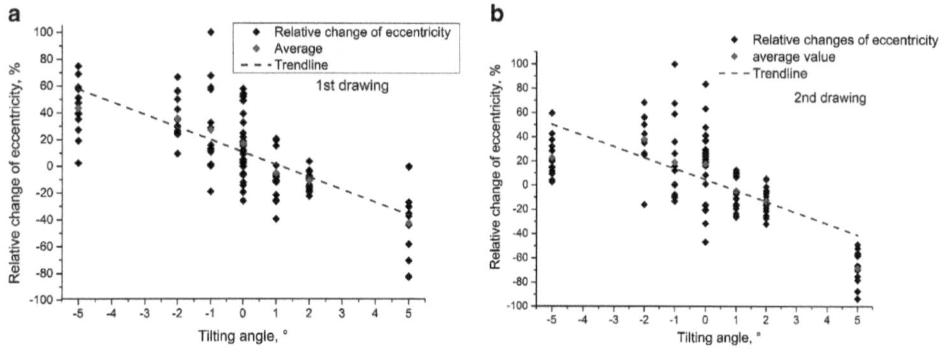

Fig. 2 Relative changes of the eccentricity by tilting for the first (**a**) and second (**b**) drawing step

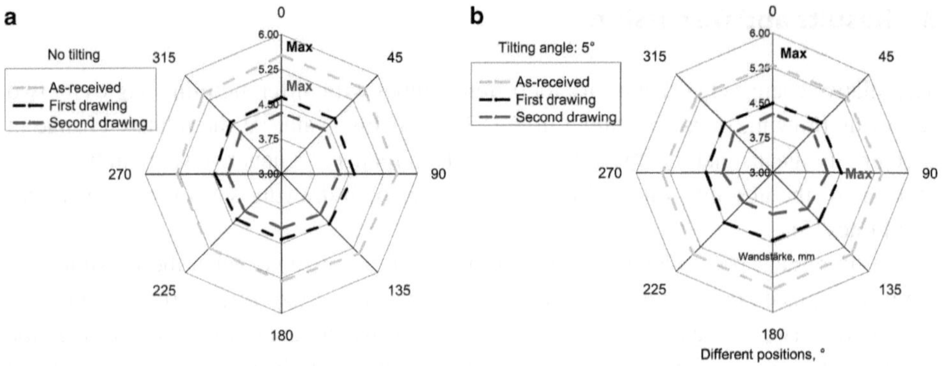

Fig. 3 Position of the maximum wall thickness before and after drawing for drawn tube with a tilting angle of 0° (**a**) and −5° (**b**)

Fig. 4 Relative changes of the eccentricity in tube drawing with shifting the dies

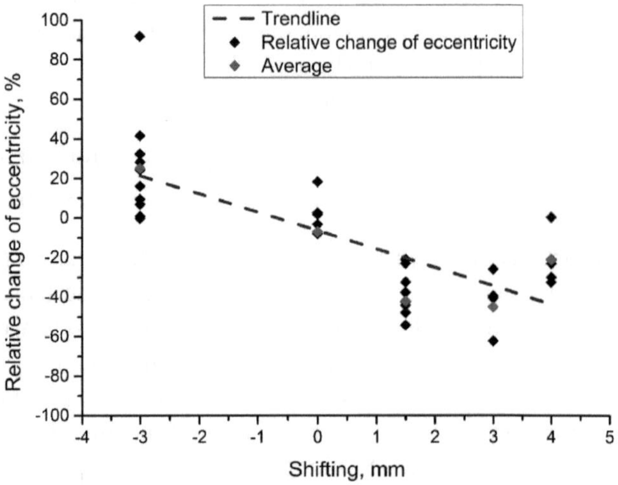

tilting angle of −5° led to a shifting of the maximum into a 90°-rotated position. Before drawing, the positions of the maximum and minimum wall thickness were 180° and 0°, respectively. After drawing, the maximum shifted approximately into the 90° position and the minimum into the 270° one.

The results for the tube drawing with shifted dies are shown in Fig. 4. For a positive value of eccentricity change there is an improvement in eccentricity and the absolute value of eccentricity is reduced and vice versa. As can be seen, the same effects concerning eccentricity can be achieved with tilting and shifting.

4 Potential

Tube drawing with tilted and shifted dies offers the possibility to

Reduce the eccentricity In applications where a homogenous wall thickness is desired, this method can be used to reduce the wall thickness difference in drawn tubes strongly and controlled.

Increase the eccentricity If a locally thicker wall thickness is desired, e. g. cases with a side contact of tubes or tubes for IHU forming processes.

Easy to build Since in this method the main parameters were applied using a die holder and a guiding die, it is very easy to utilize this method in the industry.

Influence the residual stresses It could be shown [5] that tube drawing with tilted or shifted dies influence the residual stresses as well characteristically. The tensile residual stress at the outer surface of a tube can be strongly diminished or even shifted into the compression state.

5 Conclusion

In this short paper, new possibilities to produce tubes with controlled eccentricity by tube drawing are introduced. For this reason tilted and shifted dies were used and new set-ups for drawing tubes with tilted and shifted dies were developed. In order to tilt the die, different die holders with different tilting angles were designed. For shifting, a guiding die was used. The way the tube (min/max) enters the die plays an important role in this process. To avoid the necessity to move the dies following the min/max positions of the tubes, for the laboratory investigations two different set-ups were defined: positive and negative. In positive tilting, the maximum wall thickness of the tube was in the direction of the tilting angle. Positive shifting, however, was defined the minimum wall thickness of the tube to be in the direction of shifting. Results show that the positive tilting set-up leads to an increase in eccentricity and the negative tilting leads to a decrease in the eccentricity. Tilting the tube about $-5°$ (minimum wall thickness in the direction of tilting) results in ca. 40 % improvement in the reduction of eccentricity (normal drawing gives approximately 18 %). Shifting the tube about -3 mm (putting the maximum wall thickness on the direction of shifting) improves the eccentricity about 30 %.

6 Acknowledgment

We acknowledge the support of the AiF (Arbeitsgemeinschaft Industrieller Forschungsvereinigungen – Community of Industrial Research) for the project titled, "Influence of production parameters on residual stresses and dimension tolerances for drawing of tubes" (AiF 17263 N) and Kabel Metal Europe for supporting us with the copper tubes.

References

1. J. R. Davis, S. L. Semiatin, Ed, ASM Metals Handbook: Forming and Forging: American Society for Metals, 1989.
2. T. Pirling, A. Carradò, S. Brück, and H. Palkowski, "Neutron Stress Imaging of Drawn Copper Tube: Comparison with Finite-Element Model", Metall and Mat Trans A, vol. 39, no. 13, pp. 3149–3154, 2008.
3. De Pari, Jr., Luigi and W. Z. Misiolek, "Numerical Modeling of Copper Tube Extrusion: Process and Eccentricity Analysis," Journal of Manufacturing Science and Engineering, vol. 134, no. 5, p. 51005, http://dx.doi.org/10.1115/1.4007283, 2012.
4. Y. Xin, Z. Li-hau, and L. M. Victor, "Finite Element Analysis of Residual Stress and Distortion in an Eccentric Ring Induced by Quenching," Transactions of Materials and Heat Treatment, vol. 25, no. 5, pp. 746–751, 2004.
5. F. Foadian, H. Palkowski, "Einfluss der Vorrohr- und Auslaufführung auf die Exzentrizität und Eigenspannungen von Rohren beim Gleitziehen: AiF-17263 N Final Report", Institute of Metallurgy, Metal Forming and Processing, Clausthal University of Technology, 2014.

Erratum to: 60 Excellent Inventions in Metal Forming

A. Erman Tekkaya

Erratum to:
Chapter: Hidekazu Hayashi "Extrusion of Scroll Against Counter Pressure" in:
A.E. Tekkaya et al. (eds.), *60 Excellent Inventions in Metal Forming*
DOI 10.1007/978-3-662-46312-3

Chapter: "Extrusion of Scroll Against Counter Pressure" unfortunately contains the wrong author name.

The correct author name is: Hidekazu Hayashi.

The online version of the original chapter can be found under
DOI 10.1007/978-3-662-46312-3_51

A. Erman Tekkaya ✉
TU Dortmund, Dortmund, Germany
e-mail: Erman.Tekkaya@iul.tu-dortmund.de

A. E. Tekkaya et al. (eds.), *60 Excellent Inventions in Metal Forming*,
DOI 10.1007/978-3-662-46312-3_68

MIX
Papier aus verantwortungsvollen Quellen
Paper from responsible sources
FSC® C105338

If you have any concerns about our products,
you can contact us on
ProductSafety@springernature.com

In case Publisher is established outside the EU,
the EU authorized representative is:
Springer Nature Customer Service Center GmbH
Europaplatz 3, 69115 Heidelberg, Germany

Printed by Libri Plureos GmbH
in Hamburg, Germany